BUILDING HONG KONG

ENVIRONMENTAL CONSIDERATIONS

The contributors to this book are:

Architectural Services Department
Elvis Wai-kwong AU
John BURNETT
Daniel Wai-tin CHAN
Edwin Hon-wan CHAN
Ada Yin-suen FUNG
Phillip JONES
Andrew Siu-lo LAM
Stephen Siu-yu LAU
Man-kit LEUNG
Xiang-dong LI
Bernard Vincent LIM
Chi-sun POON
Ted PRYOR
Dariusz SADOWSKI
Peter Cookson SMITH
Bo-sin TANG
Kam-sing WONG
WONG Wah Sang

BUILDING HONG KONG

ENVIRONMENTAL CONSIDERATIONS

Edited by

WONG Wah Sang *and* Edwin Hon-wan CHAN

Hong Kong University Press
14/F Hing Wai Centre
7 Tin Wan Praya Road
Aberdeen
Hong Kong

ISBN 962 209 502 X

Printed in Hong Kong by United League Graphic & Printing Co. Ltd.

Contents

Foreword

The emphasis on environmental awareness in architectural design is crucial – now and for the new millennium. I am glad that on this issue there are so many participants from different areas, both in the private and in the public sectors, who are concerned with the environment. One sector sells the goods while the other wants to convince the public that it is doing the right thing! Very often, neither the person in the street nor the architects fully grasp the problems before them! It is therefore important that these commentators do not talk at cross purposes but maintain a particular vision for Hong Kong and the region.

Academics have their own axe to grind as well. As an architect, I am handicapped by who to believe. I am prepared to listen, but I wish more symposia could be held and the environmental issues openly discussed to enable architects to make decisions for themselves. I am saddened by the fact that the poorer countries in the region cannot afford much in the area of environmental protection, and richer countries do little when it appears to affect their economies.

Finally, however, I am happy that in Hong Kong we are talking about these issues more openly than ever before. I would be even happier to see a group that would tackle environmental problems here. Such a body is desperately needed because we are the ones choking from the air we breathe. I do not believe awareness is sufficient. Concerted action is required.

Eric Kum-chew Lye
Chair Professor
Department of Architecture
The University of Hong Kong

Preface

Environmental concern is a global issue that is not confined by physical boundaries, let alone professional barriers. Such a concept is clearly confirmed by the publication of this book involving academics and practitioners from various fields of the built environment. With reference to the environment and the institutional framework for planning and pollution controls in Hong Kong, contributors proffer their philosophies and experiences about environmental considerations in the process of building Hong Kong. Contributions cover the aspects of sustainability, policy implementation, design strategies, environmental design constraints and legal framework. The book, as it is now, had been improved with the benefit of comments on the manuscript from an anonymous reviewer during the reviewing process. We are delighted that the reviewer recognizes that *'[t]he manuscript is based on research, as well as . . . the author's own involvement or practice. The study is a useful contribution to the field.'* We gratefully thank the reviewer for his/her patience in reviewing the manuscript and for his/her encouragement with insight on the subject.

This book comprises a collection of articles, compiled since 1996, under the theme of 'Building Hong Kong: Environmental Considerations'. It is grouped into four parts, namely:

Part A: Broad Issues on the Environment;
Part B: Environmental Design Strategies;
Part C: Environmental Factors; and
Part D: Environmental Legislation.

There is an abstract provided by us as an introduction to each part. The articles are based on the contributors' academic research and practical experiences. The book will be a useful reference for university students and practitioners in the subject of the built environment. It is also hoped that this book will generate worthy discussions and awareness on environmental issues among the general public in Hong Kong in the new millennium.

We would like to take this opportunity to thank all the contributors and reviewers of this book. Without their enthusiasm and support, it would not have been possible to publish a collection of papers covering such a wide range of important environmental issues related to development in Hong Kong. Thanks are also due to the publishers, in particular Mr Dennis Cheung, who saw the potential in a book of this topic and arranged for its reviewing and publication. We are also grateful to all those who helped in various ways during the editing process of this book. Lastly, we hope this book will contribute to achieving the aim, 'Quality People, Quality Home: Positioning Hong Kong for the Twenty-first Century', of the 1999 Policy Address by the Chief Executive, Mr Tung Chee-hwa.

WONG Wah Sang
Department of Architecture
Faculty of Architecture
The University of Hong Kong

Edwin Hon-wan CHAN
Department of Building and Real Estate
Faculty of Construction and Land Use
The Hong Kong Polytechnic University

January 2000
Hong Kong

Contributors

Architectural Services Department
The Hong Kong Special Administrative Region

Elvis Wai-kwong Au
Assistant Director
Environmental Protection Department
The Hong Kong Special Administrative Region

Professor John Burnett
Head and Chair
Department of Building Services Engineering
Faculty of Construction and Land Use
The Hong Kong Polytechnic University

Daniel Wai-tin Chan
Associate Professor
Department of Building Services Engineering
Faculty of Construction and Land Use
The Hong Kong Polytechnic University

Edwin Hon-wan Chan
Associate Professor
Department of Building and Real Estate
Faculty of Construction and Land Use
The Hong Kong Polytechnic University

Ada Yin-suen Fung
Project Manager
Hong Kong Housing Authority
The Hong Kong Special Administrative Region

Professor Phillip Jones
Professor
Welsh School of Architecture
University of Wales in Cardiff, UK

Andrew Siu-lo Lam
Director
City Planning Consultants Limited

Stephen Siu-yu Lau
Associate Professor
Department of Architecture
The University of Hong Kong

Man-kit Leung
Senior Architect
P & T Architects and Engineers Ltd.

Xiang-dong Li
Assistant Professor
Department of Civil and Structural Engineering
Faculty of Construction and Land Use
The Hong Kong Polytechnic University

Bernard Vincent Lim
Associate Professor
Department of Architecture
The Chinese University of Hong Kong

Chi-sun Poon
Associate Professor
Department of Civil and Structural Engineering
Faculty of Construction and Land Use
The Hong Kong Polytechnic University

Ted Pryor
Chairman
Hong Kong Civic Design Association

Dariusz Sadowski
Architectural student for AA Dipl
Architectural Association
UK

Peter Cookson Smith
Director
Urbis Limited

Bo-sin Tang
Assistant Professor
Department of Building and Real Estate
Faculty of Construction and Land Use
The Hong Kong Polytechnic University

Kam-sing Wong
Project Director
Anthony Ng Architects Ltd.

Wong Wah Sang
Associate Professor
Department of Architecture
The University of Hong Kong

Introduction: Living With The Environment In An Urban Context

Wong Wah Sang

INTRODUCTION

Charles Darwin's Theory of Evolution assumed nature's development through a process of natural selection, whereby stronger animals eat weaker animals and only the fittest could survive. A modern society like Hong Kong would agree to this principle as we witness people supplanting each other. There is keen competition all around. And the reward is materialistic wealth for a few successful people.

However, a society does not belong to a few people but to all who live in it and contribute to it. The Theory of Evolution ignored the aspect of mutual cooperation among living things to maintain a liveable Earth in equilibrium and harmony. For instance, the composition of air in the atmosphere has remained constant throughout hundreds of years despite the Earth being an open system. At every moment, energy enters and leaves the Earth. To insulate

Earth's mass is not an easy job for the atmosphere. Living things all work together to create a balance in the environment. The stable salt content in the composition of sea water is yet another example. A stable condition imparts comfort and health for living things. Any deviation will lead to instability and eventually the extinction of certain species.

Now, the importance of biodiversity is recognized as performing a vital role in the functioning of ecosystems. To allow for biodiversity, the strategy of shielding all genes, species and ecosystems from human influence is not practical. Instead, an approach is required to look at the planning of the entire ecosystem for controlled, environmentally dynamic policies so as to afford positive adaptation with minimum adverse impacts on biodiversity. This method of maintaining harmony between human beings and the environment can also be applied in the urban context.

▮ PROBLEMS OF THE URBAN ENVIRONMENT

More than half of the world's population now live in urban areas, and by 2020, the projected level is 60%. This implies that the urban environment is becoming increasingly important in a global context. To generate economic activities, cities consume a lot of natural resources while creating and leaving behind a lot of waste and pollution. Cities have become a major cause of degradation of the environment. Urban dwellers are living in generally healthier spaces with higher incomes, but at the expense of the rural environment. Human pursuits and efforts in certain economic and social developments have upset the balance of the natural ecology.

Cities consume natural resources and produce waste inside and outside the city boundaries. The environmental problems generated by cities range from those on a household level, a building level, a city level to those on a global level. The impact of these problems produces effects on human health, economies and ecosystems.

Specific urban problems vary and depend on a city's size, population, growth, topography, climate and government. Threats to human health include those from drinking water and sanitation, waste water disposal, indoor and urban air pollution, as well as solid and hazardous wastes. For large cities, environmental management is often complicated. Income level is also a factor in the creation of environmental problems. Combined with the natural features of a city and its surroundings, the type of environmental problems can be predicted. For example, air pollution increases with an increase in income level due to higher levels of car use, industrial production and fuel consumption associated with wealthier cities.

High energy consumption is a common phenomenon associated with cities. Studies have revealed that global energy use will rise considerably in coming years. Increases are in the range of 34% to 44% by 2010, and 54% to 98% by 2020. Most of these energy sources still depend on traditional coal, oil and natural gases. Renewable energy sources such as solar, wind and farm-grown energy crops are predicted to supply only 2% to 4% of the global energy for the coming decade.

Environmental problems bear both direct and indirect economic costs. Direct costs include medical costs for treatment of pollution-related diseases. Indirect costs could be reduction of productivity through lost workdays, loss of educational opportunities and shorter working lives. After cities have exhausted the natural resources in their surroundings, resources from further afield have to be brought in at greater costs. However, human health and life affected by environmental problems cannot be compensated purely by monetary means.

Creating one's own environment in the high-density urban fabric of Hong Kong.

URBAN IMPACTS ON NATURAL RESOURCES IN HONG KONG

With 11 600 persons per km^2 in its most dense area, Hong Kong is the most densely populated urban centre in the world. To gain flat land from the original hilly topography, reclamation is necessary and has built up over 25% of the total urban land area. Extensive reclamation is still ongoing. And combined with densely populated shoreline habitats and heavy fishing pressures, the shoreline ecology is being continuously threatened and altered.

Raw sewage flows into the Victoria Harbour together with toxic industrial wastes. Animal wastes and agricultural chemicals add to the pollution. The heavy shipping traffic causes hydrocarbon pollution. To improve the situation, an integrated coastal zone management is required. Laws have been passed to restrict effluent discharges from industry and ships. An urban sewer with treatment facility is under construction.

Other environmental problems in Hong Kong are associated with air, noise and energy. The source of air pollution comes mainly from vehicular traffic, especially diesel engines, construction sites and open quarries. Noise pollution is a result of the high-density urban way of life, which is a complex situation of activities, materials and traffic. As a compact city, Hong Kong has fewer energy problems. The main consideration for energy lies in the ventilation and lighting of buildings by artificial means and the use of the building enclosure to separate the natural external climate from the controlled interior.

The high density also induces an overlapping of environmental problems, as the populace is normally subjected to more than one form of pollution. Noise, air and energy problems occur together in many cases. So these environmental problems cannot be dealt with in isolation. An example is infrastructure failure causing traffic congestion. Not only is the efficiency of work decreased by unproductive waiting, but inefficient fuel use and worsened air pollution also result. Besides, workers' stress and aggravation are increased. So an overall view should be held to confront the urban environmental challenge.

TO LIVE WITH THE ENVIRONMENT

In ancient times, people respected the climate and their surroundings when designing their first shelters. For instance, in 5000 BC, around the Yellow River in China (黃土地穴居), people dug a hole in the ground and the ground was burnt to harden the earth, forming a better enclosure against water and the cold. The timber for the roof cover was covered with clay to achieve fire

Reclamation as an intrusion into one of the best natural resources in Hong Kong — the Harbour.

protection. Such was the adaptation to respect and live with the natural elements.

Modern environmental challenges are vast in scale, affecting human health, resources of nature and economic productivity. On a global scale, people now are concerned for the poor, especially the urban poor who are suffering from a degrading urban environment. The poor should be allowed to recognize their environmental risks and to determine their priorities and needs through community initiatives. Job opportunities may also be created from environmental challenges, such as in waste recycling.

Another challenge for cities is to develop strategies for economic activities with concern for environmental protection. Rapidly industrializing cities in developing countries are creating most of the worst environmental degradation. Demand on natural resources has to be worked out in ecologically sound methods without long-term harmful effects. Strategies should aim for sustainability.

Environmental management for cities is often complicated by governmental issues. National and local governments have to work together to achieve multi-goal successes. However, an informed citizenry is as important as a determined top-level management or government. Indeed, many innovative approaches to improving the environment are emerging from the bottom up – from individuals to neighbourhoods to communities – to create more human, liveable and ecologically sound cities.

In Hong Kong, the Environmental Protection Unit was established in 1977 to formulate environmental protection policy and to coordinate environmentally related activities of other government departments. This Unit was replaced in 1981 by the Environmental Protection Agency, which had developed comprehensive programmes of environmental protection measures, geared to local conditions. In 1986, the Environmental Protection Department was established in its present form with a more powerful and more resourceful set-up. The government's efforts in environmental protection are summarized in Table 1.

The balance of issues with opposing values in the city — the poor versus the rich, the disorder versus the discipline, the identity versus the non-identity...

▮ PLANNING CITIES WITH URBAN ECOLOGY

Town planning deals with the layout of districts on a large scale. Land use and resource management are planned to aim for sustainable utilization for the economic and social well-being of the present as well as future generations.

Table 1 The government's efforts in environmental protection

Government department/ unit	Functions related to environmental control
Environmental Protection	1. enforcement of environmental protection legislation; environmental monitoring; development of sewage and waste disposal programmes; policy development; planning against pollution 2. provision of waste treatment and disposal services
Agriculture and Fisheries	livestock farm licensing scheme; management of agricultural weirs
Architectural Services	water and air pollution control; noise mitigation measures; solid waste treatment and disposal; environmental review and impact assessment
Buildings	establish regulations and practice notes regarding environmental protection for buildings; control of overall thermal transfer value for certain buildings
Civil Engineering	design, construction, operation and restoration of landfills; dredging and management of contaminated mud pits; consultancy studies for disposal of marine mud
Drainage Services	design, construction and operation of waste water collection, treatment and disposal facilities
Education	noise abatement programme in schools; residential ecology/geography courses for students; in-service training courses for teachers and technicians
Electrical and Mechanical Services	operation and maintenance of refuse incineration plants; maintenance of refuse collection vehicles; phasing out the use of CFCs and halons in government buildings
Government Laboratory	laboratory services for air and water pollution control and waste management
Highways	provision of noise barriers; laying noise reduction surfacing material on highways; noise insulation work studies; waste management; environmental impact assessment/moitoring for raod projects
Marine	combat oil pollution; collect vessel-generated refuse and floating refuse; regular water monitoring
Planning	environmental impact assessment; in-house planning studies; consultancy studies
Regional Services	air and water pollution control; waste management; environmental assessment and planning
Territory Development	design and construction of sewerage network, treatment facilities and drainage channels; environmental impact assessment for highways routing
Urban Services	air, noise and water pollution control; waste management; environmental assessment and planning; education and publicity

To allow environmental planning, strategies and policies are set and implemented to provide a satisfactory and balanced environment with minimal adverse impacts on nature.

The basis of environmental planning for cities is urban ecology – the recognition of the dependence of one life process on another, the interconnected development of living and physical processes, and the recycling of living and non-living materials as a self-perpetuating biosphere. The environment is best served when town planning is perceived to cope with natural processes rather than utopian ideals. Environmental costs are also considered rather than considering just 'functional' design or good aesthetics. Cities are recognized as centres with highly concentrated nutrient energy. Biological solutions rather than just engineering solutions can be used for solving cities' infrastructure problems. Humanity and nature are considered as integrated issues.

Confucius (孔子) interpreted the *Book of Changes* (易經) as a sequence of changes with nature at work. Natural processes are dynamic. No one stage can persist for ever. Cities have the same analogy. Urban form is a result of an evolutionary process driven by physical, social, economic and political changes. Modern buildings replace old ones. Urban decay and renewal are constantly witnessed in cities.

When human beings are considered as part of the natural process, changes can be fine-tuned to afford constructive opportunities. Even destructive changes can produce benefits to the environment. Land is constantly being changed by artificial or natural processes. Planning can thus be considered as initiating purposeful and positive changes. Humankind and urban ecology are the basis for planning.

Besides the concept of change or processes, the principle of 'the least effort for the maximum gain' can be applied to urban ecology. Spending the least amount of resources and energy to achieve the best results is the principle of economics applying to cities. A city can be both the supplier and the consumer of products, like a recycling procedure. An example of a constantly recycled city is Rome, where many of its buildings are constructed out of materials and components from previous, old buildings.

Another principle for designing cities is diversity. From an ecological point of view, diversity implies health. In an urban context, diversity gives social order. Through diversity, choice is offered to meet the requirements of a diverse urban society and its lifestyles. In a larger context, diversity is related to culture and identity.

On the other hand, the different urban elements and systems are interconnected and interdependent. Any piece of land is affected by its hinterland and bio-region. So a local area has to be related to its larger context for a balanced ecological planning. Recognizing the interdependence of people and nature will create new landscapes and urbanscapes that are a mixture of the natural and the artificial, possessing ecological, economic and social values that are more adaptable to changes which occur in life.

In Hong Kong, high density is a unique factor in planning considerations. Handled positively, the urbanscape would be compact and efficient; but if mishandled, overcrowding can result, with damage to the environment. In the land use planning process, environmental impact assessment (EIA) is used as an analytical tool to allow a more dynamic approach to planning by taking account of changes in and interrelationships among different systems.

The coexistence of different functions — residential, shopping and leisure — may not strike a balance of harmony in the urban environment.

DESIGNING THE BUILDING ENCLOSURE WITH ENVIRONMENTAL CONSIDERATIONS

Buildings form the major part of the urbanscape, and the enclosures of buildings act as the interphase between natural and artificial environments. In this respect, the fundamental functions of a building enclosure include ventilation, daylighting, noise control, heat transfer as well as visual contact. The internal environment is controlled by artificial means to achieve a constant condition regardless of the external environments. Architectural design can respond to various needs, but proper building management is as important as the initial provisions.

Energy enters and discharges from a building through the enclosure in various forms, some of which are interchangeable. Solar heat, light, sound and wind are environmental factors to be tackled. Design strategies range from massing with favoured orientation, choice of materials, detailing, to the provision of special features to self-regulating 'intelligent' façades.

The impact of solar radiation is affected by the sunpath and also the location of the building. In Hong Kong, the problem is with solar heat gain in summer, especially with tall buildings. Arranging buildings with the wider façades facing north or south can reduce solar energy absorption. Service cores, including lift lobbies, stairways and toilets, are a good device to face other orientations that have more heat gain. Shading devices such as fins, overhangs, balconies or projecting eaves are effective in cutting off sunlight. Various types of glass can be used for solar control. More sophisticated methods include a double building skin and mechanically ventilated cavity façades. Vegetation can also help to cool down buildings. Alternatively, a photovoltaic envelope can be installed on the façade or roof to convert solar radiation into electricity to supply part of the energy used by the building.

Planning with environmental considerations can impart more varied urban forms.

Wind can be harnessed in different ways to benefit a building. The simplest way is to allow natural ventilation. Cross-ventilation or stack effect are possible means that can be introduced to the architectural design. Another possibility is to absorb the energy from the wind through turbines and convert this into other usable forms of energy.

Noise in Hong Kong is a problem associated with its high density, and especially with traffic. To deal with the noise problem, different design strategies involving an energy aspect can be used but will produce different outcomes.

Environmental problems can be interrelated, so an overall view has to be held during the architectural design stage. Integration with building services can also introduce a good solution to the building enclosure to make the internal and external environments harmonious.

■ ENVIRONMENTAL EFFECTS OF BUILDING MATERIALS

Cities are built up of urbanscapes that are composed of mainly buildings which are constructed from components and building materials. The extraction of materials, for building and construction, can create damage to the natural landscape, which is often difficult to recover from ecologically. Besides extraction, energy is also required for the transport of raw materials, with it also come pollution problems. Production can involve a lot of energy and may generate by-products or unusable waste. Some building materials allow recycling or adaptation to other use.

Materials are used with the aim to integrate their life cycles as much as possible with the building to achieve sustainability. This means a minimum of waste and an efficient use of materials, lengthening the life span of building components. Materials and products are encouraged to be recycled after the demolition of a building, so that natural resources can be conserved. In this respect, buildings designed for future dismantling have obvious advantages.

During the life cycle of building materials, environmental impacts in various forms and magnitudes are produced. Issues in considering the choice of materials include extraction of the raw materials, consumed energy, by-products, waste, renewability, maintenance, and lifespan.

Concrete as a common building material has low energy content, but it can be increased by the addition of steel reinforcement. And much energy is used during transport and construction. Demolition will create rubble that occupies space but can be reused as filler. Stonework and brickwork have similar environmental impacts.

Glass needs a large amount of energy for processing the raw materials required. Some pollution arises during the manufacturing process. Glass can be recycled through remelting, so the problem is that of contamination.

Ceramic is made from clay. A high energy content is involved in firing the clay and glazing. The energy content for the layer of adhesive or cement has to be taken into account too. Ceramic waste is mostly used as a filler material.

Metals are extracted and transformed into building products by using a lot of energy. Harmful by-products may be produced during manufacture. Reusability of metals is an environmental benefit and is often economically feasible.

Synthetics like polyethylene, PVC, bitumen, rubber, polyurethane and polystyrene have petroleum as the basic raw material. During extraction, harmful emissions and waste could be introduced to the environment. During the life cycle, contamination often leads to problems in recycling.

Wood is a very important renewable raw material. Little energy is involved during its production. Sustainable forestry management allows a sustainable supply of wood. Preservatives are often required. The transport of wood also requires energy. Plywood, fibreboard and chipboard, etc., are also produced mainly from renewable raw materials and offer a good choice and variety for use.

Correct choices of materials and detailing for a building are positive attributes for environmental improvement.

There are numerous paints with various compositions. Many paints release organic hydrocarbons during application and may harm the health of painters and occupants. Additives are usually also harmful. During the dumping of painted materials, harmful elements may be released.

CONCLUSION: AN ECOLOGICAL APPROACH

From the broad perspectives of building products and of town planning and urbanscapes to the selection of building products and materials, we learn that modern technology developed to shelter human beings can be unfriendly and polluting – a hard and forceful gesture in the process of civilization. Due to human activities, the recent increased carbon dioxide content in the atmosphere has increased the insulation value and hence the temperature of the atmosphere. A glazier that is formed of hard ice crushes into the others and cuts into the surrounding land, but only moves slowly at one metre per day. However, a river is smooth and meandering, flowing down cliffs and nourishing a vast landscape. And most importantly, it contains life. A considerate, lively response is much more rewarding than a forceful, unfriendly act. Environmental strategies vary by responding to different contexts, and approaches can be through legislation, the economy or technology. Environmentally responsive use of artificial creations can produce lively buildings and create cities of the best architectural and urban quality.

To conclude this introduction, I believe a broader approach is essential to cultivate environmental consciousness. When we choose to be friendly to the environment, the environment will be friendly to us. We can be part of the natural environment and vice versa. Through education, architecture and everyday life with the appreciation of changes imparted by the environment, we can learn to live harmoniously with the environment.

The correct positioning of development with environmental awareness can lead Hong Kong into a truly sustainable future.

▌ REFERENCES

Anink, David, Chiel Boonstra and John Mark. 1996. *Handbook of sustainable building.* London: James & James (Science Publishers) Limited.

Environmental Protection Department, Hong Kong. *Environment Hong Kong 1996.*

Hough, Michael. 1995. *Cities and natural process.* London: Routledge.

Wong, David O.Y. 1996. *Environmental conservation and planning: Hong Kong and overseas encounters.* Hong Kong: Woods Age Publishers.

Wong, Wah Sang. 1998. *Building enclosure in Hong Kong.* Hong Kong: Hong Kong University Press.

——. 1998. *Building materials and technology in Hong Kong* (third edition). Hong Kong: All Arts Limited.

The World Resources Institute, The United Nations Environment Programme, The United Nations Development Programme and The World Bank. 1996. *World resources 1996–97.* Oxford University Press.

Part A: Broad Issues On The Environment

This part deals with the broad issues of the environment in relation to land use and sustainability in Hong Kong. It discusses the relationships between the environment and town planning, and between sustainability and urban design. It focuses on those issues relating to the strategies, uses and sustainability of Hong Kong's Victoria Harbour.

In his chapter 'Sustainability and Urban Design', Peter Cookson Smith addresses the relationship between the two. In applying the concept of sustainable urban design in the context of Hong Kong, he discusses five key dimensions: spatial form, energy efficiency, urban ecology, tall building design and the environmental protection process. He claims that urban sustainability is ultimately a question of value judgement and a choice of improved environmental quality vis-à-vis other economic and social objectives. The achievement of environmental objectives requires coordinated efforts and the involvement of the whole community in deliberating and implementing an unambiguous and proactive agenda. In this connection, the author concludes that there is a need for genuine public consultation that aims at establishing what sort of place and urban character the people of Hong Kong want.

In 'Town Planning and the Environment: Role and Tools of Private Consulting Planners', Bo-sin Tang and Andrew Siu-lo Lam address the relationship between town planning and environmental protection in the context of private property development. They assert at the outset that property development and environmental quality are not two opposing goals. Planning is an important stage in property development projects prior to the commencement of irrevocable construction works. Their chapter discusses the role of private consulting town planners in handling the relevant environmental issues arising from land development. In essence, planners seek to manipulate three components of property development – land use, development intensity and built form – with a view to mitigating the possible environmental damages. Examples of some commonly used land use planning solutions, including preservation, conservation, segregation, buffering, integration, compensation and sustainability, are discussed.

In 'What Kind of Harbour City Do We Want?' Ted Pryor and Peter Cookson Smith focus their discussion of sustainability on one strategic asset in Hong Kong – the Victoria Harbour – which is now subject to various sorts of environmental threat. A product of human enterprise, the Harbour carries both aesthetic and economic functions for the local community. To relieve pressure on urban land, there is a need to achieve an appropriate scale, shape and location of harbour reclamation. With sensitive and innovative urban design ideas, harbour reclamation can be made into a great opportunity to revitalize and regenerate the urban centre. Considerations proposed by the authors include an upgrade of existing waterfront uses to support greater public access and enjoyment, development of a 'gateway park' to integrate recreation and tourist attraction spots with pedestrian links, careful and sensitive building design on the reclamation to enhance the harbour and new city image, and so on. Finally, they propose to set up a Harbour City Development Commission to coordinate and handle the institutional matters associated with development around the Harbour.

1

Sustainability And Urban Design

Peter Cookson Smith

INTRODUCTION

In a development sense, 'sustainability' is about maintaining and enhancing the quality of life while respecting the carrying capacity of the biosphere, the supporting ecosystem and resource base. At a more detailed and specific level, spatial planning and urban design need to be set within the urban and city planning context of natural resource conservation and local environmental quality.

Global initiatives reflecting sustainable concerns for the planet have, in recent years, helped to develop a clear ecological dimension for strategic and local planning. It is recognized that the formulation of responsive urban design parameters must also relate to the creation of lasting environments. The UK Urban Design Group's 'Agenda for Urban Design'[1] includes the statement that 'urban design is concerned with the careful stewardship of the resources of the built environment in the creation and maintenance of the public realm.' This requires a strong direction for change that allows comprehensive and realistic schemes to be produced. It involves the integration of economic, environmental and community interests in order to implement tangible projects that achieve the objectives of 'Agenda 21', which followed the 'Earth Summit' in Rio under the auspices of the United Nations in 1992.

The Hong Kong government is currently carrying out a study called 'Sustainable Development for the 21st Century'. The intention of this is to provide a basis for an improved system of corporate decision-making on how best to achieve the primary goal of creating a sustainable, urban-based way of living. This will need to take into account the need for economic vitality, along with the capacity of Hong Kong's uniquely constrained environment, infrastructure systems and land use resources.

Improving the environmental performance of cities is possibly the most pressing issue to be addressed in the twenty-first century. As Hong Kong faces the challenge of accommodating up to 2 million people over the next 15 years, the achievement of sustainable human settlements must be matched by the environmental performance of the city itself as it expands from 6.5 million people to around 8.3 million. At first glance, this might not seem such a severe task by world standards where 100 cities now accommodate 550 million people, with the 20 largest conurbations each having in excess of 20 million.[2] Urban populations in general are now increasing three times faster than overall population growth. However, the Hong Kong Special Administrative Region (SAR) lies at the seaward end of a sprawling regional growth corridor around 150km in length extending through the Pearl River Delta to Guangzhou, where continued economic growth is leading to long-term industrial-led urbanization with immense demographic pressure, lifting its population to perhaps over 40 million people by 2011 – way beyond the carrying capacity of its regional resource base.

Most concerns regarding sustainability are related to global ecological factors, and the somewhat simplistic objective of limiting the consumption of renewable natural resources to within the capacity of their replenishment. The key consideration at a planning and urban design level is to interpret these broad principles into practical design policies. The challenge for engendering conditions for sustainability in the Hong Kong SAR is not necessarily in making sense of various disparate elements, but in evolving a greater degree of collaboration among various interests, in particular government departments and agencies at different levels to ensure that sustainability works at a practical level.

Several key dimensions can be identified: spatial form, energy efficiency and pollution control, urban ecology, the tall building design environment, and the environmental protection process. The achievement of sustainable urban design must recognize all these factors within the context of established procedures and constraints that prevail in Hong Kong. While different dimensions might overlap to some extent, an analysis in the Hong Kong context nonetheless helps to define a set of issues to inform urban design theory and policy prescription. These should effectively form a planning agenda for subregional as well as local plan policy. In general, the most implementable policies are those with a clear government mandate and some form of statutory or regulatory control system, particularly in relation to environmental matters. Where responsive and sustainable policies are merely a broad aim conceived in terms of general guidance, these are often difficult to implement alongside other lands issues and priorities. One of the underlying problems in Hong Kong is that urban design itself, while being acknowledged within government planning statements as something to be broadly achieved, is not seen as an objective in itself. Sustainability cannot therefore be successfully infused within it through a proper agenda, established across all scales of design operation, and through necessary interventions in the overall development system.

▮ SPATIAL FORM AND THE CITY

In terms of city form, Hong Kong has one of the highest concentrations of people and development densities in the world. The tall building itself represents both an essential component and a major determinant of urban design, reflecting both high land values and development pressures. This, in theory, should have two advantages for sustainability: first, the economics of scale this represents for the provision of infrastructure, public transport, housing and jobs; and second, the advantages that high urban densities should embody to assist with the ecological conservation of rural environments.

While most Western cities have congested cores with large low-density suburbs adjacent to regional and national highway networks, around 50% of Hong Kong's urban population has simply been decanted to new towns, and their citizens housed within easy commuting distance of the urban area. Its overall metabolism therefore represents one of the most efficient world cities. At the same time, the metro area itself has thrived on its mix of uses and diversity, which has traditionally sustained a framework for a multiplicity of urban uses. The older tenements have, until comparatively recent times, comprised much of the urban fabric and have been able to provide a robust and adaptable mix of workplaces, retail, entertainment and restaurant uses, as well as simple forms of urban housing. This, in turn, has led to Hong Kong's familiar spirit of urbanity and helped to consolidate the economic and social importance of the inner city itself. However, as in other cities, diversity is gradually being replaced by uniformity as large parts of the metro area representing new comprehensive development or redevelopment areas are zoned for single-use purposes or single tenure estates. A combination of high land values, major housing shortages and poor environmental conditions associated with many older areas are leading to urban redevelopment on a grand scale, with little in the way of genuine urban revitalization.

The economic imperative under which the Land Development Corporation undertakes its 'renewal' operations demands that it undertakes joint ventures with property developers, and must therefore extract the maximum economic gain from land development. As this is being written, the 'Four Streets' Area, comprising part of a tight matrix of urban streets that form part of the core tourist area of urban Tsim Sha Tsui, is proposed for clearance and redevelopment, and with it a mix of uses that not only gives the area its diversity and attraction but sustains the essentially vibrant urban character and sense of history in Kowloon. It is being replaced by a 60-storey commercial building set in a landscaped plaza – one form of urbanity replaced by another – which requires a far greater level of supporting infrastructure and energy reliance. Not too far away, Hong Lok Street, until recently known as Bird Street (a uniquely urban attraction as much for its context as for its facilities), has been redeveloped and replaced by a bird 'garden' some considerable distance away.

These examples bring into focus questions and criteria about where the emphasis on urban renewal should be directed. With little in terms of architectural quality and a decaying building fabric, there is an unambiguous case for upgrading the environment. At the same time, there are less tangible aspects of the local environment that are equally relevant to the city – the multiplicity of small businesses, established trading patterns, local identity, colour and vitality – that are almost inevitably lost in the process. To recognize and retain at least some of these aspects requires careful economic restructuring, responsive lands mechanisms, and a sensitive approach to urban design that should ideally reflect the characteristics of robustness, permeability and mix of uses associated with the traditional street environment. Wider forms of community gain need to be recognized as part of the publicly sanctioned redevelopment process, in terms of moulding and reinforcing city fabric, introducing aspects that provide city vitality and interest at street level, and helping to sustain the character of the city for locals and tourists alike.

At the same time, new development possibilities in the metro area provide opportunities to consolidate its compact environment and high density. The first of these is the government's ambitious reclamation programme, incorporated within Metroplan and the Territorial Development Strategy Review and approved by Exco in 1998. This presents an unparalleled opportunity to examine large parts of the urban area as representing the nucleus of a better quality of life, with both high-density residential nodes and coordinated expansions to existing commercial districts integrating new extensions to the mass transit system. Properly designed reclamations also hold out the possibility of new and continuous pedestrian systems incorporating new urban 'places', links with adjoining older districts, and promenades. There is, however, a need in these new areas to direct the emphasis on urban design from a preoccupation with engineering and architectural standards to ways in which we can satisfy community concerns for environmental quality. There is also a need to replicate, as far as possible, the inherent complexity, multi-use and overall efficiency of the city itself. Places are more important than buildings to the public at large, and good design, linkage and continuity of these means added value and a high level of pedestrian use – all essential characteristics of urban sustainability.

The second aspect relates to the continued development of public transport. Hong Kong's compact environment and high-density new town conurbations overcome, to a large extent, the normal inconsistency between environmental criteria and personal choice. Car reliance is substantially reduced by efficient multi-mode public transport services and people movers. This places large concentrations of people within easy reach of rapid transit. These services are optimized in a cost-effective way, linking together areas having a 'critical mass' of housing and employment, which make the system viable. At the same time, property development in effect pays for railway construction, while high densities ensure an operating profit – a combination of factors almost unique in world cities. Since its inception, the MTRC has constructed a total of 18 rail-related projects, including over 30 000 flats, 200 000m^2 of office space and 250 000m^2 of retail space. At present, there is in excess of HK$200 billion associated with committed new rail links within the urban area. The overall liveability of the city habitat would, however, be markedly increased by reinforcing the high use of public transport by pollution-free vehicles and the integration of pleasurable, pedestrian-friendly environments that accentuate the attractiveness of walking.

▮ ENERGY EFFICIENCY AND POLLUTION CONTROL

The urban design approach must be oriented around several key dimensions concerned with the husbanding of key resources. This calls for a responsive design process that equates built form with energy efficiency. There is both an urgent need and an opportunity to develop environmentally sound technologies, including energy-efficient and pollution-free forms of transport, efficient energy supply systems, and the means to improve the 'clean' energy performance of urban buildings. Similarly, recycling technologies can be made more efficient. At present, less than 10% of Hong Kong's waste is recycled, and the three strategic landfill areas in the New Territories are expected to last only to 2012.

At a territory-wide level, the new towns and large residential conurbations on the fringe of the urban area comprise a sustainable model, with high-density nodes generally linked by efficient public transport corridors with easy access to 'green' country parks and large areas of coastline. This combination of compact environment and high density to a large extent overcomes the normal inconsistency between environmental criteria and personal choice. People are able to live, work and participate in leisure and recreational activities without having to travel too far. Car reliance is reduced by efficient multi-mode public transport services with rapid commuting times. This is probably just as well, as there is only around 8.4m of road lane length per vehicle in Hong Kong. This acts to justify a large investment in territory-wide transit and highway systems as Hong Kong continues to grow and develop. It also demands a new level of concern for an even more efficient and resourceful city.

A fundamental issue is the need to overcome energy wastage and the pollution that usually arises from it. Hong Kong's Environmental Protection Department generally takes a pragmatic stance on pollution control. This reflects a concern to maintain industrial growth and the needs of business interests when issuing new regulations. Compromises tend to be found through discussions with trade associations and the Federation of Industry. However, the essential principle is that the polluter pays, and Victoria Harbour, which represents around 2% of the territory's waters, is now designated as a water control zone – the last of ten such zones. This was prepared for in advance by chemical waste regulations and the establishment of the privately operated Chemical Waste Treatment Centre in 1993, the first in Southeast Asia, that provides an alternative to dumping waste into the sea. The Centre charges for its services from around 8 000 registered factories.

Under the Pollution Control Ordinance,[3] it is obligatory for developments to connect to a sewerage system and to ensure that all water treatment facilities are operated properly. Licensing now covers all major waste treatment and disposal facilities. While the government bears all the capital costs for sewerage, sewage treatment and disposal facilities, users of sewage services contribute by paying the operational and maintenance costs. Environmental protection is, however, still a relatively new notion in Hong Kong, and the current aim is generally to contain pollution before attempting to reduce it. New environmental ordinances require remedial, restoration and clean-up work to be carried out. Since April 1998, private waste collectors can dispose of commercial waste at refuse transfer stations at a nominal rate.

▮ AN URBAN ECOLOGY

The urban planning approach must have built into it a set of urban quality considerations related to policies that underpin the fundamental values of the city. In planning for sustainable development, effective environmental management should extend the values that are currently placed on protection of country parks, where some of the strongest development control powers are lavished, both to maintain countryside environment as a recreational resource and to conserve ecological habitats. It is important to differentiate between different landscape elements and ecosystems. Natural landscape and open space need to be seen as vital elements of urban infrastructure, in effect an urban ecology that should extend the values that are currently placed on valuable rural habitats. This should ensure quality for the built environment and add value to it through the provision of active and passive recreation facilities. It needs to comprise a 'Green and Blue Plan' in response to existing settlement patterns, microclimate and ecological consideration based on interlinked principles of continuity, individuality and diversity. Tree planting can, for instance, remove up to 75% of particulate pollution in the atmosphere.

One of the most despoiling environmental factors in Hong Kong is the proliferation of urban sprawl through 'temporary' uses in former rural areas, which has led to the inhibition of farming activities and has detracted from new development. Wetlands, scattered across the New Territories and islands, are being continually reduced and threatened by redevelopment and pollution. Illegal village encroachment threatens lowland areas and marshland that are valuable breeding grounds for rare bird species. Wetlands are at particular risk as they generally comprise old fish farms or abandoned paddy fields, which are bought up by developers. Direct enforcement provisions against unauthorized development have been recently introduced through the extension of areas under statutory planning control. However, new strategic growth areas, for example in the northwest New Territories, will put these areas under further pressure, and it will be important to incorporate existing environmental 'qualities' into the broad development framework, including sites of special scientific interest, areas of *feng shui* importance, village boundaries and areas of landscape relief and vegetation. In this way, natural elements can be extended through new development areas to help establish a framework for urban design.

In terms of recycling urban infrastructure, Hong Kong's traditional procedure is to redevelop entirely. Buildings rarely change their use unless they are designated as historic monuments and converted to museums. Historically, older structures have been redeveloped to higher densities allowed under successive changes to the Buildings Ordinance. Older parts of the urban area are still characterized by intensive forms of development with densities of up to 4 500 persons per hectare. Urban renewal must now be facilitated by the assembly of sites of sufficient size to allow viable development. Ideally, urban renewal should encompass various measures of improvement, rehabilitation and upgrading at a scale sufficient to incorporate comprehensive new layouts, but with conservation or re-creation of essential urban elements along with necessary environmental improvements. If this process were to be carried out successfully, some of the older communities would be left substantially intact within a rejuvenated urban structure that would reinforce their sense of identity. This calls for public spaces to be integrated properly, i.e. streets, plazas and pedestrian precincts defined by robust multi-activity edges that would stimulate private investment in quality buildings around them.

The process needs to re-establish the traditional sense of street vibrancy and multi-use that could help knit together urban districts. In this sense, it should be a fundamental objective to regenerate and revitalize rather than simply redevelop. However, the accelerating pace of large-scale urban renewal is based largely on market imperatives, which are inevitably leading to a depersonalized economic culture reflected by grand architectural symbolism rather than environmental values and sense of place. 'Ladder' streets, market streets and areas of special or even unique trades characterize certain areas and help to induce a strong sense of local identity. It is important to recognize and accommodate such places of interest within the new planning framework. These are in tune with a pedestrian-oriented city, and serve to indicate that a blanket application of type-cast solutions for large redevelopment areas is inappropriate and will gradually produce not just an undifferentiated city form, but one which has little diversity or individuality.

At the same time, new reclamations provide a means to apply more sustainable solutions and better integrate city and harbour. An integral part of the programme for Central/Wan Chai, West Kowloon, Southeast Kowloon and Green Island is the potential to create up to 35km of new waterfront for pedestrian activity, tied back into the core urban areas by continuous open space corridors comprising a sequence of linked public 'places'. There is an opportunity to make these into welcoming, people-friendly places that meet current public aspirations and act to continue Hong Kong's dramatic identity through its association with the central harbour. This is where a substantial amount of human contact and interaction will take place, as and when the public are able to enjoy increased access to it. Urban sustainability must be a fundamental objective of the planning approach, the implementation process and the overall development criteria. It must also extend the urban design process in a visionary way.

Social and environmental objectives must essentially be seen in tandem. Improvement of air pollution and noise levels improves comfort, while energy-efficient buildings and transport cut costs. An improvement in local environment provides an impetus to pedestrianization, civic upgrading and better landscape, as well as providing for more community interaction. Caring for places will ultimately impel us to embrace resilient environments.

▮ SUSTAINABILITY IN A TALL BUILDING ENVIRONMENT

The tall building is both an essential component and a major determinant of urban design form in Hong Kong. There are no immutable rules to its design, which is largely determined by function and context as well as the regulatory framework. This raises the issue of how contemporary urban environment should integrate the cultural and sustainable objectives of the tall building design. The tall building relates to the city in a number of ways: at a broad level, its collective visual image relates to the skyline; at a city-block level, its form and massing relate to local environmental factors, and the way in which podia are linked can influence the dynamics of street environment and microclimate; and lastly, it affects the way in which pedestrian movement systems are integrated at ground and elevated levels.

New technologies are liberating established aesthetic traditions in terms of external architectural image. It is now possible to open up interiors by climatically controlled atria running through entire buildings, together with skycourts, internal and terrace gardens, soaring lobbies and galleria that, in many cases, have replaced the wider urban precinct by dense and secure cores. The new metropolitan experience in Hong Kong is one of integrated and stacked facilities, both recreational and cultural, that have come to personify a new kind of vertical urbanism. However, as tall buildings will continue to be a mainstay of city growth, more attention must be paid to designing them in an ecologically responsive way. While Hong Kong's high urban density and high public transport use make average energy consumption per inhabitant relatively low, around 50% of all energy use is normally associated with tall buildings. A meaningful step forward should therefore be taken to develop design strategies for 'greening' these major elements. The air-conditioned glass tower has become a symbol of profligacy, increasing energy consumption through the cooling load demand. There is a need to relate sustainable energy policy to what is technologically possible and what is economically practicable.

New industrial techniques and technological advances offer creative potential and have, inherent in them, ways to enrich the city. However, there is now a social and environmental dimension that extends design responsibilities, with opportunities to transmit a new aesthetic order and interact with the public realm in a more expressive way. They must, however, be defined by different forces than those in the past – a greater respect for environmental context with low energy consumption, minimal operating costs and a generally 'green' vocabulary. On the wider front, these factors must be reflected by means of a greater synergy between high-rise structures and the urban realm.

The Hong Kong developer is primarily concerned with the intensification of building area over relatively small sites, with floorplate configurations that must satisfy optimum letting and working conditions. Personalization comes through individual fit-outs, often at high cost to the user, but the buildings themselves are rarely user-friendly, and are rarely and expressly related to culture or place. The key is to apply climatic responsiveness with high quality and, in the process, to engender a legitimate starting point for new urban design expression. This involves the creation of 'places in the sky' together with public realms and linkages that involve contact with the wider environment, not dissimilar to those at street level.

While there is no single solution to the design of environmentally responsible buildings, they must relate to low-energy design and call for bioclimatic solutions which, as far as possible, provide for maximum use of natural ventilation, solar energy and daylighting. The architect Ken Yeang, who has produced innovative bioclimatic buildings such as the Menara Mesiniaga Tower in Kuala Lumpur and the MBF Tower in Penang, sets out certain ecological design principles based on a consideration of the use of energy and materials in terms of various interactions and interdependencies, both external and internal.[4] This implies, first, a response to the siting, orientation and configuration of the building in relation to ambient and environmental characteristics; second, that built form should encourage energy conservation in its mechanical and electrical systems; and third, that this should be reflected in the management of the building's energy efficiency. This means that ideally the building should be configured to minimize the use of mechanical and electrical engineering systems, taking whatever advantage possible of climatic benefits – the bioclimatic design approach. In addition to these factors, the building's outputs should be both non-polluting and, in terms of its construction materials, able to be recycled or reused.

The incorporation of soft landscaping through planting and through introduction of organic matter associated with façade elements, skycourts and balconies provides both ecological value and visual advantages. This can be incorporated through direct integration or in juxtaposition with major building elements. Hong Kong's microclimate, in particular, calls for energy savings in mechanical cooling and artificial lighting. This implies an initial diagnosis of sites, for example, orientation in relation to the sun, shadow effects, the direction of prevailing noise, wind and airborne pollutants.

Planting can provide shade to internal spaces, minimize heat reflection and glare, while evaporation can help the cooling process on the façade and can significantly reduce summer ambient temperatures. Studies have shown that plants help to process internally generated carbon dioxide, release oxygen into the air, and remove some airborne pollutants and chemicals. Rooms or offices overlooking 'internalized' sky gardens and atria, for example, would receive air oxygenated by plants. Landscape at podium level in the form of roof gardens is an accepted part of urban design and open space provision in Hong Kong. Most plants can grow in 0.6m of soil with 0.2m for drainage, and trees of up to 7m can be grown in tree pits of around 1.2m depth, centred on the prevailing support structure. This also helps to reduce heat absorption on floors below. In a similar way, planting can be used in skycourts – atrium spaces recessed into the façade that provide transition areas between internal working areas and the outside. Such areas can also serve as visual amenities and landscape zones for social meeting places, providing recreation and leisure facilities associated with the external environment.

New urban buildings might very well utilize a range of high and low technologies. 'Nerve centres' within buildings that continually gather and transmit information and programme their own energy-saving response can be complemented by low-tech elements such as roof canopies and sun-screens.

These would act to reduce glare and 'stretch' the façade while providing a new design aesthetic of climatic moderators, which control light and heat energy. Outer building skins can also be designed to respond to changing climatic conditions by means of photovoltaics, photochromic glass and shading devices linked to special sensors so that wall filters and blinds could be automatically inclined to reduce excessive sun penetration, reflect sunlight or react to typhoon conditions. Devices such as wind deflection shields or sun-scoops activated by hidden neural networks within the building walls will generate their own aesthetics. In combination, these would combine to provide an expressive design vocabulary that is recognizably modern, reflecting energy-efficient solutions for a sustainable high-rise urban environment.

THE ENVIRONMENTAL PROTECTION PROCESS

The government's Environmental Protection Department (EPD) has set out some of the groundwork for a sustainable urban environment by introducing legislation to support anti-pollution strategies. This has been met by a willingness on the part of the business community to support projects with an environmental agenda – there is now, for example, an environmental rating scheme for new building designs that can be used by developers in the marketing process.

The Technical Memorandum on the Environmental Impact Assessment Process[5] sets out criteria and guidelines for assessment of air quality, noise, water pollution, waste management, ecology, fisheries, landscape and visual factors, and sites of cultural heritage. In general, these are positive measures aimed at reducing or eliminating the impact of development actions. A full environmental impact assessment now forms part of all urban planning and development studies for the government. This comprises a detailed assessment in quantitative terms, wherever possible, and in addition an assessment in qualitative terms of the likely environmental impacts and benefits of the project. The general principles involve a description of the environmental characteristics, impact prediction, evaluation of anticipated impact, and proposed mitigation measures.

This comprehensive process underscores a serious approach to environmental protection. It provides control mechanisms over most environmental matters and determines conditions to be imposed in an environmental permit, including monitoring and audit requirements. These necessary but somewhat belated measures now have a far-reaching effect. The first and arguably most important benefit is that environmental matters are now high on the development agenda, and controls can be rigorously enforced. In terms of its impact on urban design, however, there is a case for stating that if it is not a matter of 'too little too late', it is rather 'too much too soon'. Given the almost complete lack of environmental controls until comparatively recent times, coupled with what might be termed a somewhat nonchalant disposition towards environmental discomfort on the part of the public at large, the rigid and somewhat bureaucratic approach to certain types of control is questionable. The constant juxtaposition of incompatible elements throughout the SAR through a combination of expedient development policies, poor regulation and often by sheer economic necessity, is difficult to overcome quickly. In many cases, it requires not merely environmental controls but stricter planning enforcement. Many aspects of environment are unacceptable, from land despoliation through container storage in the North-west New Territories, to quarries in vulnerable locations around the urban fringe. The needs of road and rail infrastructure are paramount and the most ambitious mitigation measures cannot possibly overcome all adverse effects. There has therefore been a tendency to apply an overly heavy-handed approach where an environmental ordinance can be applied to the latter, for example, the quantifiable aspects of noise impacts, often leading to convoluted design solutions and ugly mitigation measures, such as noise barriers, that tend to resolve one environmental problem at the expense of others.

The introduction of environmental impact assessment procedures, although not always integrated into the land use system *per se*, has nonetheless assisted in strengthening treatment of crucial land development – environment interface issues. Professor Peter Hills has stated that particular attention needs to be given to the cumulative impacts of development in Hong Kong, where there will always be potential conflicts between the need for growth and development, and the natural assimilative capacity of the environment.[6] A major aspect of this is the need for a more holistic approach whereby 'environment' is not merely a residual issue, but achieves a sustainable balance with development. A clear impact such as a harbour reclamation should, for example, be set against opportunities for a more genuinely sustainable urban area. While quantifiable aspects, such as problems of water quality and contaminated mud, can be tackled in various ways, landscape and visual impacts are more difficult to determine and therefore open to a more emotive public response. The answer here is surely that environmental and sustainable gains need to be integrated into the planning process. A piece of harbour reclamation that is perceived as destroying a significant historical resource in order to house elements that could be integrated elsewhere is clearly not sustainable. If, however, it is shown to meet public aspirations for urban improvement and a better environment, developed and sustained for the benefit of the community and as representing the nucleus of a better quality of life, then this is altogether a different picture. It is not a 'black and white' issue, but one with many shades of grey.

Genuine urban sustainability is most likely to be achieved when environmental interests and good planning come together to ensure the integration of a whole range of design aspects meaningful to the wider environment. In *Sustainable Settlements*, a guide for planners, designers and development in the UK,[7] it is stated that established statutory planning processes should provide the proper means by which public interests are defended, although there is clearly a need for meaningful and goal-oriented public participation to help achieve some of these ends. There is also, arguably, a need to encourage new and innovative financial and institutional mechanisms in certain situations that might deal specifically with sustainable urban design and civic considerations rather than the somewhat single-minded attitude prevailing at present. Among other things, it is necessary to ensure the following:

- proper mechanisms should be put in place, through a combination of environmental initiatives and the established statutory planning process, to ensure the best creative care for the physical quality of the urban environment;
- a proper level of investment should be made in public 'places', whether these are promenades, gateway parks, urban squares, plazas, water-based features or conservation initiatives. These are more important than grand architectural schemes or even some environmental controls to the public at large. This is less to do with the achievement of 'standards' than quality, related to the pedestrian environment;
- less rigid demarcation between different arms of the government, which deals with aspects of design and control of the public realm. This implies a greater level of synergy among those responsible for lands and financial matters, open space, cultural provision, planning and environmental control; and
- generation of community gain through investment in the public realm in terms of a variety of measures that contribute to a wider level of urban richness and vitality. In this sense, urban design should be utilized as a creative discipline to create lasting and sustainable urban forms in keeping with the wider environment.

▮ COMMUNITY COMMITMENT

The starting point for sustainable policies is not merely acknowledging a need but in establishing the state of the current environment in all its forms. Monitoring of air pollution and water quality, which is carried out at present by the EPD, needs to be extended to wider aspects of habitat through environmental audits and indicators. The *South China Morning Post* reported on 21 June 1998 that, while the EPD had been measuring ambient air quality since 1995 at stations between 17m and 25m high, the first roadside monitoring stations had revealed that air quality was up to 171% more polluted than previously indicated.

While scientific monitoring might inform the process, sustainable forms of design are essentially about urban values – that is to say, how much one aspect of environment is valued against others, and to what extent the creation and sustenance of environment is valued against other economic and social criteria. This should involve information from a wide range of sources, including planners, urban 'thinkers' and community groups, and would then be a resource not only for environmentalists and government agencies but also for development bodies. Better design and planning will promote resource efficiency. Hong Kong could in fact become a facilitator for environmental information, recycling renewable energy resources and technology transfer across the border. This in turn requires the setting up of urban development objectives and explicit systems for meeting these and for assessing their effectiveness. We must recognize that the implementation of sustainable policies, particularly those relating to the built environment, can never be achieved by reliance on government alone. In recognizing this, the 1992 Earth Summit recommended a coordination of interests through public, private and voluntary sectors working towards the production of integrated plans.

There is undoubtedly a high cost to growth. For decades, the annual increase in Hong Kong's GDP has been matched by a deterioration in the local environment and wastage of scarce resources. In the coming years, the way in which population growth is achieved is likely to be the most critical indicator of environmental health. In a recent survey of 6 000 people interviewed by the Social Research Department of the University of Hong Kong, almost 60% viewed the environment as worse than two years previously.

There is a need to set out an unambiguous and proactive environmental and planning agenda geared to the achievement of sustainability in the context of Hong Kong's future development. This represents an opportunity for some genuine public consultation aimed not so much at specifics, but at establishing what sort of place and character people want. There might be certain common goals with more detailed agendas for economic growth, culture, recreation etc. This would involve giving people the right sort of comprehensive information so that they can fully understand the implications of controls and procedures related to the built environment. This should act to inform the planning and development process itself, and in so doing generate forms of urban sustainability most suitable to Hong Kong's particular needs.

▮ CONCLUSION

A research team at the Centre of Urban Planning and Environmental Management recently completed a report[8] which shows that Hong Kong was gradually moving further away from a sustainable development path. With the pressures from a projected population growth that might rise from the current 6.6 million people to 8.3 million and beyond by 2016, there is an urgent need to provide an acceptable level of environmental quality through fundamental changes in the way we meet needs for land, transport and energy. New problems also tend to overtake old ones – no sooner has progress been made in cleaning up waterways than air pollution reaches new and formidable levels, largely due to industrial pollution drifting across from Guangdong and the continued use of diesel fuel for around 155 000 commercial vehicles on our roads. Although these constitute only 32% of the total number of vehicles in Hong Kong, they account for 62% of total vehicle travel and for around 98% of respirable suspended particulate emissions from vehicular sources. This is exacerbated by the continued use of low-grade diesel that can be purchased across the border or illegally within Hong Kong for around one-third of the price of high-grade diesel.

It is intended that a White Paper, 'Agenda 21' dealing with sustainability, will be issued in the year 2000, but establishing what can and cannot be resolved while ensuring economic growth and social development is a central issue. It is therefore about planning for the future. The government's commitment to construct up to 85 000 flats a year will necessitate significant new areas for development. Every year, Hong Kong will have to produce about 100 hectares of land just for high-density housing, associated uses and infrastructure. In developing such areas, there is also an opportunity to rationalize the building process, and produce a better standard of urban design and environmentally friendly building process and environmentally friendly buildings that are both safely designed and energy-efficient. Quality and efficiency of energy and materials are the key to ecologically sensitive design. This requires proper incentives to reduce capital and operational costs. At the same time, there is a need for the use of lease conditions to specify better building requirements and the use of the voluntary Hong Kong Building Environment Assessment Methods rating system intended to improve the environmental performance of commercial buildings. It might also be possible for the government to reduce land premiums for buildings that meet certain specified environmental criteria.

Finally, there is a need for a more considered approach to the planning and urban design opportunities presented. Hong Kong's urban renewal programme is formidable – around 950 hectares of existing development fail to meet reasonable standards of design and environmental quality. However, the 'market drive' approach clearly has a downside in that older mixed-use areas, which have imbued the city with both their character and sense of activity, are being swept away to be replaced by somewhat undifferentiated urban forms and single types of use. There is a need in the future urban renewal programme to cater for more of a revitalization approach, based on more incremental development to generate both physical and economic diversity, colour and vitality. Similarly, there must be a more adventurous and innovative approach to urban design in special areas to strengthen and enhance the attributes of the city – recreationally oriented new waterfronts, better improved public access to the harbour, good landscape and continuity of open space elements, and pedestrianization schemes for both new and existing areas. The opportunities for innovation need to reflect the many special processes that make up the urban environment. We now need to embrace a culture of quality and a design vision applied to the wider environment and the building industry for the benefit of the community as a whole.

▮ NOTES

1. *Urban Design Quarterly* 57 (January 1996) published by the Urban Design Group.
2. Girardet, Herbert. 1996. *The Gaia atlas of cities*. London: Gaia Books Limited.
3. Planning, Environment and Lands Branch, Hong Kong. 1996. *Heading towards sustainability: the third review of progress on the 1989 White Paper: pollution in Hong Kong – a time to act*. Hong Kong: Government Printer.
4. Yeang, Ken. 1996. *The skyscraper, bioclimatically considered – a design primer*. London: Academy Editions.
5. Environmental Protection Department, Hong Kong. September 1997. *Technical memorandum on environmental impact assessment process* (Environmental Impact

Assessment Ordinance, cap 499, s 16).
6. Hills, Peter. 1997. The environmental implications of land development. *Building Journal Hong Kong, China* (January 1997).
7. University of the West of England and the Local Government Management Board. April 1995. *Sustainable settlements – a guide for planners, designers and developers.*
8. Barron, William and Nils Steinbrecher, eds. 1999. *Heading towards sustainability? Practical indicators of environmental sustainability for Hong Kong.* Hong Kong: Centre of Urban Planning and Environmental Management (CUPEM), the University of Hong Kong.

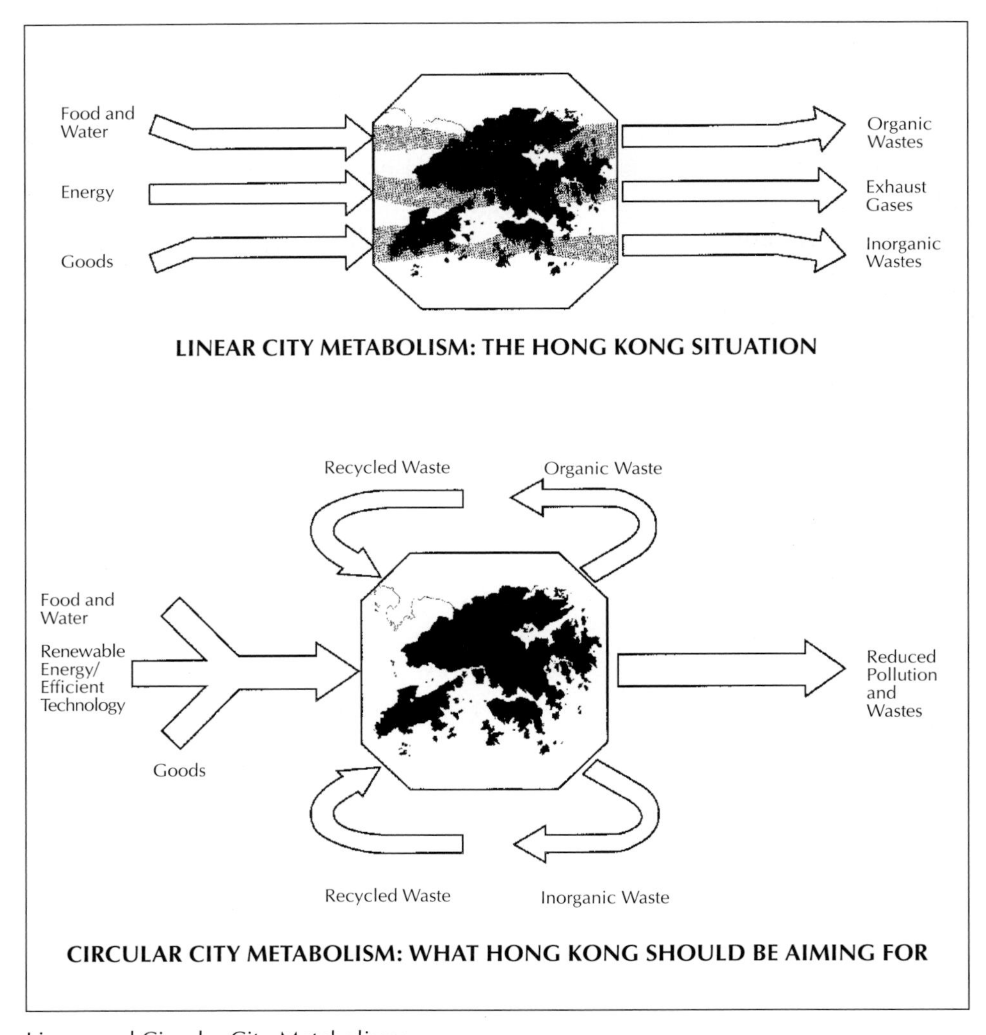

Linear and Circular City Metabolisms

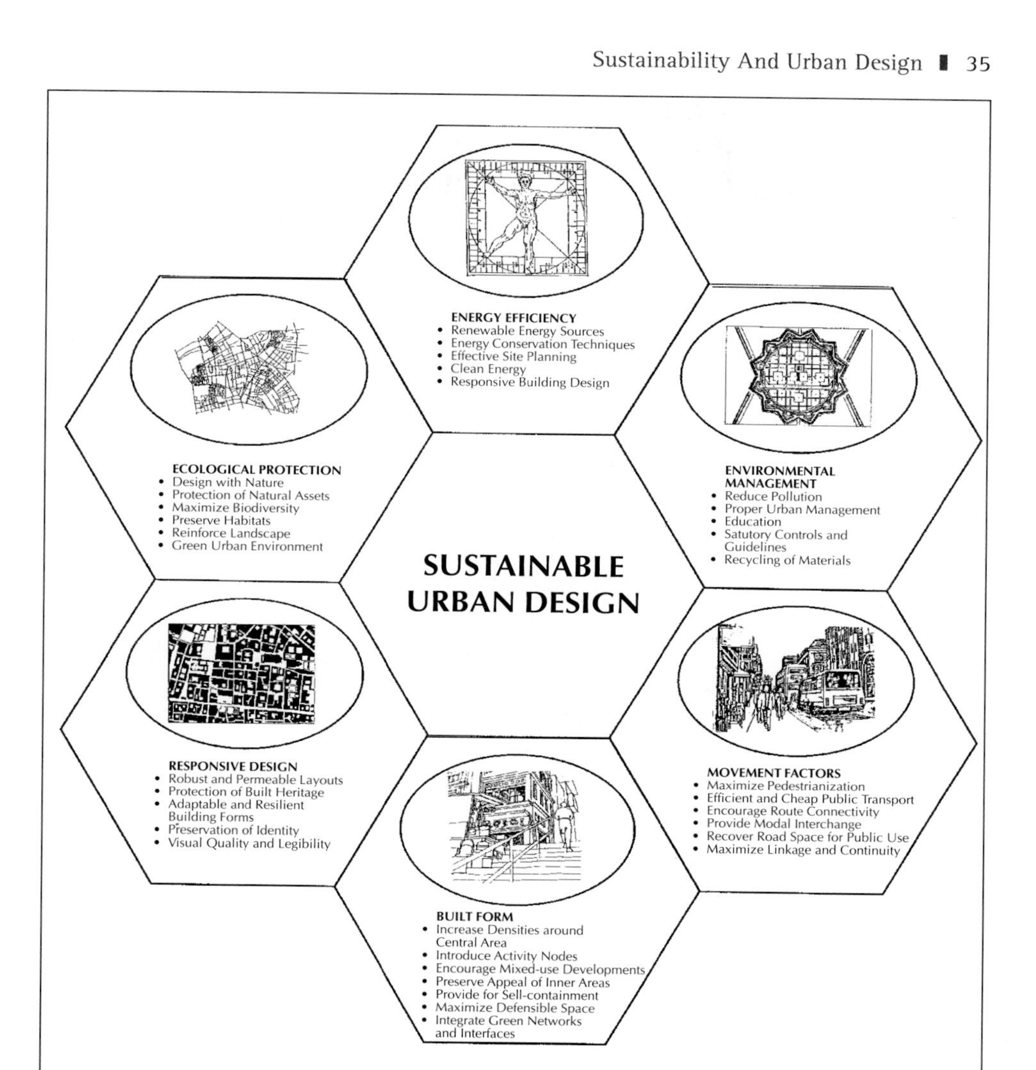

Sustainable Urban Design Components

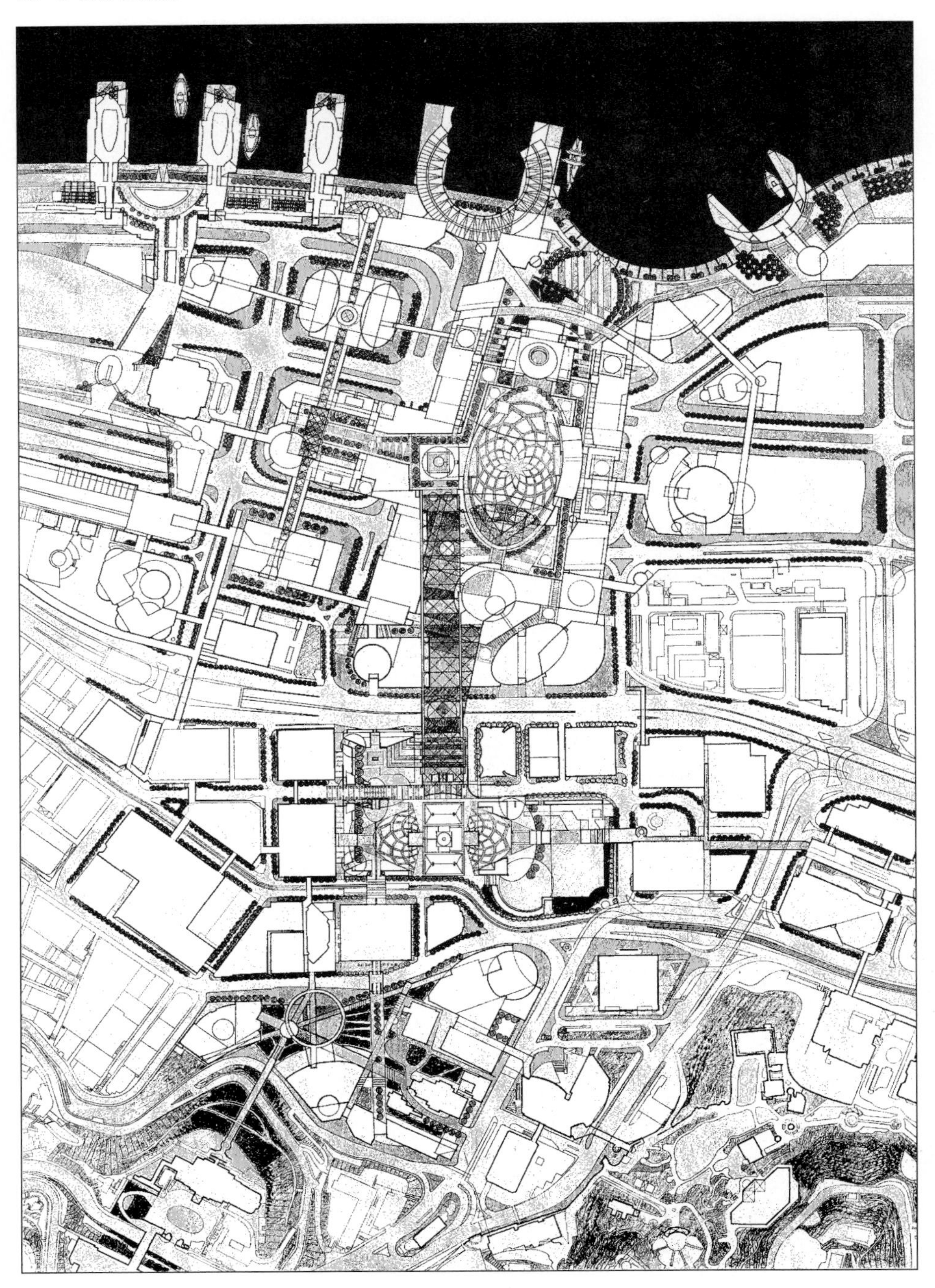

Proposed extension of Statue Square as a corridor of civic buildings and open spaces. This represents part of urban design proposals and parameters for the wider reclamation area.

Proposal for tourism and waterfront development in Causeway Bay.

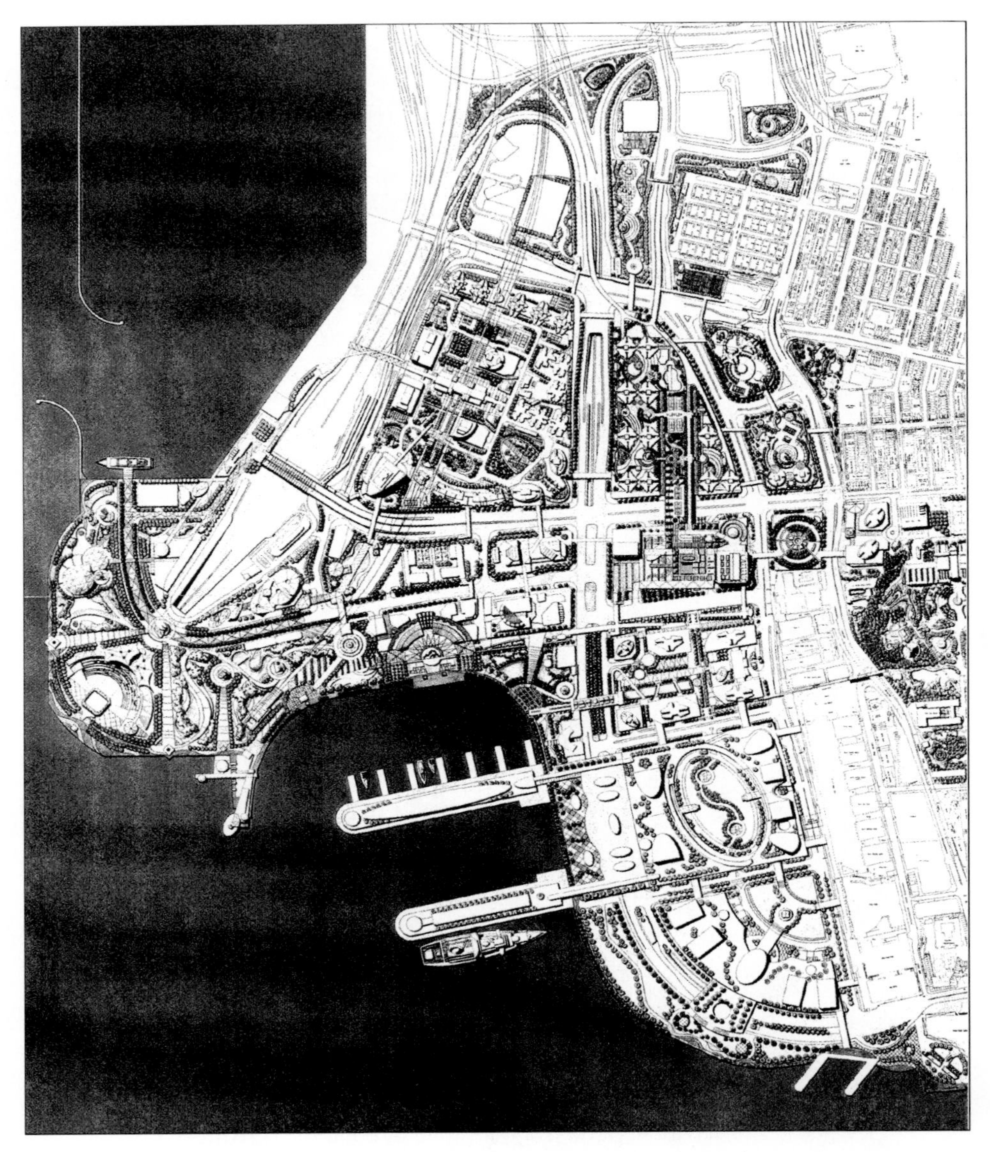

An urban design framework for West Kowloon, as a means of coordinating various major waterfront open spaces, station developments, cruise terminals and pedestrian links with Kowloon Park and core Tsim Sha Tsui.

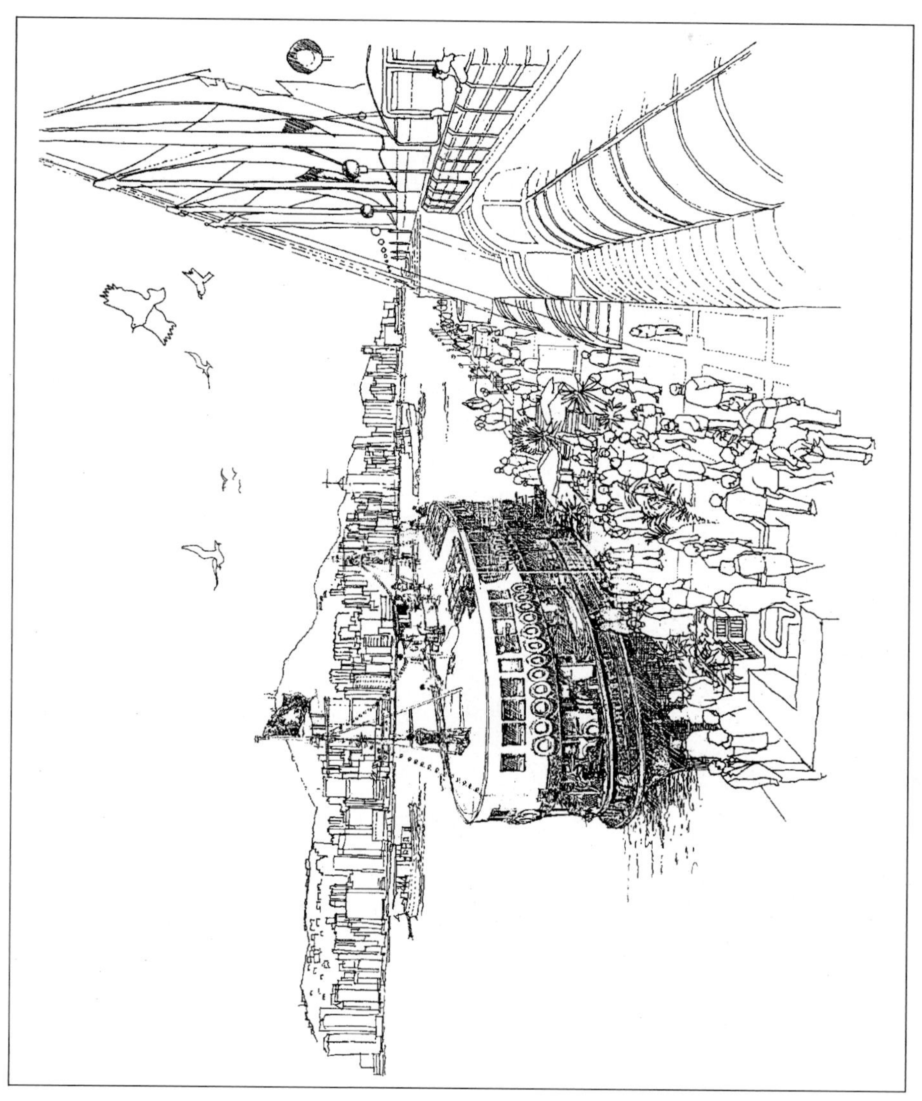

Proposed Promenade Design, West Kowloon

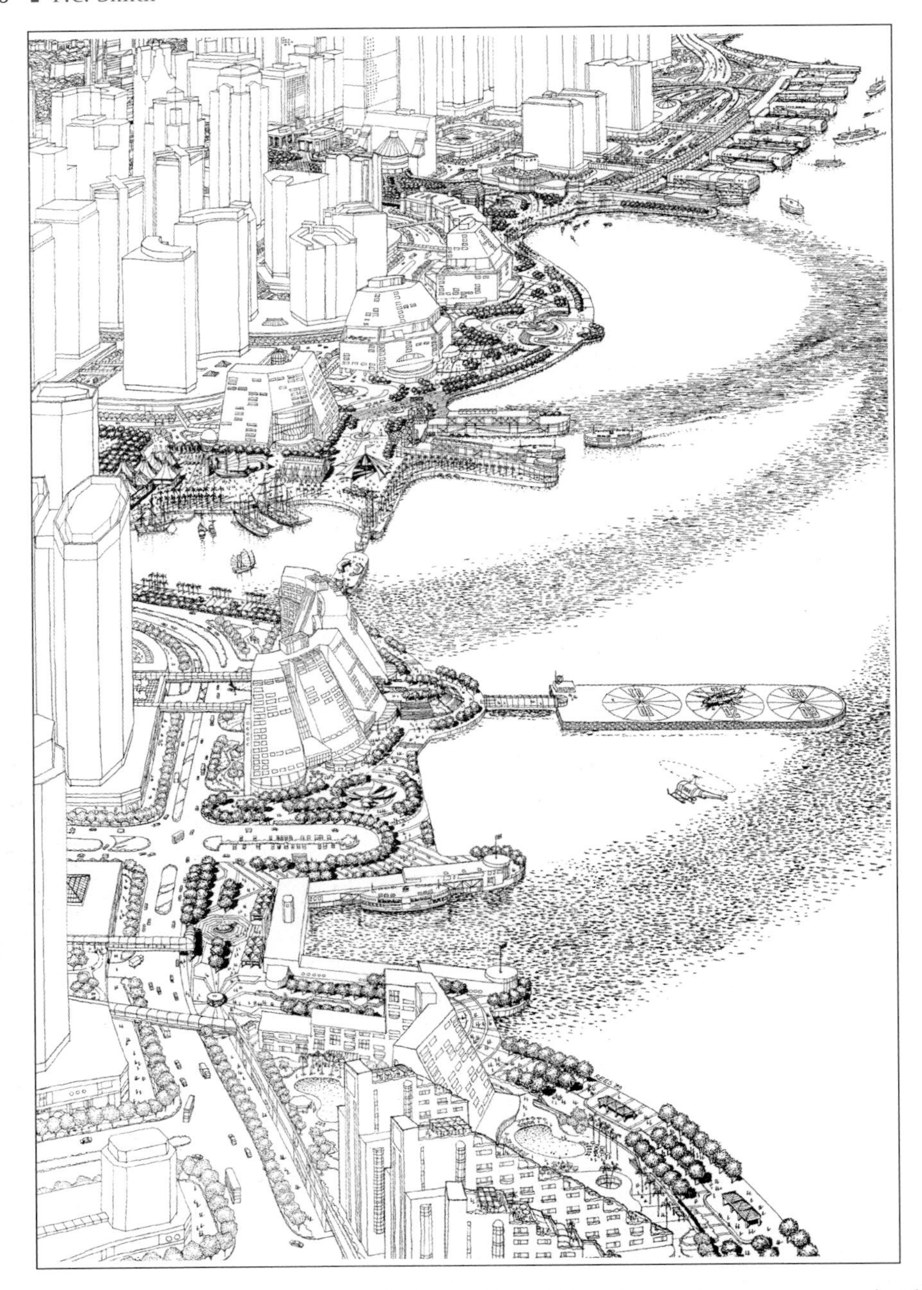

Sketch of Central/Wanchai Waterfront for initial Central/Wanchai Reclamation proposals. The layout is based on linkage and continuity of pedestrian elements and spaces tied back into the existing urban area. The waterfront itself is intended as a continuous animated sequence of recreational activities, water basins and facilities, such as festival market attractions, with points of reference related to the harbour.

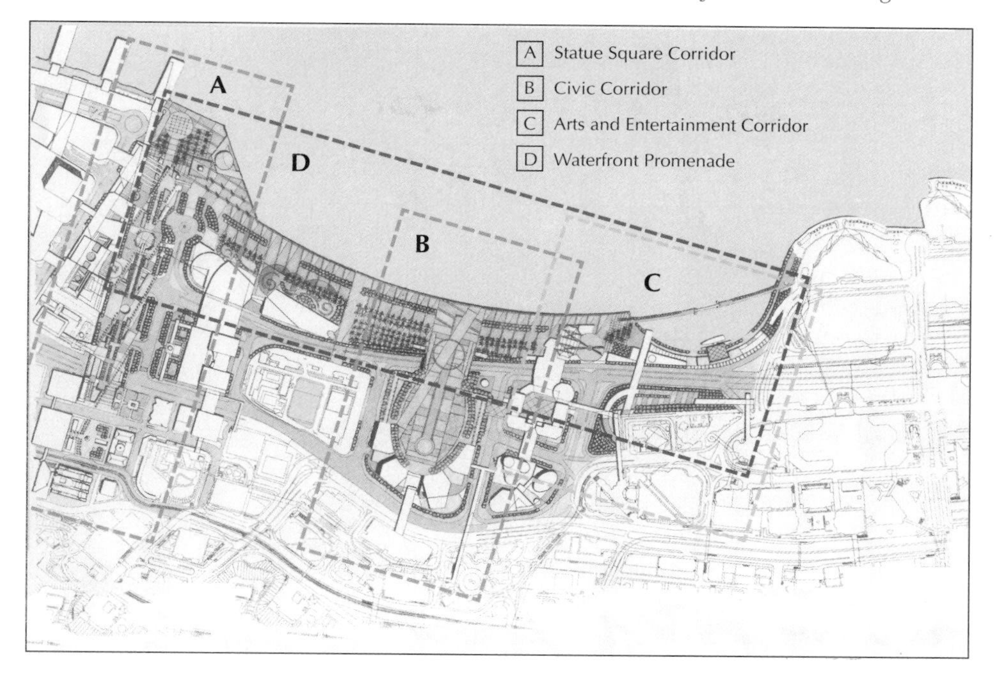

The New Central Waterfront is composed of four main elements. These comprise the Statue Square Corridor, the Civic Corridor, the Arts and Entertainment Corridor and the Waterfront Promenade. These will form distinctive components of the New Central Waterfront and will each imbue an individual identity and function which will enhance visual interest and variety.

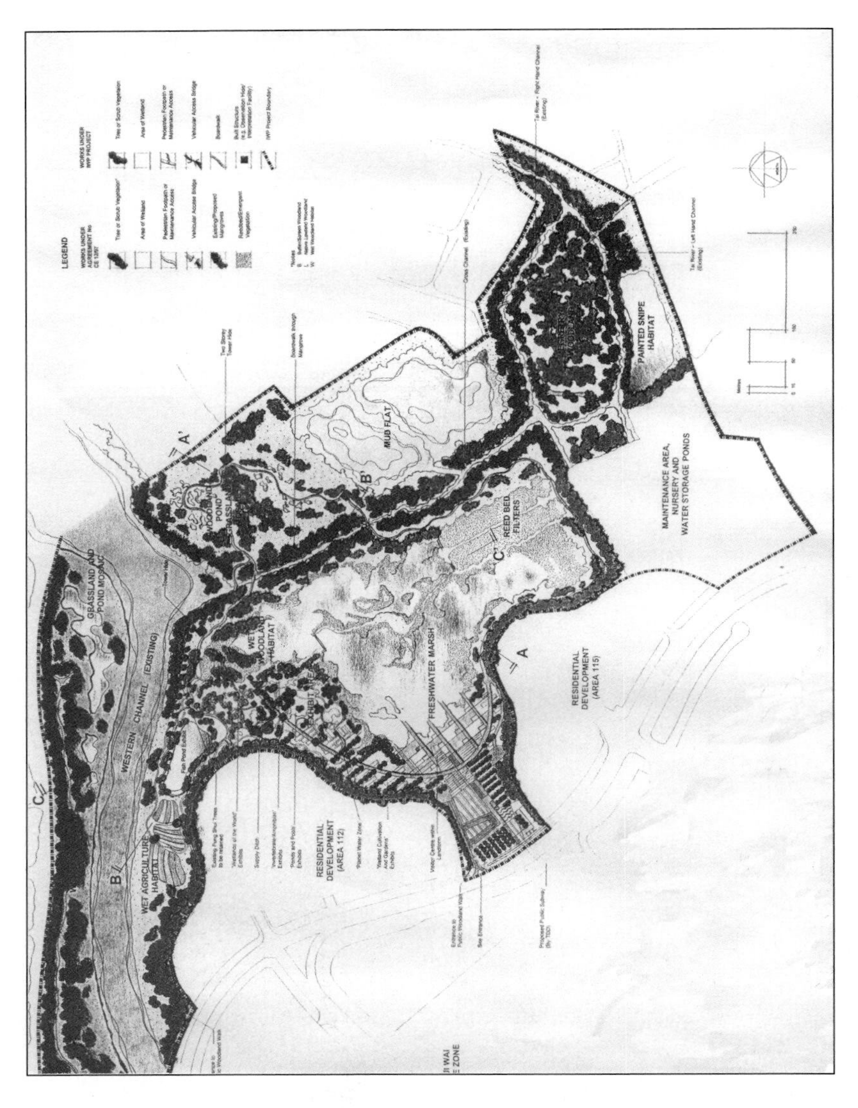

Proposal for wetland restoration, International Wetland Park and Visitor Centre, on the northern edge of Tin Shui Wai. The masterplan is divided into salt water and fresh water wetlands. The fresh water wetland will include a lake and fresh water marsh, wet woodland, reed bed, habitat for painted snipe, egretry, and small ponds for amphibians and dragonflies. The salt water wetland, which is rather smaller, will comprise mangrove, tidal creek and intertidal mud flats.

2

Town Planning And The Environment: Role And Tools Of Private Consulting Planners

Bo-sin Tang and Andrew Siu-lo Lam

INTRODUCTION

Hong Kong is both a beneficiary and victim of its incessant development. Rapid economic growth brings with it significant improvement in material wealth that is enviable by many developing countries. However, our remarkable success also leaves noticeable traits of environmental damages (see Table 1). Some people say that, as there is no such thing as a free lunch, poor environmental quality is the price we pay for our actively growing economy. This is definitely incorrect. Development and environment are not necessarily at two opposing ends. Specifically, the relationship between property development and environment can be in the form of a virtuous cycle (see Cairncross, 1993). Property development creates value to society. Such wealth can then be fed back into various environmental improvement provisions that further enhance the property development values. As such, it is not justified to regard environmental measures as simply a waste of resources that does not generate returns in the end.

Table 1 Pollution complaint cases in 1996

Type of pollution	Number of complaints
Air	6 160
Noise	7 441
Liquid waste	1 164
Solid waste	535
Water	83
Miscellaneous	9

Source: Information Services Department, Hong Kong Special Administrative Region (SAR) government

This chapter will briefly illustrate the role of private consulting town planners in handling environmental issues embodied in private sector development. Planning is the very first task in the inception stage of any property development project. In Hong Kong, town planning control is at the forefront of the government's development control system. Most often than not, there is a need to first obtain planning approval from relevant public authorities before the development scheme can proceed further. In many of these schemes, private town planners play an important function, albeit few people may notice, in the application process.

▌ TOWN PLANNING AND DEVELOPMENT

Private property development usually starts off with this question: How to make the best use of the subject site to obtain the highest possible returns? Different professionals may approach this question from varying perspectives and come up with different answers. For instance, an architect may be most interested in formulating an aesthetically attractive building design scheme. On the other hand, a property valuer may tell how much the space is currently worth and advise whether it is appropriate to sell, retain or develop the site. As far as town planners are concerned, they are most interested in the land use, density and layout design of the development of the site.

Within the profession, town planning *(chengshi guihua)*, land use planning *(tudi liyong guihua)* or simply planning *(guihua)* are often used as equivalent terms. Nevertheless, it must be pointed out that planning can perform very diverse functions in different modes of society. Planners in a socialist economy would obviously have a greater authority in many aspects of social life than their counterparts working in a capitalist society.

Hong Kong is often praised as a *laissez-faire* economy in which market decisions dominate. Within this context, town planning is primarily concerned with managing the use of land. The purpose of planning is to optimize land and property development so as to 'promote the health, safety, convenience and general welfare of the community' (Town Planning Ordinance, Chapter 131, Laws of Hong Kong). The ultimate goal is to achieve social and economic progress of the whole society.

Despite its emphasis on land and property, planning is less physically-oriented as it appears. It is often said that a planner's role is to promote the right development at the right place at the right time. This task calls forth a lot of value judgement on what kinds and forms of development are right, less right or wrong. There is no simple and clear-cut answer. Quite often than not, land use planning has to take into consideration various social, political and economic factors, in addition to the physical and technical matters that include the environment. It is indeed about identifying reasonable solutions and making choices rather than reaching an absolute perfection.

Town planning is always regarded as a government activity. This is valid because only the government has the authority and mandate to look after the public interest by planning ahead for the entire city. In Hong Kong, a significant proportion of the planners work within the public sector. Their functions cover a wide range of activities between forward planning and development control at strategic, district and site-specific levels. Planners, on the one hand, are involved in the preparation of territorial and sectoral development strategies, and formulation of general planning standards, guidelines and policies with long-term and extensive impacts. On the other hand, they participate in establishing detailed planning parameters for various development sites.

Despite this, planning is not a monopoly of the government. Hong Kong, as a market economy, relies much on the development initiatives of the private sector. The built environment is as much shaped by private sector development forces as by government planning. What government planners try to do in this respect is to guide private sector resources to planned development areas and to control undesirable development.

Undesirable development, however, has a definition governed by the passage of time, location and, most importantly, the changing values of the society. Government planning, no matter how well it is formulated, does not necessarily reflect all the private market considerations and the best welfare of the community. Although government planners regularly review town plans in the light of prevailing circumstances, private consulting planners still have a role to play in the property development process. Our development control system provides the flexibility for possible deviations of property development from the planned provisions in town plans. It contains provisions for various applications from the private sector to seek changes (see Planning Department, 1995). Through involvement in development applications, private consulting planners help to articulate and achieve what is the 'right' development at the 'right' place at the 'right' time.

▮ LAND USE PLANNING AND THE ENVIRONMENT

In a broad sense, property development is about manipulation of the environment to suit human needs. Most people, perhaps, may not realize how valuable our environmental asset is until, one day, it becomes scarce. Indeed, the quality of some of the free gifts from nature is not as good as before. To some extent, the entire community is paying for the damages it has indicted. Supported by increasing affluence, society is now demanding cleaner air and water, protection of wildlife habitat and natural woodland, and elimination of various kinds of environmental pollution. Environmentalism has become a social movement with the emergence of major green interest groups in many Western countries. The establishment of the separate Environmental Protection Department in 1986 to specifically handle environmental control and management was a positive response of the then Hong Kong government (now the Hong Kong SAR government) to this worldwide trend (see Tang, 1994).

Arguments for and against development do not subside with the setting up of a government environmental watchdog. Some people say we have to consume less and build less in order to sustain our environment. Others say we should keep developing and make whatever remedies we need in order to sustain our economy. In a pro-development society like Hong Kong, the latter argument is particularly appealing. Strong development pressure from the private sector in this small territory implies that the government is often forced into a defensive, guarding position. Under the auspices of town planning, for instance, government planners always take up the role as regulators of private development. In determining the overall development pattern, land use planners perform a key role as 'environmental goalkeepers' in safeguarding the last frontier, and implement necessary measures to channel development to less environmentally sensitive areas. Similarly, environmental planners in the government perform the function of controlling the environmental impact of the proposed land projects.

There is nothing wrong for the government to undertake the regulatory function. However, the over-dominance of this role often creates an undesirable impression that private development and environment are two opposing goals. A related ideology is that private developers must be greedy, unscrupulous money-grabbing animals that will seek every opportunity to exploit negatively the natural environment to fulfil their selfish motives. This is, of course, a biased view. There is no reason why property development initiated by the private sector cannot be environmentally friendly. As pointed out in the earlier section, sensible environmental measures can surely add value to the development projects which work to the ultimate benefit of society. Property developers are often willing to incur the costs to meet this end. To encourage this outcome, there is a need for greater flexibility within our planning permission system and for more effective communication between the government and the private development sector to work towards the common goal.

Within the property development process, planning provides an ideal stage at which environmental considerations can be incorporated in property projects. There is always more room for manoeuvre in a less constraining scenario. As such, land use planners will always find it easier to plan ahead and improve the environment for development of a greenfield site than within a developed context.

How does land use planning help to achieve environmental improvement? Consider a place where there are only three types of land use illustrated by different colours. The area covered by water is painted blue. The undeveloped area covered by countryside, woodland, farmland and the likes is coloured green, and the built-up area is coloured grey. Suppose over 60% of the plan is painted blue, only 6% is grey while the remaining is coloured green. The planner faces this situation: how to accommodate more development? What are the options? Could we have more grey by taking a bit out of the green or the blue? Or should we make the grey area a bit darker by increasing its density? Or should we have both?

Actual planning exercises are more complicated than the above illustration. However, it suffices to point out the critical elements in a planning process: formulation of choices and evaluation of these options. Different options will have varying implications. There can be no perfect solutions. A rational, comprehensive approach is often said to include a series of tasks such as problem identification, comprehensive investigation, option formulation, evaluation and recommendation (Figure 1). In essence, the land use planning process is an analytical course through which the costs and benefits of actions taken on land can be examined comprehensively to facilitate rational decision-making (see Meyerson and Banfield, 1955, p. 314).

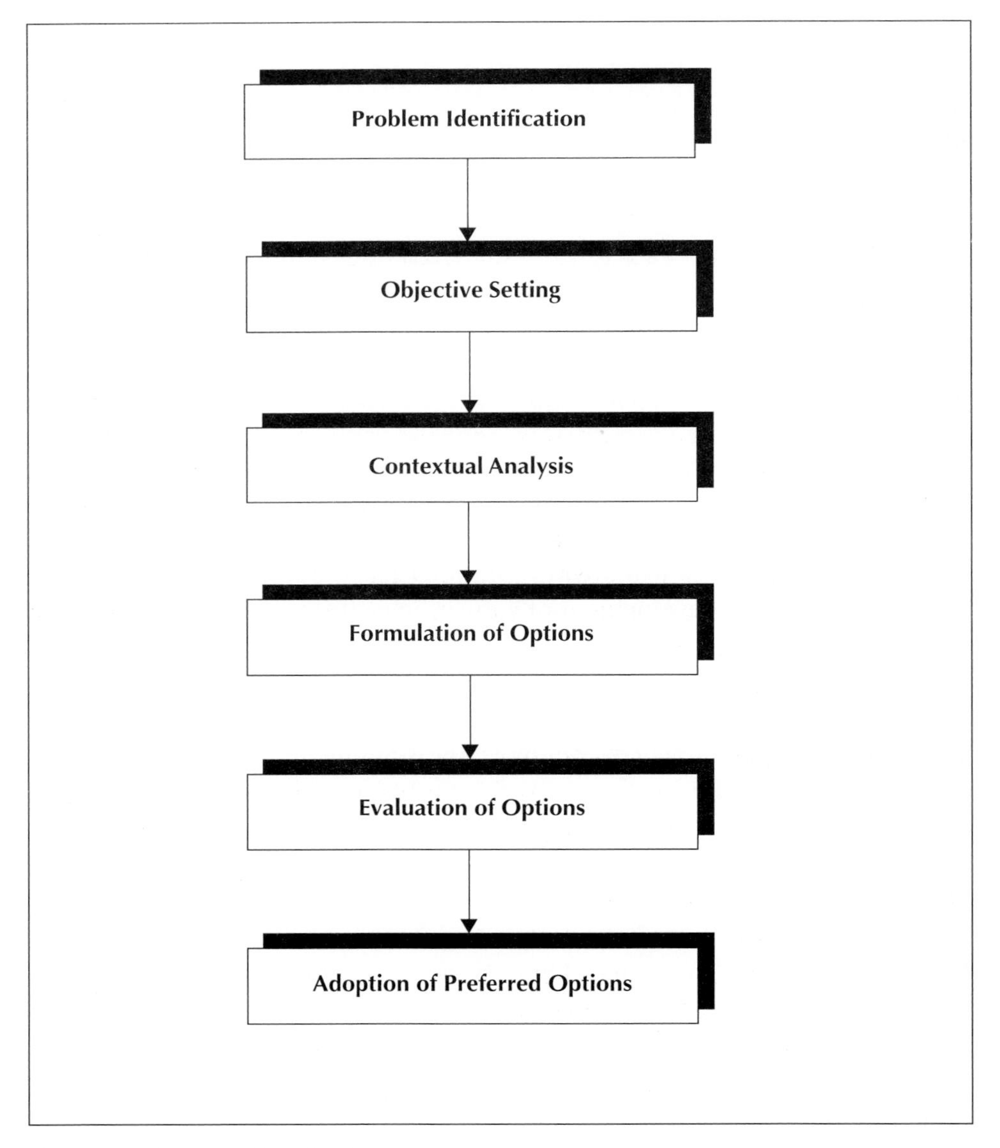

Figure 1 A Rational Planning Process

Land use planning has to be a multi-disciplinary activity. Environment is one of the technical aspects to be considered in property development. Other relevant sectors include transport, drainage, sewerage, civil engineering, community impact, and so on. After all, environmental standards and policies are derived from the basis of both subjective and objective criteria. It is a collective expectation that establishes policies and then standards. Planners work within the framework set by these policies and standards but are not entirely bounded by them. Planners are supposed to be forward-looking and visionary in the discharge of their professional duties. Policies and standards are subject to alterations facing the ever-changing social, political and economic circumstances. So are planning solutions.

▮ WHAT PRIVATE CONSULTING PLANNERS CAN OFFER

Private consulting planners are particularly sensitive to market changes because of their daily business encounters in the private sector. In Hong Kong, private professional planners often perform the role of consultants to property developers, land owners, interest groups and public bodies. They provide a wide range of land-related services including planning surveys and studies, strategic planning, master layout planning, planning advocacy service, development advice, and so on.

When private planners are invited to engage in property development proposals, they are always serving two constituencies. First the client, and second the professional body, and these two constituencies are of equal importance. On the one hand, the planners are appointed to represent the client's private interest for development. As such, they are undertaking the position as planning advocates in support of the client's development scheme. On the other hand, most planners are often members of the professional town planning institutes. In Hong Kong, these include the Hong Kong Institute of Planners and the Planners Registration Board. Hence, they have to adhere to the professional ethics requirement imposed by the relevant professional bodies. This sort of dual role is common for professional practitioners such as lawyers, surveyors, etc., and is therefore not unique to the planning profession.

Private planners' dual role as both advocates and professionals requires them to strike a balance between self-interests and collective requirements. In the process of planning for private property development, they look for preferred solutions out of many feasible options.

Environmental impact is definitely one major aspect for consideration in the course of property development planning. Nonetheless, planners are not omnipotent. Most planners are not environmental scientists who can master the technical aspects of environmental problems. As far as property project planning is concerned, planners are often most interested in manipulating three key elements in development schemes with a view to influencing the extent of environmental impacts so arising. These three elements are land use, development intensity and built form (Table 2).

Table 2 Land use planning considerations

Planning elements	Key sample considerations
Land use	• What sorts of activities should be accommodated within the site? • How do these activities relate to one another in a spatial and temporal sense? • How do the proposed uses of the land relate to other sites within the planning region, etc.?
Development intensity	• What is the preferred plot ratio of the development? • How much gross floor space is provided? • How much of the site is roofed over, etc.?
Built form	• What should the building height be? • What is the layout design of the development? • How does the built form take care of its surrounding environment, etc.?

Many existing environmental problems are caused by inadequate considerations on these elements during the initial planning of the property development. For instance, air and noise pollution are likely outcomes due to development of incompatible land uses in adjoining sites, for example when a residential building is developed within an old industrial area. Environmental pollution will also arise if the development intensity exceeds the capacity of the sewage treatment plant in handling the volume of effluents generated by the property. Inappropriate built form of a real property in a conspicuous, exposed location may create visual intrusion. Many of these problems could be avoided in the first instance if due care is exercised at the time the project was initially planned.

What specifically can professional planners contribute to a property development project? Regarding environmental protection, consulting planners can provide the following services.

Comprehensive Appraisal Of Environmental Impacts

Most land use planners are not environmental specialists; generally they are not trained to be capable of carrying out very technical, scientific assessment of environmental problems. However, their professional training and working experience equip them with sufficient knowledge to provide a preliminary, yet comprehensive appreciation of all the possible environmental issues arising from a development proposal. These issues relate to the nature of land use under consideration, the surrounding environment and development context of the subject sites. Planners can manipulate the contents of development proposals in order to reduce harmful environmental impacts (Figure 2) or propose possible remedial measures to enable the development to proceed. A preliminary review of possible environmental problems is definitely valuable. This is because, depending on the extent of investigations, detailed environmental impact assessment (EIA) studies take time to complete and can be costly to private developers. A planning appraisal therefore helps to define the scope for more detailed assessment and economize costs of development. It can sometimes give very constructive recommendations at the outset of project inception, especially when a development proposal is in apparent conflict with the prevailing planning intention.

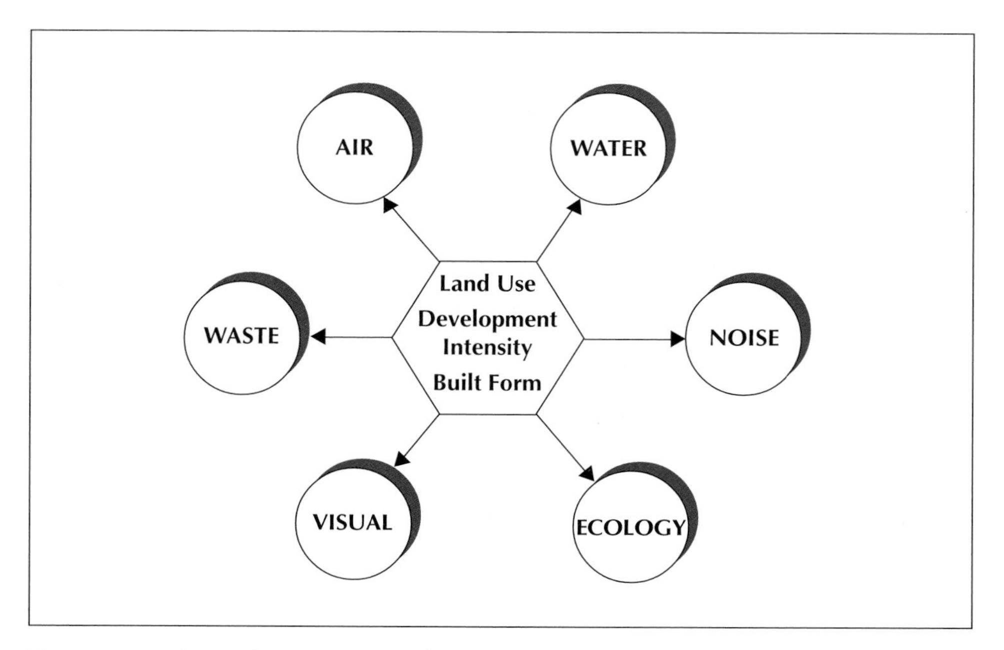

Figure 2 Land Use Planning Appraisal on Environmental Elements

Identification Of Environmental Consultants, And Teamwork

This aspect of work follows the previous task. A preliminary planning appraisal helps to identify the key environmental concerns that warrant further study. Professional planners can assist in identifying the appropriate environmental specialists to conduct the necessary investigations. Nowadays, environmental concerns have a very wide coverage and there is also a general trend in the profession towards more finite specialization. For instance, visual impact, terrestrial and marine ecology, archaeological, historical and cultural heritage are emerging environmental areas in Hong Kong. Specialist consultants are required to handle these areas. Planners can perform the role as project coordinator in the team.

The ultimate objective is to ascertain that all the environmental problems are not insurmountable. Their technical studies will help to systematically quantify the possible environmental impacts and prepare viable mitigation measures. As development planning is an iterative process, there is bound to be a lot of interaction among planners, environmental specialists and other professionals in this stage of work. In a team, they work closely together to identify the critical environmental issues, establish design principles and guidelines, try out different development options, testing environmental implications, revising options, and so on.

Formulation Of Planning Solutions

Planners can help to mitigate adverse environmental impacts by modifying the nature, scale and design of development proposals. These three components are related, and numerous planning options can be generated depending upon the actual circumstances. Solutions for a particular technical issue may create problems in other technical aspects. Thus solutions embodied in development planning are often in the form of a package rather than as independent means. It is beyond the scope of this chapter to outline all the planning methods and solutions. From our experience, these methods can be grouped under a number of themes as follows.

Preservation

To preserve is to maintain the status quo of the environment. During the course of development appraisal, planners may in certain situations identify some environmental assets within the subject site that should best be retained. These assets, for example, may refer to vegetation covers with significant ecological values. If development were to proceed, these environmental assets might be lost. It appears that the logical outcome, under these circumstances, is to avoid development at all. Nevertheless, few environmental assets are entirely free from interference. Even if left untouched, their environmental quality may deteriorate and the status quo will still degrade over time. Thus the act of preservation necessarily requires some deliberate human action to intervene in the environment, just like preserving an Egyptian mummy necessitates the use of preservatives.

In many cases, town plans and other relevant planning documents show clearly where the environmentally sensitive areas are. These include country parks, marine parks, nature reserves, sites of specific scientific interests, coastal protection areas, countryside protection areas, sites of archaeological, historical and architectural interests, and so on. The value of these designated areas is widely accepted, although they need to be further substantiated. There are far more isolated valuable spots yet to be identified. A seemingly simple task, such as the retention of a mature tree within a development site, could sometimes present a major challenge to site layout planning of a development.

Conservation

To conserve is to achieve a meaningful retention and protection of key environmental features of a site. To some extent, it represents a lesser degree of preservation, and should be considered when there are clear indications of natural degradation if neglected further. Certain important natural environmental assets can be deliberately taken care of or even improved to enable sensible human uses. Within a large development site, for instance, consideration may be given to retain rather than level off a greenery knoll to make it a visual feature. Further landscaping to blend with minor foot trails could be planned on the knoll to enable passive recreation uses. This is probably more desirable than leaving the knoll undeveloped.

Conservation is also appropriate for historical monuments, of which routine management and regular maintenance are absolutely essential for their upkeep. Conservation, if used on its own, can be quite costly. One associated means to recoup the costs is the transfer of development potential. Such a transfer can be exercised in the form of land swap or intensification of development scale on another piece of site held by the relevant parties (Figure 3). However, flexible land administration, planning and building control systems would be required to facilitate application of this approach.

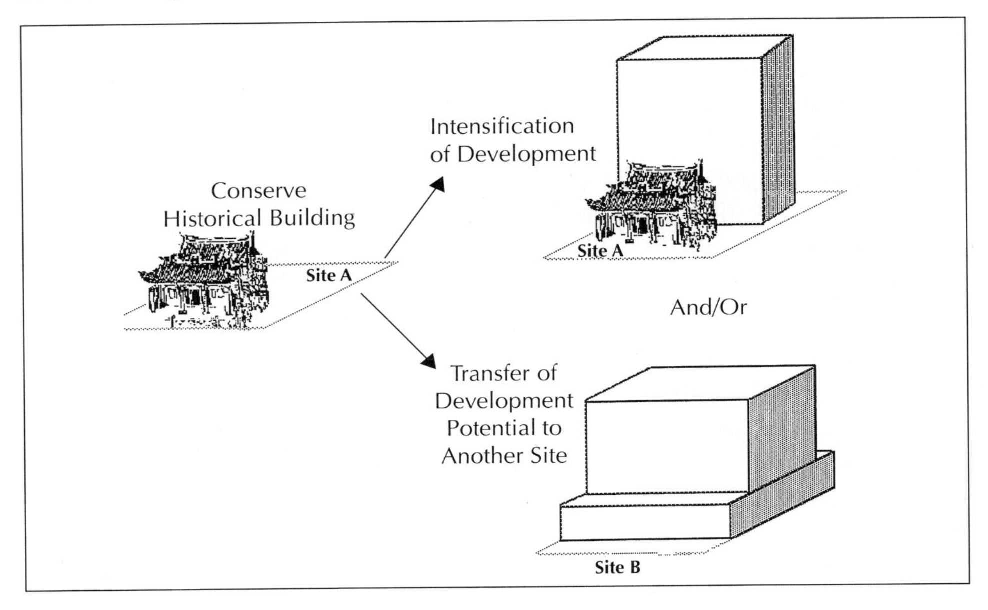

Figure 3 Conservation

Segregation

To segregate is to separate incompatible activities in a spatial and temporal sense so as to minimize negative interference. Land use zoning is the most common method used by planners. By grouping similar land uses in one zone and by segregating different zones geographically, planners can achieve environmental harmony in developments (Figure 4).

A more popular feature in urban Hong Kong is the adoption of vertical segregation (Figure 5). A more intensive use of airspace and underground areas is a widely accepted means of land utilization in a multi-layer city. Vertical land use zoning provides the benefits of both segregation and integration of a variety of land uses. Although its application is not without drawbacks, it represents an effective solution to minimize possible conflicts among different land uses in a compact city environment. Obvious examples include the use of lower floors of residential blocks for commercial purposes and the extensive use of podium decks to segregate flows of pedestrians and road traffic.

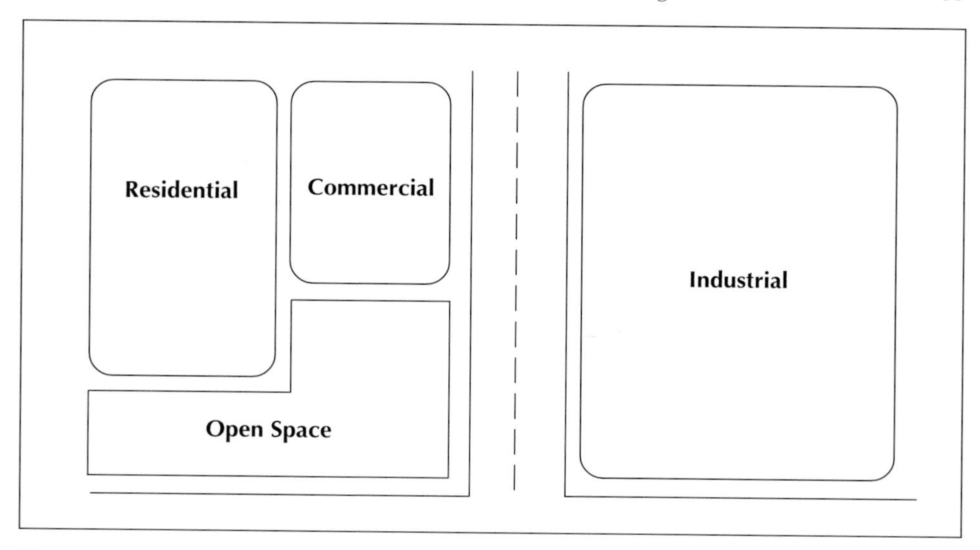

Figure 4 Horizontal Segregation

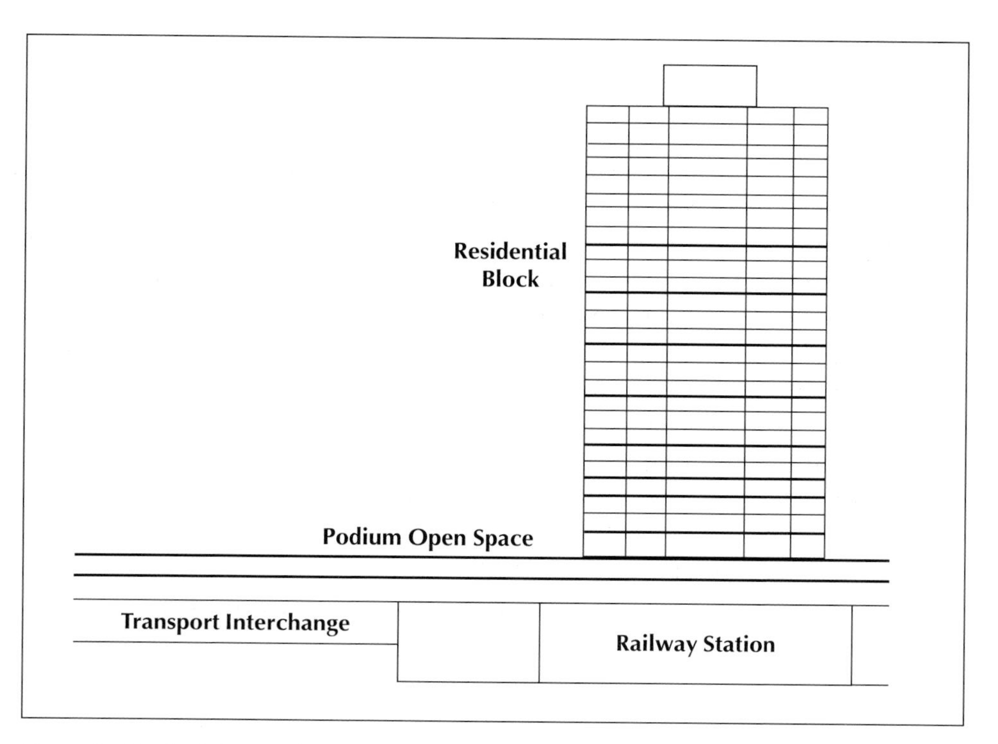

Figure 5 Vertical Segregation

Buffering

This is also another tool used in managing land use and built form. To buffer is to put something in between incompatible spatial activities. A buffer can be in the form of physical structures such as visual screens, noise barriers, etc. It can also be land uses, such as an open space strip separating a flyover and housing development (Figure 6).

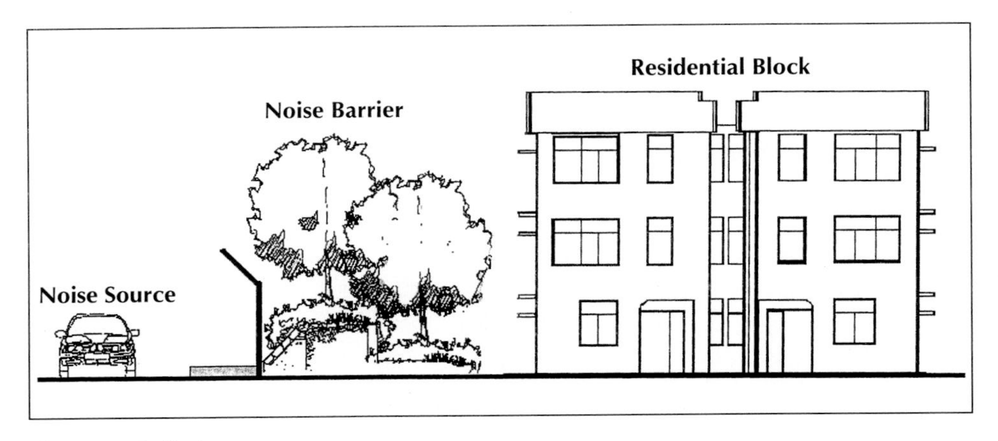

Figure 6 Buffering

The application of this tool depends very much on the spatial setting. Where land space is abundant, *buffering* together with *segregation* can be used to achieve a desirable land use pattern. Planning of greenfield rather than brownfield sites is always easier. Conversely, buffering could result in sterilization of valuable land resources, and thus significant economic loss to the community if no beneficial uses are identified on the buffer land. In some cases, the concept of buffering is applied on a large scale. The Deep Bay Buffer Zones in Hong Kong is a good example. These designated buffer zones are development bands within which a restricted range of land use activities are allowed in order to protect the ecological integrity of the Deep Bay area.

Integration

To integrate is to maximize the synergy effects and enhance mutual benefits by a spatial juxtaposition of compatible uses. This requires a careful study of activity patterns and requirements generated from different land uses. Modern property development of a reasonable scale normally contains more than one single land use component (Figure 7). Vertical segregation of land uses in a sense can be regarded as a form of horizontal integration within a development site. Examples of successful integration of land uses may include kindergartens within major housing developments, a car park within commercial blocks, etc. Successful projects sometimes require imagination. In the UK, for instance, a sewage treatment facility has been turned into a visitor attraction spot under a planned 'edu-tainment' (education-cum-entertainment) project. This shows how integration can be achieved with the application of creative thinking.

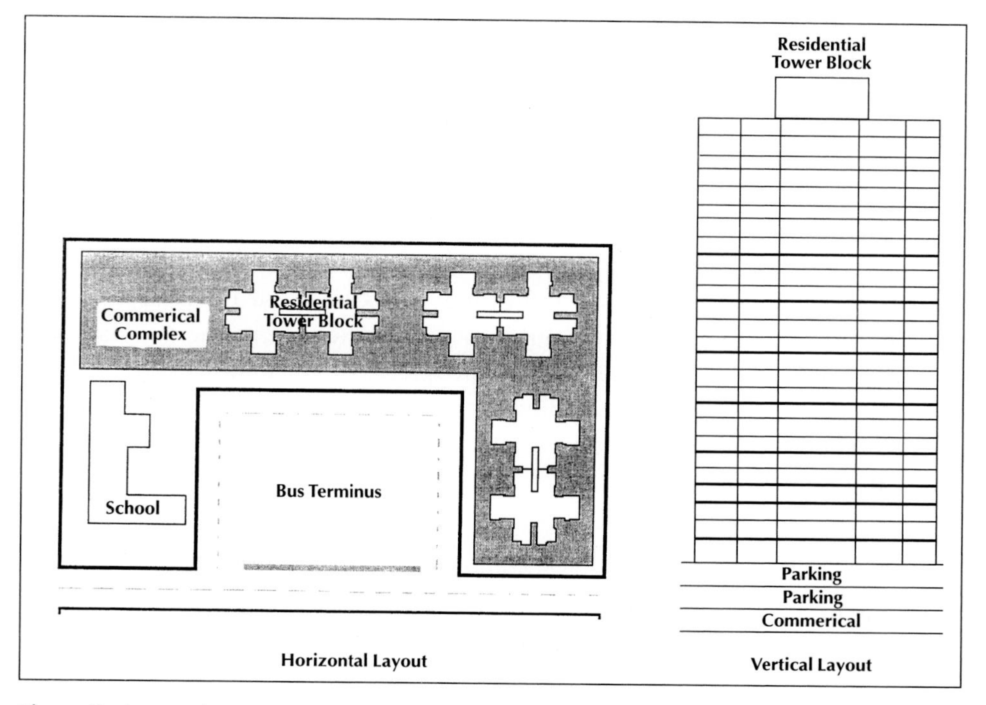

Figure 7 Integration

Compensation

Planners may sometimes run out of methods to eliminate or minimize adverse environmental impacts arising from development proposals. The existing environmental assets have to be sacrificed if the development proposals are to proceed. Provided that sufficient justifications are made to support implementation of the development, compensation may be a suitable measure to restore and replace the environmental assets affected. For instance, tree compensation is a rather popular measure in local development projects when cutting of existing trees is required. The same principle may also be applicable to development within ecologically sensitive areas. However, compensation is a remedial action. It should be used only as a last resort as loss of some environmental assets could be permanent and hence irreplaceable. Detailed assessments must be undertaken before implementation of compensatory action.

Sustainability

Sustainability has become the buzzword for development in the 1990s. The United Nations' definition of sustainable development is 'development that meets with the needs of the present without compromising the ability of the future generations to meet their own needs' (PELB, 1993). This tends to give a strong notion of environmental protection and resource-saving. An emphasis on sustainability of developments implies a heightened awareness about 'capacity' – a threshold level of the environment beyond which more development would become excessive and harmful. Environmental impact generally increases with development intensity. Hence, reducing development density is a means to lessen the environmental stress. For instance, a higher level of resident population may put a strain on sewage treatment capacity in an area. Planners cannot direct where people live, but planners can control the intensity of development. By lowering the total floor space or the number of flats in a development, planners help to exert their influence on the design population level.

Communicate Outcomes Of Assessment And Feasible Solutions

It is essential to explain clearly to interested parties the EIA results and the preferred solutions prepared by the team. These parties may include the relevant public authorities responsible for development approval and other groups affected by the proposed development scheme. As most of these people are not environmental specialists, the information must be conveyed in layperson's terms. Professional planners are often well suited to take up this task.

▌ CONCLUSION

This chapter seeks to demonstrate how private consulting town planners can contribute to enhance the environment in private sector property development projects. Planning in general is a conscious, systematic process to identify problems, find out solutions, study alternative proposals and recommend preferred options. When this is applied to the town planning profession, the objective is to achieve the right property development to satisfy human needs. This appears straightforward, but in fact the process involves a lot of subjective judgements and conflicting arguments about what is right or not right. Another problem is that planners in a market economy like Hong Kong do not possess all the authority to control other people's activities. In other words, planners' ability to plan is seriously constrained. Compared with their government counterparts, the power of private consulting planners to plan is less.

Nonetheless, planning provides an ideal stage for serious thinking before action. This is particularly needed in property development projects, in that all the environmental issues can be addressed before construction starts. Professional training of planners makes them particularly sensitive to environmental impacts generated from property development. When they engage in these projects, they can help to strike a balance between environmental protection for nothing and development for profit. There is no reason to view environment and development as necessarily mutually exclusive. By carefully deliberating on land use, intensity and built form of development schemes, planners can surely come up with environmentally friendly developments. It is against this background that our society will progress further in both economic and environmental terms.

▌ REFERENCES

Cairncross, F. 1993. *Costing the earth.* Boston: Harvard Business School Press.

Environmental Protection Department, Hong Kong. 1990. *Environmental guidelines for planning in Hong Kong.* Hong Kong: Government Printer.

Meyerson, Martin and Edward C. Banfield. 1955. *Politics, planning, and the public interest.* Glencoe, Illinois: Free Press.

Planning Department, Hong Kong. 1995. *Town planning in Hong Kong: a quick reference.* Hong Kong: Government Printer.

Planning, Environment and Lands Branch (PELB), Hong Kong. 1993. *Hong Kong environment: a green challenge for the community.* Hong Kong: Government Printer.

Tang, B.S. 1994. Environmental protection in Hong Kong: past, present and future. *Shui On Construction Review 1994.* Hong Kong: Shui On Group.

Town Planning Ordinance (Chapter 131) of Laws of Hong Kong.

3

What Kind Of Harbour City Do We Want?

Ted Pryor and Peter Cookson Smith

INTRODUCTION

For some time, there has been a growing ground swell of public concern about how our harbour city is changing character and, in the process, is increasingly portraying an erosion of its historical vibrancy and sense of place. In particular, there continues to be a never-ending game of ping-pong played between those for and those against harbour reclamations. There seems to be no clear way forward for resolving a number of key issues that keep returning to the public arena whenever new plans are unveiled by the government. The latest proposals for the redevelopment of the former airport site at Kai Tak and the associated reclamation of the highly polluted Kowloon Bay are a case in point, having been recently the subject of animated discussion by the Planning, Lands and Works Panel of the Legislative Council. Because of the complexity of the issues involved and the polarization of views, many concerned people in our community must now find it difficult to 'see the wood for the trees'.

A question that must therefore now be asked is, 'what kind of harbour city do we want?' While there is no easy answer to offer in return, it is an issue that should not be allowed to resolve itself by default. At the heart of the matter is the unarguable need for more urban land and its disposition; the need for the restructuring of obsolete parts of our harbour-based city; and the need to improve the city by better three-dimensional design and by the greater provision of better planned and integrated open spaces, leisure activities and waterfront amenities for the benefit of city dwellers and tourists alike. Related to these issues is the process by which plans for new urban development schemes are produced.

The following thoughts are presented in an endeavour to point a new way forward, while recognizing, however, that there are many other views that also need to be heard.

▮ SOME POINTS OF PRINCIPLE

To state the obvious, the urban environment in which we live is mostly the product of human enterprise. On that account, the present configuration of our harbour, particularly the central part, is not solely the product of being a 'natural asset'. To a very large extent, the shape of the harbour is the result of human endeavours over many years, arising from a rather ad hoc series of reclamations. In many ways, it represents the dramatic image of the city as a whole, and might perhaps be seen as a testimony to the creative abilities of the Hong Kong people.

The historical roots of Hong Kong have been firmly linked to the harbour, which initially provided a safe anchorage for trading vessels. The central part of the harbour between Hong Kong Island and Kowloon was, for many years, the hub of port activities with wharves, godowns, trading houses and ship repair yards along many stretches of the coastline. From the early days of British rule, reclamations had to be undertaken to provide new land for development. Such reclamations were, by and large, relatively small and dispersed. Changes to the profile of the shorelines had little overall impact on the aesthetics of the harbour.

In recent years, we have faced a radically different situation largely due to the industrialization of the Pearl River Delta and the consequential growth of Hong Kong as an entrepôt. This coincided with the advent of containerization in the early 1970s and the use of deep-draught ships, leading to the development of mega container terminals at Kwai Chung. Coastal and river vessels from

China also increased, requiring the opening up of new buoyage areas in the western harbour between Hong Kong Island and Lantau. All these, in turn, led to the central harbour relinquishing its traditional role as the heart of our port. At the same time, however, rapid economic development, the growth of population from 3.1 million in 1961 to 6.8 million by mid-1998, the growing obsolescence of inner city districts, and the need to upgrade infrastructure required major efforts to be directed at new land formation, including harbour reclamations. This led to substantial changes to the shape of our harbour which, by good fortune rather than by good planning, is an aesthetic asset that can be well appreciated from high-level vantage points.

Equally, however, the 'public asset' that has been created in certain areas can be seen in a more negative light, especially at close quarters. The sad fact is that the great majority of the city-harbour coastline is inaccessible to the public due to the uncoordinated development of waterfront sites for godowns, boat yards, power stations, wholesale markets, sewage treatment works, water pumping stations, major highways and other infrastructure facilities. We surely can and must do better than this!

Yet, there are advocates for permitting new reclamations just for new infrastructure works. While reclamations for some works of this kind may be necessary, reclamation of harbour areas solely for that purpose would compound an already unsatisfactory situation, given that infrastructure works by themselves alone would not be conducive to enhancing or safeguarding the harbour as a public asset and natural heritage.

There are also views, on the one hand, that favour reducing the extent of reclamations so as to minimize the further narrowing of the central harbour but, on the other, advocate increasing the densities and consequently the height of building development so as to 'optimize' the use of scarce land resources. Against such contentions are appeals for a more people-friendly urban form, with places and spaces to which the person in the street can identify comfortably and appreciate at first hand a more traditional, street-oriented urban form.

One important aspect at the heart of the matter is our need for more land for urban use, which brings into question the potential sources of supply.

▌ THE NEED FOR MORE URBAN LAND

An obvious and fundamental issue which our community has to face is that the population of Hong Kong has, over the past 40 years, increased at about 1 million people every decade. Our population now stands at about 6.8 million people living in confined territorial limits. There are predictions that, by 2011, we will have to accommodate 8.1 million – an increase of more than 100 000 people per annum. In broad terms, this means that every year we will have to produce about 100 hectares of land just for high-density housing and associated uses and infrastructure. Over a period of 12 years, we would need to produce an area the size of our new airport at Chek Lap Kok. Without a steady supply of land, there would be diminishing prospects for achieving the ambitious production targets for the supply of both public and private housing. We do not have the luxury of being able to sit on our hands, hoping that the problem will somehow melt away.

Is the answer, then, simply to develop remaining reservoirs of land in the New Territories? This has been done over many years, and additional new town development projects are in the pipeline. Other lowland rural areas, mostly in private ownership, are fraught with a multitude of problems that would be ruinously expensive to resolve and take an inordinate length of time to sort out. The further urban transition of lowland rural areas in the New Territories is undoubtedly inevitable, but meeting the very significant land needs of our community in a timely way requires a broader look at other supplementary means of land production. We must surely acknowledge that our precious country parks, which provide green lungs for city dwellers and which occupy 40% of Hong Kong's total land area, are off limits.

An inescapable conclusion is that, in addition to whatever new land for urban use can be provided in a well planned way in the ecologically and environmentally sensitive land-based areas of the New Territories, supplementary provision also needs to be made by means of reclamation. Not only will most further new town areas need to resort to such measures, but the metro area, in which almost 60% of our population now lives, often in crowded conditions, also requires additional reclaimed land in carefully chosen locations to provide space for living, leisure and work. Into this equation has to be injected perceptions of the future configuration of the harbour itself, not as a by-product of new reclamations but as a key element of city design.

▮ RESTRUCTURING THE CITY

It has been increasingly recognized in recent years that there are many parts of the metro area around the harbour that are badly in need of restructuring. Hong Kong's urban renewal programme is formidable. Around 950 hectares of housing and industrial areas were identified by the government's 1991

Metroplan Strategy as failing to meet reasonable standards of design and environmental quality, and hence falling considerably short of the needs and expectations of occupiers.

The private sector has taken the initiative over many years to redevelop obsolete properties. There are a number of notable, comprehensively designed schemes, such as at Taikoo Shing and Laguna City, that have essentially created new urban communities. Mostly, however, private developers have replaced relatively low-rise tenement blocks with high-rise apartments on small sites, thus adding to the congested character of the inner districts.

In the public sector, the Housing Society and the Housing Authority have wiped the slate clean of very obsolete housing blocks and, in their place, have created more spacious estates that provide higher standards of accommodation and a full range of community facilities. Since 1988, the Land Development Corporation (LDC) also has made inroads in tackling the redevelopment of decaying pockets of old buildings that, for various reasons, private developers have been unable or unwilling to tackle.

The main problems in implementing urban renewal projects have been the assembling of sites of sufficient size to allow viable and worthwhile development, along with the difficulties of rehousing affected households and the removal of long-established non-domestic uses. Over the past ten years, urban renewal projects in the private sector have cost something in the order of HK$50 billion without burdening the public purse. This has involved the participation of private developers as joint-venture partners with the LDC. However, the resulting property-led and 'market-driven' approach to renewal clearly has marked ramifications for the urban environment. While relatively large areas of technically obsolete buildings have been comprehensively redeveloped, the former mixed-use and multi-activity character of the affected localities has inevitably been swept aside. There is a growing view that this process has failed to integrate elements that might bring about genuine revitalization, diversity and economic regeneration in the longer term.

The inception of any major renewal scheme first requires the provision of new land to rehouse displaced households and to provide sites for new community facilities and infrastructure – not in the more distant parts of the New Territories but on the doorstep of those areas that need to be restructured. The considerable experience of the Housing Authority, the Housing Society and the LDC has established that people affected by redevelopment schemes have strong preferences for being rehoused in the same general locality.

Since the provision of new land in already highly congested areas is, by self-definition, very difficult to achieve, and since the scope for forming land in the metro area by the terracing of hillsides is limited, we need other adjoining areas of land to enable the city to be restructured in a comprehensive and logical way.

The former airport site at Kai Tak embodied within the plan for South-east Kowloon provides new development opportunities. However, even that relatively large site has a limited potential by itself and its present configuration, and coupled with its designation for predominantly public housing to an extremely high density, places limits over urban design and the provision of essential new infrastructure required to enhance the functional efficiency of the city as a whole. Also, of course, other parts of the city need new land to facilitate programmes of urban improvement.

Hence, there will continue to be a need for reclamations of an appropriate scale and which need to be planned and developed in a sensitive and cost-effective way. Achieving that objective will not be easy and not without compromise as the land, engineering, environmental, aesthetic, social, economic and institutional issues interact in highly complex ways.

▌ FOUNDATIONS OF DESIGN FOR THE CITY

It is thus seen that comprehensive and well-planned reclamations of the 'right' scale and shape offer a unique opportunity for our community to revitalize and strengthen the aesthetic and recreational values of the harbour. To leave the harbour and waterfront areas in their current state would only perpetuate the present legacy of human-created problems.

The harbour is an asset that needs to be appreciated by people, both from the hills around and, more interactively, along its extensive shorelines. This, in turn, also requires a fresh look at the three-dimensional form of the city as a whole, with plans designed to enhance the integration between urban places in the heart of the city and the waterfront.

We also need to integrate areas of mixed use, colour and vitality not only in new development areas but also in urban renewal projects, which can otherwise change lively, street-oriented environments into standardized, inward-looking and hermetically sealed environments of mostly single-type uses in commercial complexes. As the city has grown both upwards and outwards in response to economic and social pressures, we have witnessed the irrevocable loss of traditional street markets and mixed-use shopping streets, the random visual quality and informality of which gave the heartland of our city a unique appeal.

We need to become more concerned with 'place making' and not merely the creation of standardized, high-rise projects, which the current system of planning and building regulations tend to perpetuate. We need to create coordinated extensions to the urban area that strengthen and enhance the aesthetic attributes of city living. This means, among other things, developing an emphasis on pedestrian spaces: squares, precincts and plazas, together with urban streets that become enjoyable pedestrian corridors, and establishing good linkages with older adjoining districts and waterfront areas.

We should not be afraid of creating new waterfront development, and there are generally few other options but to do this on carefully selected areas of new reclamation. Other cities around the world are doing exactly this and are finding that, if done sensitively and with creative design, new waterfronts come to be the most attractive and vibrant parts of the city – witness the renowned success of Darling Harbour in Sydney, Fishermen's Wharf in San Francisco, Boston Harbour, and Clark Quay in Singapore. By contrast, in Hong Kong, opportunities to create exciting harbour fronts and new urban communities are being compromised almost at the outset, both by an inhibited and expedient approach that seeks to equate minimal reclamations with maximum densities and also by plans that place new building forms within layouts which offer little flexibility for innovative design on a district-wide basis. The cruel irony in this for all who are concerned about the visual sanctity of the harbour itself is that the existing aesthetic drama of the harbour, originally defined and offset by spectacular mountain peaks and ridgelines, is being compromised by tall building massing that overwhelms the visual quality of open areas of water. What, then, is the gain to the public?

Faced with ongoing opposition from certain quarters to any new harbour reclamations, there appears to be a growing inclination for the plan-makers to keep any new schemes to the very minimum of limits, with shorelines set along alignments that offer few concessions to the dictum 'nature abhors a straight line'. While this may be seen to offer a pragmatic compromise, opportunities could be foregone to create attractively designed marine basins of appropriate scale providing new amenities that can add interest, colour and a sense of liveliness at focal locations where the harbour and city meet. Such basins can also provide attractive settings for public uses and tourist attractions that could offer viable opportunities for private sector investment. Some could also be designed in a sensitively conceived way to incorporate new residential communities and other amenities around their edges; good examples of which can be found in other harbour cities.

In common with proactive measures taken by other cities, we need to make up the deficiencies in public access to the harbour, integrate major, colourful, recreational leisure and visitor-oriented facilities (such as festival markets, 'fisherman's wharf', marine-related uses such as maritime exhibition and museum facilities, tall ship moorings, etc.) that will add to the drama and visual interest of the harbour itself. We also need to seek out and pursue the development of continuous but sinuously configured promenades along as many stretches of the inner harbour as possible, linking together points of recreational attractions, tourist facilities and restaurant areas within a landscaped framework. These are not new ideas – they were proposed several years ago as part of a 3km waterfront between Central and Causeway Bay, the scale, configuration and design of which are now proposed to be redefined to more conservative limits with a view to minimizing reclamation. Are we now in danger of 'throwing out the baby with the bath water'?

A fresh view is also needed in respect of operational facilities needed to service the port and other marine-based activities. The general aim should be for certain existing or proposed cargo handling facilities, marine workshops and a mixed bag of short-term tenancies scattered along various stretches of coastline, and adjoining residential areas to be relocated from the central harbour to outer perimeter sites designed for such purposes. This would open up opportunities to promote the use of the central harbour for marine recreational activities, and also make available key waterfront sites that could be developed for a wide range of compatible purposes which would inject new life into the city. For example, there now seems to be no sound reason for current plans to incorporate a major cargo working area at Causeway Bay where there would be scope for developing a new and exciting harbour community.

As a backdrop to the harbour and the waterfront, more thought needs to be given to the challenge of how to create a better configured city profile that would, on the one hand, preserve wide views of the harbour from the surrounding hills and, on the other, enable the distinctive features of the same surrounding hills to be appreciated from lower waterfront zones. However, there are clear signs that, in many parts of the city where large-scale development has been undertaken in recent years and in certain areas still at the planning stage, the use of standard, high-rise blocks is resulting in a monolithic and substantially undifferentiated urban form that overpowers the natural landscape. Current plans, produced partly in response to high population growth scenarios, call for more of the same. This may provide an essential catalyst for the rapid development of our new towns. However, for 'sensitive' areas of design at key locations within the metro area itself and around the harbour, a new perception is required. This, in turn, calls into question the way in which our plans are conceived and implemented.

▮ THE PROCESSES OF URBAN DESIGN AND DEVELOPMENT

What we see each day of our city form and functions is increasingly the product of the many and diverse processes that contribute to urban development in Hong Kong. To achieve a better modulated city form requires a clearly articulated urban design framework and a close review of the institutional measures required to achieve compliance with defined parameters that are aimed at achieving visionary design. Such measures should be capable of being set in place at least for key design areas on new reclamations, with enforcement through lease conditions and the more explicit provisions of statutory outline zoning plans, which take precedent over other statutory building controls.

The remarkable achievements of Hong Kong in producing urban communities of a massive scale are a response to a critical and ongoing need to provide decent housing for our burgeoning population. The use of standard designs for housing blocks, schools, clinics and other key community facilities has played a major role in the coordinated and speedy development of the new towns. But, has the time not now come when we need to think more deeply how the attributes of our harbour city could be enhanced, by means of a more integrated and comprehensive process of city design that has fuller regard to the three-dimensional and architectural aspects of urban forms? This would require a re-evaluation of what we would like our city to look like as a place of world renown befitting the new millennium, encompassing new areas of reclamation, major urban renewal projects, the use of as yet undeveloped land-based sites, the harbour itself and landforms that provide a natural backdrop. By contrast, while there have been well produced feasibility studies that have adopted a comprehensive design approach, in some cases with extensive urban design guidelines and parameters, these usually come before the public as broad statutory zoning plans with no meaningful provisions relating to three-dimensional aspects.

Such plans essentially provide a general framework within which private developers are sold sites to be developed, with the limits of height, site coverage and plot ratios defined under the Building (Planning) Regulations, or within which certain public bodies are allocated sites for the development of standard designs, especially with respect to housing and G/IC facilities. The end result is a city form that is a collection of largely independently conceived urban forms that bear a strong imprint of uniformity. In defence of this 'bottom-up' approach, however, it has to be said that it provides one way of producing relatively quick results to dependable building standards.

And yet, the Town Planning Ordinance permits outline zoning plans to include provision for 'any matter whatsoever' by means of 'such diagrams, illustrations, notes or descriptive matter' as the Town Planning Board thinks appropriate. As noted above, such provisions have the force of law and take precedent over other related provisions within other statutes and regulations. The door surely is open to make our outline zoning plans more purposeful in the field of city design and, at the same time, give adequate scope to members of the architectural profession to exercise their innovative skills.

Besides a 'bottom-up' approach relying heavily on standard designs and building planning regulations that offer relatively little leeway for innovation, there should be a 'top-down' planning process. Such an approach was advocated under 'Metroplan' (a strategic plan approved by the Executive Council in 1991) to provide an overall land use-transport-environmental concept for restructuring the city. Within that framework there should be tailor-made layout plans for specific areas and the definition of the key design parameters to provide a better integrated framework for project design. Enforcement of such tailor-made parameters through lease conditions would be an effective mechanism. Such an approach has been applied successfully by the Urban Renewal Authority in Singapore, which generally decides what is most appropriate for an area in terms of form, massing, conservation, new building, etc., and then circumscribes the planning means and urban design controls to make this happen. In the case of Hong Kong, despite earlier efforts to follow such a path in the context of Metroplan, innovative detailed development plans under subsequent Development Statements have largely failed to materialize through the current statutory planning system.

It is also perceived that the planning process needs to provide more opportunities for professional bodies, business associations and community groups to be given an early opportunity to contribute their ideas on the way in which our city should be developed. At present, the statutory planning process provides only for the lodging of objections to plans that are of a very general nature. This seems to be a rather confrontational way of 'public consultation', and somewhat late in the day to expect the authors of the gazetted plan to keep an open mind about their proposals, which are often based on detailed feasibility studies, of which the general public seldom has a chance to be involved in or to see.

The city belongs to people from all walks of life, and it is advocated that there should be a two-step process whereby, at an early stage of plan-making, a range of conceptual options is made public for comment and discussion. Unless a 'driving concept' is formulated and agreed, the prospects of producing coherent proposals for individual projects would be diminished. Following the assumption of an agreed concept, more detailed studies can be undertaken at a second stage with a greater level of confidence that they will be acceptable to the community at large. The results of such detailed studies should be made openly available for public reference and, where appropriate, the key elements of design and associated parameters should be incorporated in the statutory outline zoning plan. There should be no impediment for either the government or the Town Planning Board to pursue such a path as part of the process of plan formulation.

If such a modified approach is to be launched, it would also be necessary to review critically the processes by which planning studies are undertaken. In particular, for areas identified as deserving innovative design, such as waterfronts, town centres and around major transport nodes, the very best of professional skills (which are abundantly available in Hong Kong) need to be assembled and the appointed designers given sufficient freedom to generate new ideas that stretch beyond conventional standard solutions. However, current government procedures require planning study briefs to be cast in a rather rigid mould, better suited for engineering projects, and are at times over-specific as to the end product required.

Study briefs relating to major urban design projects need to be more open-ended and investigative, leaving adequate leeway for the iterative generation of progressively more detailed proposals. There should be scope to allow examination of the merits of new development concepts that extend beyond currently prevailing standard designs. The initial exciting designs for the redevelopment of the Central Ferry Piers are a case in point. For such studies, there should be adequate budgets, a realistic time frame that is adhered to, and a consultancy selection process that does not allow rock-bottom fee bids to over-compromise the quality of design inputs. In short, we need to recognize that the product hoped to be produced should be reflected in the processes used for consultancy appointments – not the other way round. Feedback from consultants in this field indicates that the present system for the appointment of specialists for planning and urban design studies is in need of review.

▮ DOING GOOD WITH WHAT WE HAVE

It should be abundantly clear from the foregoing that the issues associated with the design of our harbour city are complex indeed. Also, if it is accepted that there is a prima facie case to warrant a rethink of the processes which mould the shape of the city and its harbour, it has to be expected that any such review would take some considerable time to come to fruition to the reasonable satisfaction of the key stakeholders in both the public and private sectors. As a portent of that view, over 25 years have passed by without seeing much come to fruition from the efforts of the government's planners and the Hong Kong Institute of Planners to comprehensively review the Town Planning Ordinance, first passed into law in 1938.

In the meantime, we have no choice but to adhere to current systems of planning and decision-making, which in many ways have enabled Hong Kong to satisfy its economic and social needs in a bold and pragmatic way. It is thus held that while continuing to work towards a fresh look at the processes of planning and urban design, initiatives should be taken for 'doing good with what we have'. The following three possibilities are presented for deeper consideration.

Upgrading Of Existing Waterfronts

One example that could offer a good opportunity to 'reclaim' a valuable stretch of already formed coastline for the wider public benefit could be the substantial area of land beneath the Eastern Island Corridor. Much of this land is government-owned, and in the absence of any firm and imaginative plans for the development of waterfront amenities, sites have been allocated by the Land Authority for such temporary uses as barging points for the storage of sand, aggregates and other building materials; for large parking areas for cars, trucks and coaches; and for USD and other government depots.

Such uses have a role to play in providing for the day-to-day running of the city. But, now that a large area of Hong Kong's former international airport at Kai Tak has become available, would there not be scope for the relocation of some of these temporary uses? For parked cars and other private vehicles, there may well be an opportunity to accommodate them in privately operated, multi-storey parking buildings under selected elevated sections of the expressway. Also, for stretches of water between the expressway and the present coastline, there should be good scope for reclamation of areas that, in their present state, otherwise offer little of aesthetic or functional value.

The way would then be opened to turn valuable stretches of coastline into areas that could provide recreational and other amenities of great benefit to the local community and also to visitors. There would be good opportunities to create landscaped waterfront gardens, festival activity places for, say, performing artists and 'flea' markets, indoor and outdoor sporting facilities for all age groups, seafood restaurants and Asian food bazaars, small- and medium-scale exhibition halls, etc. Along an extensive stretch of the expressway, it should be possible to span from pier to pier to create a unique coastal walkway and cycle-way. By such means, the acute shortfalls of open space in the adjoining high-density districts of North Point, Quarry Bay and Shau Kei Wan could partly be alleviated. Joint venture schemes between the government and private developers on the basis of design, build and operating arrangements may offer a way forward, especially if linked with adjoining sites for private residential development (such as the site of the former Government Supplies depot in Oil Street, North Point).

Development Of A Gateway Park

There is a further opportunity to create a major new waterfront-based project along the southern limits of the West Kowloon Reclamation. The concept we propose is the development of a 'gateway park' in close proximity to the MTR West Kowloon and airport railway station, and also adjacent to a terminus for the KCRC West Rail connecting the city to the North-West New Territories. Together, these two nodal points will become important points of entry and exit for tourists and other visitors to our harbour city. However, achieving such a vision would require a fairly radical rethink of current planning proposals.

The present statutory outline zoning plan for the West Kowloon Reclamation provides for a number of relatively regular-shape blocks for commercial, residential and other forms of urban development. The Mass Transit Railway Development Corporation has already led the way for the transformation of the area by the phased development of a mega project around the West Kowloon station. However, the site is somewhat removed from the city and, for some years to come, the intervening area is likely to be a 'wasteland' of undeveloped lots, many of which are likely to be allocated by the Land Authority for a mishmash of temporary uses.

Current plans also provide for the reservation of a large area of land at the south-western point of the reclamation for a future park and this, in itself, is a welcome feature. Provision is also made for a waterfront promenade connecting the park to the city. A number of sites adjoining the promenade are set aside for local open spaces, in between which are blocks of land that are zoned for high-rise commercial development and various community uses and which are designed for implementation on a site-by-site basis.

A significant part of the area was identified in a recent feasibility study commissioned by the Hong Kong Tourist Association as a site for an Expo event to mark the new millennium. While that scheme was not accepted by the government, it demonstrated well the very significant potential for the development of a 'gateway park'. Such a park could comprise a wide, landscaped pedestrian corridor connecting the existing Kowloon Park at the eastern end and the proposed new park at the south-western end of the West Kowloon Reclamation. The corridor could be designed to incorporate a wide range of open-air and indoor recreational amenities, some of which could be run on a commercial basis. Also embedded in the corridor could be harbour-front, mixed-use, new lifestyle community development schemes and innovative and well-designed tourist attractions. The various nodes of development could be linked together by an integrated system of traffic-free pedestrian links, including harbour-front promenades and civic squares.

This is precisely the type of project that might well be sponsored by the Jockey Club Charities Fund. Given the current economic downturn that continues to grip Hong Kong and Asian countries, there would appear to be no great need or hurry to commit the West Kowloon Reclamation to new commercial-led development. This, in turn, should offer a good window of opportunity to review current plans on the basis of a comprehensive and three-dimensional urban design study by the very best of professionals in the field.

Redesigning The Central/Wanchai Reclamation

The scene is now set for a change to the scale, shape and function of the balance of the Central/Wanchai Reclamation, for which detailed plans were first produced ten years ago. Changing circumstances and development needs led, in early 1998, to the gazetting for public objection of a new Outline Zoning Plan. In part, the new plan might be seen both as a response to the Harbour Protection Ordinance, which has set in place a presumption against new reclamations in the central harbour, and as a justified reduction in the amount of predominantly commercial development proposed under the original plan.

There is an increasingly pressing need to finalize a new plan, principally to open up the way for the earliest possible implementation of works for the provision of a new underground bypass to relieve traffic congestion, and an extension to the Mass Transit Railway. It is possible that a further reduced reclamation may be an outcome of the deliberations by the Town Planning Board in the light of the objections received. From a joint city-and-harbour design point of view, a truncated reclamation profile could still provide the community with a unique opportunity to create an extension of the city of outstanding design that would offset the visual waterfront drama created by the new extension to the new Convention and Exhibition Centre.

Outline zoning plans essentially indicate broad land use patterns and major road systems. They are intended to provide a framework for the production of detailed development plans and also to regulate redevelopment. They nevertheless, in practice, tend to establish from the outset quite a rigid framework for the demarcation of land uses and, of course, have the force of law. While this type of plan has been employed for many years in the development of our new towns to enable these massive developments to proceed quickly and expediently, applying the same approach to a hugely sensitive site in the central part of the harbour must be questioned. This area, above all other areas, needs to be designed in a way that deserves international acclaim. For such an important site with its inherent opportunities for stimulating design in every sense, there is a need for a planning approach geared directly to meeting important urban design objectives at the outset. We are of the opinion that it will be extremely difficult to incorporate these in retrospect as a somewhat belated response to public objections.

Achievement of design objectives not only requires imaginative layout plans, but also calls for clear-cut design and control mechanisms to best ensure that objectives are met in practice, both to enhance the harbour and to create a new city 'heritage'. This means that the planning layout itself must be responsive to a large number of planning and design factors. This would help ensure that the built framework would eventually have a sufficient level of urbanity, a stimulating sense of place, a coherent arrangement of forms and spaces, user-friendly and interlinked pedestrian ways and open spaces, exciting waterfronts, and so on.

An outline zoning plan thus also needs to be supported by well-conceived and solid urban design parameters related to building massing, spatial relationships and the creation of urban places of distinction. One requirement should clearly relate to height controls, similar to those proposed for the original Central/Wanchai Reclamation development plans in 1993. Such controls should be enforced by the Town Planning Board and/or the Building Authority to avoid having enormously tall towers that block out the Peak ridgeline backdrop and the dramatic profile of the Kowloon Hills. Such guidelines are clearly stated in the Metroplan approved by the Executive Council in 1991. Regrettably, a number of recently approved projects flagrantly breach those guidelines on the rather dubious grounds that they would serve as 'landmark' towers in a central part of the harbour.

It is difficult to see how the Town Planning Board will be able to bring together in a coherent way the diverse views of objectors, some of whom have put forward a range of new design concepts for the Central/Wanchai Reclamation. Replanning of this key area is not a process that can be easily dealt with through the current statutory planning system, under which decisions made by the Board on objections are final and beyond appeal.

Further serious thought needs to be given to launching a comprehensive detailed design review of the Central/Wanchai Reclamation. In this sense, there is a need for the government to show that everything is being done in terms of three-dimensional urban design to give Hong Kong the best possible new waterfront development, reflecting both its geographic position and Hong Kong's particular culture and context. This requires vision and the application of 'tailor-made' mechanisms to achieve the desired end-product. Such a review should be within broad shoreline limits set by the Town Planning Board and also would need to accept the vertical and horizontal alignments of the new bypass highway and MTR line, in respect of which there would seem to be little practical scope for change. There should be no compromise in a commitment to produce a new concept and detailed urban design proposals of the very highest standards for such a key area.

There may also well be a need for a similar review of the three-dimensional design basis for the Kai Tak-Kowloon Bay area, taking account of the large number of objections received to the recently published outline zoning plans. In a similar way to the Central/Wanchai Reclamation, there is a need to ensure that urban design objectives can be achieved in practice and opportunities maximized to create a stimulating new urban environment and waterfront, especially given the current objective of the plans to maximize development intensity.

▮ CONCLUSION: WHERE DO WE GO FROM HERE?

The conventional planning approach in Hong Kong has evolved over many years and, in doing so, has become increasingly oriented around 'processes' rather than innovative and stimulating city design solutions. Imaginative new concepts that are given due support need to be translated into detailed projects through effective institutional controls (including land disposal mechanisms) which may require current systems to be modified. Otherwise, compromises to expediency can effectively kill off 'hallmark ideas' at an early stage.

The time has now come to open up our minds to new visions. Despite existing constraints and practices, which range from the massive cost of land to the use of standard high-rise building forms, it can be seen that there are opportunities for bringing new visions to reality, by means of a comprehensively devised strategy for restructuring and redesigning our harbour city within the framework of modified reclamation profiles and better conceived urban renewal schemes that adopt an area-wide approach rather than just focusing on individual projects of relatively limited scale.

In seeking a way forward, we suggest that the following questions need to be addressed.

- What is the current overview of the future long-term requirements for urban land to satisfy both the social and economic needs of our society and the restructuring of the city?
- If Hong Kong is to accommodate a growth rate that could increase the existing population by up to one-third over the next 18 years, how can this be best met without massively compromising the character of the city and its harbour environment?
- What potentially available land resources in suitable areas throughout the territory could be made available to meet such needs in a timely way?
- How can potential demands for land be best met in a timely way within the limits of defined potential development areas throughout the territory – this being a way to help set in place the scale and function of harbour reclamations?
- What broad vision of the city, including the harbour, should we assume?
- What kind of plans, associated urban design controls and institutional mechanisms do we need to translate visions into tangible assets that will take us into a new era of outstanding city design?
- What consultative and statutory processes of plan-making should we employ that will enable key stakeholders to contribute, first, to the shaping of new development concepts and, second, to facilitate the preparation and implementation of detailed plans?
- What organizational structure is required to help achieve the coordinated implementation of development plans, so that the city can function more efficiently as an integrated whole rather than as a sum of individual parts?
- What meaningful and visually stimulating projects should we aim to achieve in the near future to get things moving along a new path of vision, using the planning and development systems now in place?

At first sight, such questions seem rather daunting. However, there is already a substantial reservoir of information from various official studies that should suffice to enable a broad picture to be produced, especially on the question of future land use needs and potential development areas. It is acknowledged that development and redevelopment pressures have increased in recent years and that, over the next 10 to 15 years, levels of development are likely to far exceed the population/employment quantum recommended by Metroplan. However, this should entail not just a numerical review of land use/transport forecasts and the recommendation of a revised development strategy, but a 'vision' for the city itself and its means of realization. There is a corps of well-founded professional expertise in the community from which new views and ideas can be assembled, especially on the systems of plan-making and -implementation.

Left unattended, the issues highlighted above are unlikely to go away of their own accord. Similarly, unresolved major points of contention can create delays to essential works required to provide a sustainable development framework to help maintain the future prosperity and environmental integrity of our community. A lead needs to be taken to create a new vision of the city, and the sooner the better.

Who or which body should take such a lead? Are we to assume that the Town Planning Board, the government's own 'in-house' high-level Committee for Planning, Land and Development, the Legco Planning, Lands and Works Panel, the Housing Authority, the LDC (soon to become transformed into an Urban Renewal Authority) and a plethora of government departments will be able to produce, in some consensus-led way, a collective view that makes sense? This seems unlikely, as each body has its own prescribed remit and a focus of views that may cloud perceptions of a broader vision.

Perhaps, therefore, the time has come to set up on a provisional basis a Harbour City Development Commission charged with the responsibility of reviewing the kind of issues presented in this chapter and required to produce a report for public consultation before the end of the millennium – say, by end-1999. Such a commission should be under the direction of a well-respected professional in the design or building field, of good political standing and with a sound knowledge of the current systems of planning and building development. Other members should be drawn from a diversity of relevant backgrounds, from high academic/professional levels to grass roots foundations. Funds would be required to undertake studies through a technical backup team, with a competent executive director, reporting directly to the Commission. Such a team should be of the highest possible calibre, based on a combination of public and private sector experts. The groundwork now being laid by the Planning Department for a review of the Metroplan and other parallel studies on urban design guidelines and specific harbour reclamation projects would need to be incorporated into a more open review process. By such means, we will hopefully be able to reach a consensus to answer the question posed by this chapter, 'what kind of harbour city do we want?' Once that question has been resolved, we should be able to enter the new millennium with a greater sense of direction and purpose in the broad field of planning and urban design.

Note: This chapter has been jointly prepared by Ted Pryor and Peter Cookson Smith, who are founder members of the Hong Kong Civic Design Association. They have long careers in the field of urban planning and design respectively in the public and private sectors in Hong Kong. The Hong Kong Civic Design Association was established to provide a forum for the generation of new ideas to help promote the future development of Hong Kong as a better place in which to live and work.

Part B: Environmental Design Strategies

This part discusses how environmental design strategies are considered by designers and applied to actual architectural projects. For an insight into actual practice, architects from both private and public sectors have given their views on design strategies based on experience.

In 'Building into the Environment', the Architectural Services Department (ArchSD) explains how its Green Manager Committee helps with the protection of the environment. For large projects, a preliminary environment review is made at the inception/feasibility stage. In the design stage, design studies include overall thermal transfer values, reduction of the use of tropical hardwood, minimization of waste, and resource-saving by recycling or reuse. Energy conservation is exercised in the fields of air-conditioning, lighting and management systems. Renewable energy is not commonly used but has been applied to a number of ArchSD projects. Some existing buildings with environmental problems have also been improved by the ArchSD.

In 'Use of Technology to Assist Environmental Design: A Case Study of Verbena Heights in Hong Kong', Kam-sing Wong uses Verbena Heights as an example to explain the design and research process. Research, responsiveness and review are the key headlines of study. Being an important part of the project, technology was used in relation to the local climate. The project also dealt with the elements of water, energy, sun and wind.

Ada Yin-suen Fung in 'A Public Housing Experience' describes planning and design for public housing. Environmental assessment studies were made at an early stage for the large quantity of housing required. Energy efficiency and noise control were design criteria. Ozone depletion was avoided. An automated refuse collection system helped to provide more hygienic living conditions. The construction process was also monitored to be less damaging to the environment.

The chapter 'Passive Environmental Strategies for Architectural Design', written by Bernard Vincent Lim and Man-kit Leung, states the relationship between human comfort criteria and climatic context. The particular context of Hong Kong is examined, and a list of passive environmental strategies is provided. The building envelope is also examined in respect of ventilation and daylighting.

4

Building Into The Environment

Architectural Services Department

INTRODUCTION

To ensure proper consideration is given to environmental improvements, the Architectural Services Department (ArchSD) has introduced a systematic and measured approach through an environmental management system (EMS). The ArchSD has built on its previous ISO 9001 initiatives to obtain certification in March 1998 of ISO 14001. Recognizing that in day-to-day operations, it is inevitable that the environment is impacted in many ways, the aim is to minimize the extent of this impact, wherever and whenever possible.

The EMS can be found in two divisions as it addresses both the process/organization through housekeeping/administration, and products/projects. A comprehensive range of multi-disciplinary professional and technical services for buildings and facilities is committed through a quality management system to improve the environment by:

1. undertaking the design, procurement and maintenance of community facilities in an environmentally responsible manner, such as reducing energy consumption and the use of materials which are harmful to the environment;
2. continuously developing and maintaining an EMS in accordance with ISO 14001;
3. complying with relevant environmental protection ordinances;
4. providing training for staff to increase awareness for continuous improvement in protecting the environment and preventing pollution; and
5. communicating and making available environmental policy to the construction industry and the public at large.

PROCESS/ORGANIZATION

Prior to certification, the Green Manager Committee was set up by the ArchSD in early 1994 in support of the government initiative to improve housekeeping aspects of office administration.

The targets and objectives of the committee, which focuses on the ArchSD's internal process, are shown below.

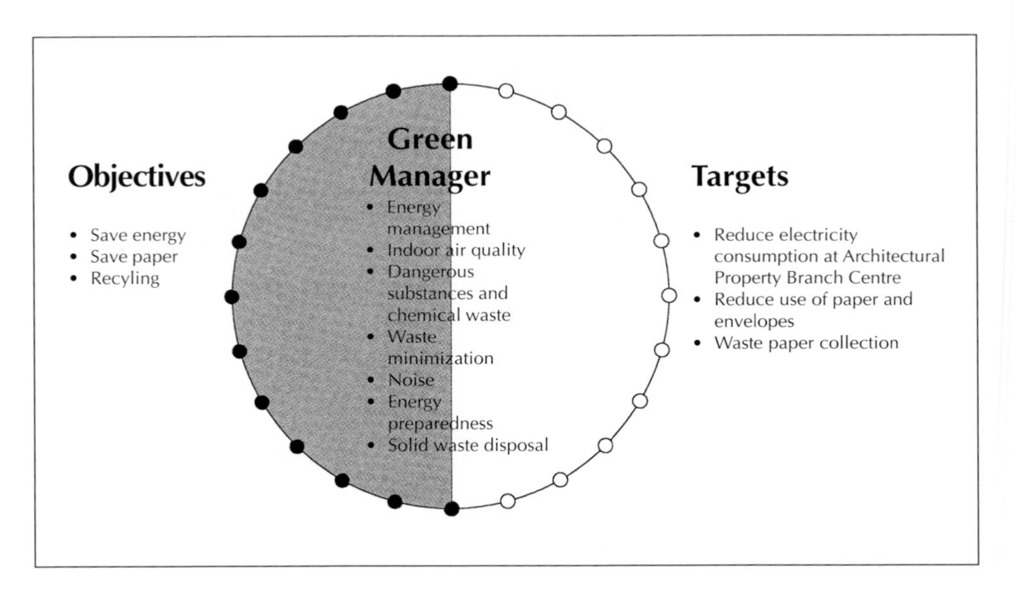

Architectural Services Department: Environmental Management System

The Green Manager Committee actively reminds ArchSD staff to consider all administrative actions that can help to protect the environment. Examples are switching off lights when out of the office for some time; using both sides of sheets of paper; providing recycling bins for waste paper; reducing the use of consumables; and reconsidering typing layouts and formats of large reports. The response has been very encouraging, and improvements continue to be made each year. A summary report of resources and waste recovery for 1996 is shown in the Appendix (page 110).

▮ PRODUCTS/PROJECTS – THE PRELIMINARY STAGE

The illustration below indicates the types of projects and the range of clients dealt with by the ArchSD.

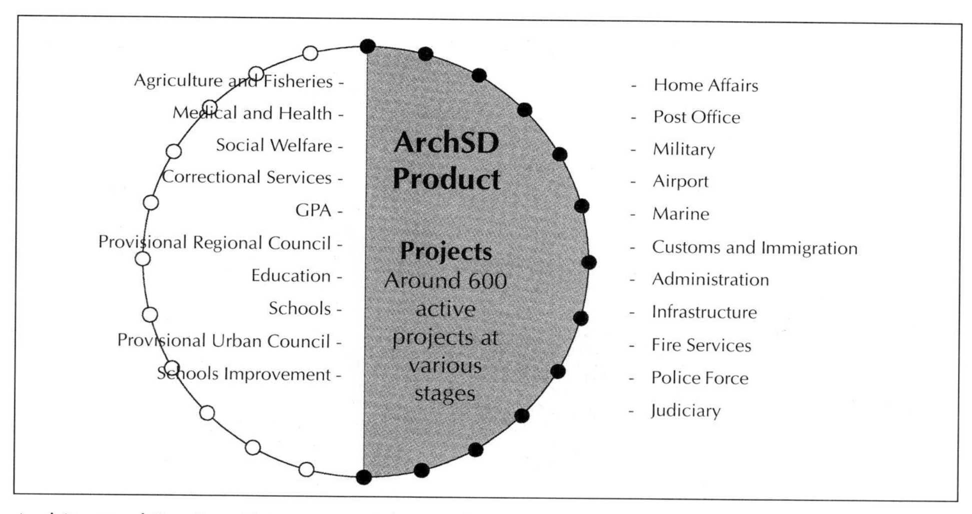

Architectural Services Department: The Product

Project Inception

At the very start of a project, all aspects and impacts on the environment resulting from our actions are addressed. How the external surroundings impact on the project and how the project will impact on the environment are considered in broad terms.

The Client's Brief

To respond to the client's needs, it is important to discuss with and advise the client during the preparation of the brief on all environmental aspects. For example, the client may envisage a curtain wall building and not be aware that the result would increase energy running costs. The aim is to highlight important issues during the briefing process to try to influence the client into introducing appropriate environmental aspects into the brief, allowing the client to select the most appropriate development option at the end of the feasibility stage.

The Type Of Project

The type of project has a large impact on the environment. School projects by the ArchSD have different and perhaps lesser impacts than a multi-storey, multi-user maintenance depot. Open space and park projects can enhance the overall environment, whereas other projects may contribute little due to their functional nature, such as buildings for storage. The strategy is to minimize the environmental impact as far as possible in accordance with the client's brief.

North District Park as an Urban Oasis

The Site

The physical surroundings of the site, including the location, orientation, size and shape of the site, are all critical to the development of the project and its environmental impact. There are many aspects to consider, such as the location of vehicular entrance/exit points, views from and of the site, existing ecology and landscape, noise sensitive receivers nearby, etc. These have to be assessed and it is often found that the ideal orientation of the building to minimize heat gain cannot be achieved as the physical or other constraints take precedence. The aspect of minimizing heat gain is then addressed in the design, through the window-to-wall ratio, choice of materials, consideration of sunscreens, insulation materials, etc.

Preliminary Environment Review (PER)

For major public works (PW) projects valued over HK$15 million, a PER is required at the inception/feasibility stage of the project. The PER will address the major environmental issues and ensure that the project will not create unacceptable environmental problems in the future. If the PER identifies a traffic noise impact, the development of the design and the details will need to take this into account. If the PER identifies that the project has a major impact on the environment, then an environmental impact assessment (EIA) will be required. The designer will need to provide effective measures to remove or reduce the adverse impacts to the satisfaction of the Environmental Protection Department, before the project can proceed with construction.

The EIA Ordinance was gazetted in February 1997 and it is now in force. In the Ordinance, a schedule is established and stipulates which categories of projects are required to carry out EIAs. It is noted that some of the projects under the control of the ArchSD fall into one or more categories of the schedule.

For non-PW projects and projects valued below HK$15 million, the ArchSD has introduced a project environmental check-list to ensure that the major environmental aspects of all projects have been considered at this preliminary stage.

▮ PRODUCTS/PROJECTS – THE DESIGN STAGE

Objectives And Targets

The objectives and targets set by the ArchSD on areas of major significant environmental impact are illustrated as below.

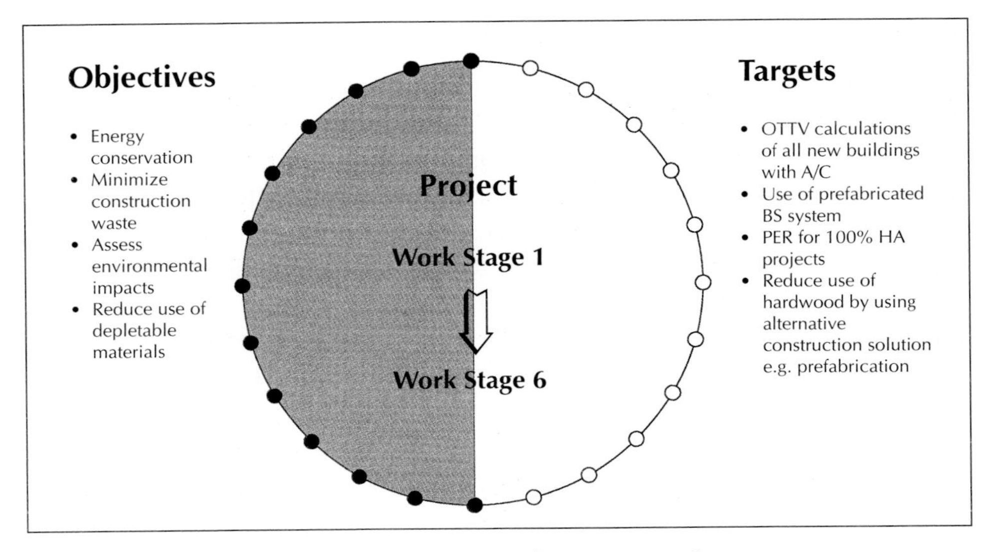

Architectural Services Department: Environmental Management System

Overall Thermal Transfer Value (OTTV)

To help reduce energy usage in maintaining the indoor comfort level of a building, it is important that the heat gain from, or heat loss to, the external environment be kept low. Heat gain/loss is mainly affected by the building orientation, type of construction, type and thickness of insulation material used, type of glazing material used, external shading provided, and colour of the external surfaces of the building.

In July 1995, the Buildings Department introduced the Building (Energy Efficiency) Regulations that require the building designer to calculate the OTTV of the building envelope if it is of an air-conditioned commercial building or hotel. The current limiting value of OTTV is $35W/m^2$ for tower blocks and $80W/m^2$ for podiums.

All ArchSD projects with air conditioning are now designed to comply with the specified limits or better. Upon evaluation of results, the future target is to reduce internal OTTV compliance figures. The use of external shading features and special glass, and reducing curtain wall areas, etc. are considered. Experience has indicated that attention to the following will help to lower the OTTV and reduce electricity consumption:

- minimize east- and west-facing glazing areas to reduce air-conditioning solar load;
- use lighter colours for the external surface of the building to reflect direct solar radiation;
- external shading based on summer solar angles to reduce direct solar radiation;
- reduce the use of curtain walls on all elevations;
- specify reflective or protective glass for curtain walls when used; and
- careful selection of external wall materials and careful construction to reduce thermal transfer.

The Legco building and the school covered walkway below illustrate the use of shaded walkways, which help to reduce OTTV.

Legco with Shaded Walkways

School with Covered Walkway

Project Example: Castle Peak Hospital, Phase I, Stage 1

The project features simple environmental measures such as main orientation on a north-south axis; landscaped gardens on thermally insulated roofs; use of light-coloured external materials; retention of existing boundary trees; use of natural air ventilation in mild weather; use of recycled cool air on extract for secondary rooms. The OTTV is 12.83W/m^2.

Roof Garden to Assist Thermal Insulation

Boundary with Trees Providing Greenery

Project Example: The Hong Kong Public Records Building

Having attained the Energy Efficiency Building Award for 1997, the project includes:

- minimal window area to control the internal storage environment;
- environmentally friendly fumigation process;
- special, climatic zones for specific documents;
- UV-resistant tinted glass in windows where provided;
- thick, insulated external structural walls, which also reduce the internal columns;
- acoustic barriers to the A/C plant on the roof to reduce noise transmission to neighbours;
- 24-hours-a-day, 7-days-a-week filtered A/C system;
- seamless, stain-resistant flooring to ease cleaning and reduce dust build-up; and
- special lighting and storage systems.

External elevation and interior of the Hong Kong Public Records Building designed with high energy efficiency considerations.

Landscape Input

The ArchSD takes up major open space and park projects of the Provisional Urban and Regional councils. Working with the client, care is taken to introduce new planting and soft landscape areas to ensure the future greening of Hong Kong.

For each building project, the landscape group is consulted to advise on the possibilities of introducing soft landscaping, but demands on available ground floor space tend to result in more podium-type gardens. Particular care is taken at site boundaries and areas of major visual impact. Where appropriate on major projects, integrated irrigation systems are used. For example, site investigation at the hillside site of the Pamela Youde Nethersole Eastern Hospital indicated a small but permanent run-off of fresh water. This was utilized and fed into storage tanks by gravity for the irrigation system. A back-up fresh water supply from the mains was needed, only as an emergency supply.

Landscape Project Example: Kowloon Walled City Park

An example of an area of urban decay that was taken away and replaced with a park is at Kowloon Walled City Park. The project, which replaces the old walled city, enhances the overall urban environment, provides open space for the public, retains historical elements and commemorates another part of the historical development of Hong Kong.

Walls of Chinese Style to Provide a Sense of the Past

Circular Moongate as a Traditional Chinese Feature

Reducing The Use Of Tropical Hardwood

Before the government's call for minimizing the use of tropical hardwood, the ArchSD issued an internal instruction in 1993 with a view to preserving the tropical rain forests by stipulating that no tropical hardwood should be used for site hoarding, covered walkways and signboards. Alternative sheet materials need to be specified instead. It is recognized that metal sheets may initially require more energy to produce, have more environmental impact and be more costly, but due to its potential for reuse it is more environmentally acceptable than unsustainable hardwood. The use of hardwood is also restricted to site accommodation, excavation supports, etc.

Hardwood is not totally restricted to use in formwork as work is often of a non-repetitive type; however, contractors are encouraged to use vertical metal propping and bracing as falsework to support formwork for concrete floor construction. Initiatives to employ system building and dry construction to minimize the use of timber shuttering and wet trades are ongoing.

Minimization Of Waste

Encouragement is made for better use of materials, including prefabricated systems and better detail design. Designing to suit exact tile modules, which results in minimum cutting to reduce construction waste, is encouraged. This also assists with buildability and the quality of workmanship.

Saving Resources By Recycling/Reuse

Further improvements are through the use of recyclable materials, reusable steel formwork, reusable concrete pavers, tiles, concrete kerbs, etc. Other areas include consideration of automatic flushing/washing sensors and smaller cistern capacities, which will help to reduce water usage.

▮ PRODUCTS/PROJECTS – ENERGY CONSERVATION

Energy conservation is one of the major areas of ArchSD policy and is very important in protecting the environment. In February 1995, the Building Services Branch of the ArchSD updated and published a booklet entitled *Energy Conservation in Buildings*. The document includes some basic design parameters and concepts to provide architects, engineers and those involved in building management with information on energy conservation and on the effective use of energy. It is available from our Building Services Branch.

It is important to note that energy conservation does not mean stopping the use of energy to meet our needs, but to use it wisely. Energy conservation is aimed at influencing the rate of increase of energy usage while at the same time maintaining or improving our standard of the living environment. The consideration of OTTV for the building envelope mentioned earlier is the main influence on energy requirements, but there are other significant areas in building servicing installations that can help to reduce energy usage.

Air-conditioning

There is no doubt that air-conditioning is a major energy consumer and contributor to the energy bill. The energy requirement for air-conditioning of a large office building is about 50% to 58% of the total energy requirements. As such, it is considered that the following points should be noted during the design of an air-conditioning system.

- Try to use chillers with higher coefficients of performance. Pay particular attention to the efficiency of the chiller during off-peak operation.
- Use independent A/C system/equipment for areas with separate requirements or different operation hours to avoid operating the main plant for a very small load.
- Reset the chilled water temperature to a higher level during off-peak hours or cooler seasons.
- Adopt variable air flow and variable water flow design.
- Use frequency inverted type of variable speed drives for pumps and fans.
- Consider heat recovery from chillers, boilers or exhaust air for other systems.
- Use heat pumps to provide warm water (for swimming pools, for example) and warm air.
- Use outdoor air instead of recirculated cool air for cooling in cooler seasons as it reduces energy usage and subsequently recurrent costs.
- Designate areas of different orientation as different A/C zones for optimum chilled water flow.

Cooling Tower with a High Coefficient of Performance on a Rooftop

Heat Recovery

Another important area in air-conditioning installation that can contribute to energy saving is heat recovery. Air-conditioning is in fact a mechanical means of transporting heat from within the building and disposing it externally, or vice versa. In projects where cooling and heating are simultaneously required in different locations, consideration should be given to recover the heat rejected from one location for use in another location. Suitable equipment to reclaim the heat rejected has to be installed. If the following conditions exist, consideration can be given to heat recovery.

- There is demand for low-grade heat, e.g. preheating of domestic hot water, space heating, etc.
- Heating and cooling loads coincide for a considerable period of time during the year.
- The difference in temperature and in latent heat content is large between the heat source and heat absorber.
- The payback period is reasonable.

Waste Heat Boiler Contributing to Energy Saving

Electrical System And Lighting

For a modern complex building, the demand for electrical power is very high. It is desirable to locate the source of power supply close to the load centres in order to minimize power loss during transmission. Consideration should be given to improving the power factor to an optimum value because having a low power factor is an inefficient way of transmitting electrical energy.

For lighting installations, choosing energy-efficient light sources is paramount to energy saving. For example, use energy-saving fluorescent tubes instead of conventional fluorescent tubes; use energy-saving miniature fluorescent lamps instead of incandescent lamp bulbs. Design for flexible switching circuitry to encourage the use of daylight when the work area is close to windows. Avoid using opal lighting diffusers as they screen out much light, especially when dirty. To summarize, the following aspects should be considered when designing an electrical and lighting system.

- Improve the power factor to an optimum value.
- Use high-efficiency lighting sources.
- Use localized lighting for the work area only and reduce the illumination of the general areas. (Note that the ratio of illumination levels between general areas and work areas should not be more than 1:4.)
- Use high-efficiency lighting control gear.
- Design for flexible switching of the lighting system to encourage the use of daylight.
- Install timer/automatic control device to switch off groups of lighting during outside operation periods.
- Provide indicator lights to alert watchmen that the lights inside unattended rooms have been left switched on.

Building Energy Management System

This type of system is useful when the mechanical and electrical systems of the building become very sophisticated and complex. It gives an integrated system for overall control and monitoring of the engineering systems installed in the building.

Building Energy Management System (BEMS) with well-developed energy conservation software programs can optimize design conditions and reduce energy consumption in the operation and maintenance of buildings. Typical operation programs can optimize the start and stop of chillers, boilers, pumps and air-handling units to cope with the actual and varying needs of the building; adjust the set comfort points with respect to ambient conditions; control electrical demand; minimize use of lighting outside operation hours by lighting switching control, etc.

Energy Conservation Features Incorporated In Projects

In the past, a number of the above building services energy conservation features had been incorporated in ArchSD projects. The following table is a summary of the application from 27 May 1985 to 15 September 1997.

Description	Application	
Air-conditioning		
Sea-water cooling	22	
Cooling tower	11	
Variable air volume for supply air	21	
Variable speed pump drives	33	
Run-around coils/heat pipes	5	
Heat pumps	10	
Thermal wheel	18	
Heat recovery chiller	13	
Natural air cooling	4	
Subtotal		137
Lighting		
Daylight utilization	244	
Low-loss control gear	154	
High-efficacy discharge lamps	344	
Energy-saving fluorescent tubes	377	
Timer/photo switch control	379	
Subtotal		1 498
Building automation		
On/off scheduling	107	
Remote resetting control points	77	
Energy-saving programme	54	
Subtotal		238
Grand total		1 873

Renewable Energy

Renewable energy from natural sources includes wind energy, hydro power, solar energy, etc. Due to geographical and natural resource constraints, solar energy is the main type of renewable energy that offers the most benefits for application in Hong Kong. However, the use of solar energy is not common due to the following reasons:

- The dense population, small site coverage, high-rise buildings obstructing solar radiation in the urban area
- Small roof areas
- Stable and economical supply of other types of energy (electricity, town gas, etc.)
- High capital cost of solar energy systems
- Low efficiency of photovoltaic cells, i.e. many cells are needed for more power

The ArchSD's development work on solar energy application is mainly confined to solar hot water installation. This includes provision of warm water to swimming-pools and preheating of water for general purposes. The following are projects that have incorporated solar energy systems into the hot water supply:

Project title	Solar collector area (m^2)
Stanley Latrine and Bathhouse (1980)	60
PWD Depot at Malaya Lines, Shek Kong (1981)	4
Lai Sun Prison (former drug addiction treatment centre), Hei Ling Chau (1981)	400
Tai Nam Street Bathhouse, Kowloon (1981)	48
Shek Pik Prison (1983)	498
Training Depot, Brigade Gurkha Swimming Pool (1984)	200
Tuen Mun Hospital Hydrotherapy Pool (1989)	96
Hot water system to Young Inmate Centre, Hei Ling Chau (1990)	30
Tsuen Wan Swimming Pool (1998)	320

Solar Panels on a Rooftop

▮ PRODUCTS/PROJECTS – IMPROVEMENTS TO EXISTING BUILDINGS

Noise Abatement Measures For Schools

Many schools, especially those located in urban areas next to noisy transport routes or those under the airport flight path, experience continuous noise levels that are above the latest recommendations in the Ordinances. On 22 October 1980, the Environmental Protection Advisory Committee's Special Committee On Noise considered the noise problem in schools and recommended that classrooms and special rooms experiencing excessive noise levels should be noise-insulated to improve the teaching and learning environment. The project 'Noise Abatement to Existing Schools' commenced, and more than 10 000 classrooms and special rooms have been treated in this ongoing work, which was originally carried out by private consultants with the ArchSD as the monitoring department. The schools were divided into four groups according to the severity of excessive noise levels. Noise abatement measures included installation of double-glazed windows and air-conditioning units for the affected rooms. One of the difficulties of the project lied with the short time span available for installation work, i.e. from mid-July to the beginning of September each year. Another difficulty the project team faced was the space requirement for additional transformer rooms, if required, when the original electrical supply had to be upgraded to accommodate additional air-conditioning load.

In 1991–92, the government decided that the programme should be advanced, and the ArchSD manage all works for the project by letting a number of Design and Build term contracts to noise-insulate about 400 additional schools. The work for the remaining schools included in the programme was still in progress at the end of 1999; however, as standards improve, further work may be required in the future.

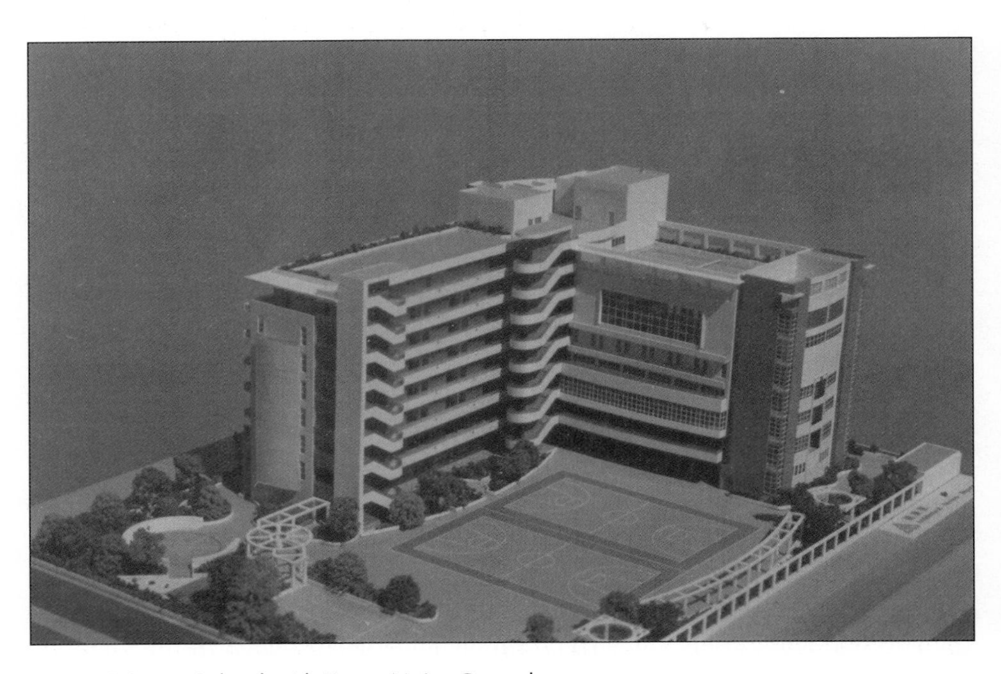

A New Primary School with Better Noise Control

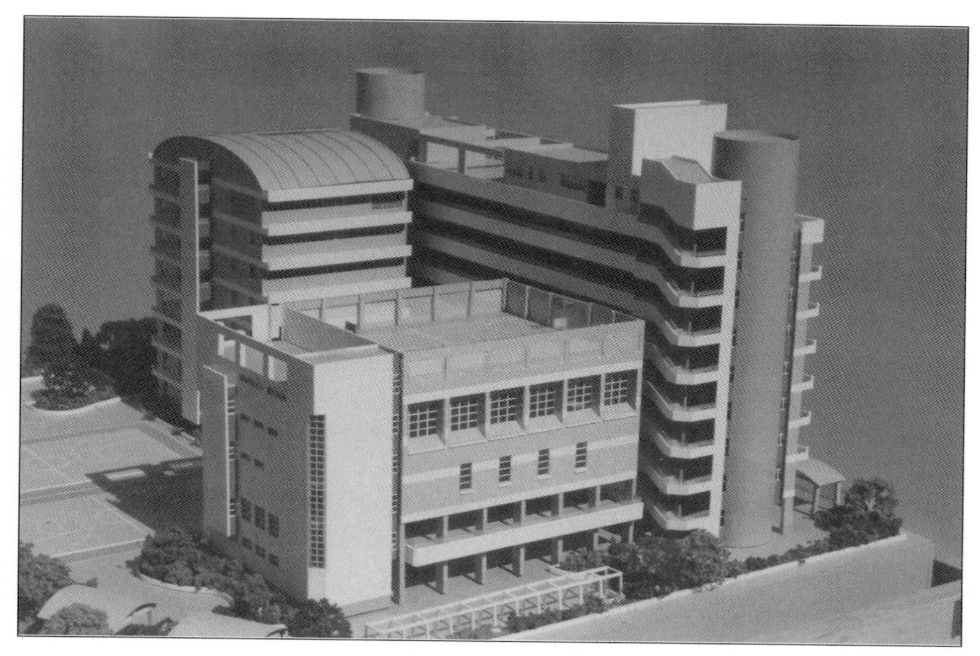

A New Secondary School with Better Noise Control

Renovation Of Sterilization Systems Of Swimming-pool Water

The ArchSD undertakes a full programme of public swimming-pool projects. One of the advancements in the service design of the swimming-pools is the transition from using chlorine gas to an on-site hypochlorite generation system for disinfection. Chlorine gas is able to sterilize water to the required standard if its residual concentration in water is maintained at 1ppm or more. Besides the undesirable side-effects of unpleasant odour, irritation to the eyes of swimmers, bleaching effect to the hair, etc., chlorine gas is classified as dangerous goods in Category 2. It poses a hazard to the swimming-pool neighbourhood if there is a leakage. Chlorine gas is an acute respiratory irritant producing a range of effects from minor irritation to death as the dose is increased. Because of the potential hazard in handling and storing chlorine gas, the ArchSD completed a conversion of all the 1000kg chlorine gas drums to 50kg cylinders in 1993 in order to reduce the risk of leakages.

In 1992, the ArchSD completed a study on alternative disinfection systems for swimming-pools, and recommended to the Urban Council and Regional Council that they should adopt an ozone sterilization system and an on-site hypochlorite generation system for public swimming-pools. The ozone sterilization system has many obvious advantages. It is the most powerful oxidizing agent and is more effective than chlorine gas in disinfecting against coliform, bacteria and other micro-organisms. For ozonated pools, the residual chlorine gas concentration is greatly reduced and its associated effects, such as eye irritation, bleaching odour, undesirable chemicals, etc., will be eliminated or reduced to an acceptable level. At the same time, the hazard associated with the handling and storage of chlorine gas is removed. The main drawback of the ozone sterilization system is its higher cost and higher space requirements for its larger plant room.

Up to 1998, the conversion work for all the scheduled Urban Council and Regional Council swimming-pools in the agreed programme had been completed. This included ten Urban Council swimming-pools: Morse Park, Kwun Tong, Pao Yue Kong, Lai Chi Kok, Kowloon Tsai, Tai Wan Shan, Shum Shui Po Park, Wanchai, Chai Wan, and Kennedy Town. It also included ten Regional Council swimming-pools: Shatin, North Kwai Chung, Tsuen King Circuit, Yan Oi Tong, Fanling, Yuen Long, Tuen Mun, Kwai Shing, Sai Kung Outdoor Centre, and Lady MacLehose Village.

Since 1992, all new swimming-pools have adopted the ozone sterilization system, including Hammer Hill, Jordan Valley, Tin Shui Wai, Tsing Yi, Tsuen Wan, Ma On Shan and other projects in the planning stages.

The improvements to sterilization systems together with other additional provisions in recent years, such as leisure and play equipment, have greatly enhanced the amenity value of the swimming-pools and provide a better and safer environment for the public.

Exterior of Hammer Hill Pool

Interior of Hammer Hill Pool

Programme Of Phasing Out Chlorofluorocarbons (CFCs) And Halons

In response to the Montreal Protocol and its amendments, the Hong Kong government enacted the Ozone Layer Protection Ordinance in 1989. As a responsible administrator, the Hong Kong government is determined to phase out CFC chemicals. A five-year programme of phasing out CFCs estimated to cost HK$310 million started on 1 April 1993. All chillers and refrigeration machines using CFCs as refrigerants would be either retrofitted to use environmentally more acceptable refrigerants or replaced with new machines that use refrigerants with zero ozone-depleting potential. A four-year programme started on 1 April 1994 to phase out halons. This programme was estimated to cost HK$127 million to replace existing fire-fighting installations with clean agent fixed fire-fighting installations.

The CFC phasing-out programme was scheduled to complete in 1998. Up to now 90 chillers and refrigeration systems totalling 69 300kg of CFCs have been replaced or retrofitted. The halon programme was also scheduled to complete in 1998 and to date, 110 halon fire-fighting installations with a total of 7 800kg of halons have been replaced.

Energy Management Study Of An Existing Office Building

Assessment of improvement measures for the energy performance of an existing office building starts with the fixed building envelope and the building being occupied as the two key components. It is then necessary to examine in depth the original design criteria, the operating modes of the services systems and the use by its occupants.

In 1991, the ArchSD appointed energy consultant Messrs Parson Brinckerhoff (Asia) Ltd. to carry out an energy management study for Wanchai Tower. A list of recommended energy management opportunities (EMOs) was presented in the Final Report issued in March 1992.

The implementation started in early November 1992. Due to operational restrictions, such as heavy usage of the courtrooms and unacceptableness of interruption of services to the public, certain recommendations could not be implemented. Nevertheless, after implementation of the remaining recommendations, savings of around HK$500 000 per annum have been achieved. The Consultant further monitored the status of the implementation of individual EMOs for Wanchai Tower and assessed the overall effectiveness. In summary, the following energy-savings can be achieved.

Saving achieved in 1994	490 000kWh
Further saving when all work completed	167 000kWh
Total saving anticipated	657 000kWh

Subsequent to the energy study conducted for Wanchai Tower, the consultancy was extended in October 1992 to cover energy studies for the Immigration Tower and the Revenue Tower, which were completed in 1994.

Project Example: Environmental Resource Centre

An example of a smaller building project that illustrates environmental awareness is the conversion of the old Wanchai Post Office into an environmental resource centre. The following measures were designed:

- Recycling the counter and post boxes as current day facilities.
- Reuse of materials from a demolished wall.
- Selection of new materials to match some existing works.
- Installation of water-saving wash-basins-cum-WC systems from Japan.
- Use of energy-saving lamps.
- Inclusion of a nature garden which introduces plants, rocks of Hong Kong to promote the concept of green landscape in a fully built environment.

Environmental Resource Centre as Transformed

Environmental Resource Centre as Transformed

Restoration Work

The ArchSD has a specialist group in its Property Services Branch responsible for the restoration of historical buildings. The group has undertaken many projects that contribute to recording history and providing visual relief from continuing new developments.

Kowloon Walled City Park Restoration

CONCLUSION

ISO 14001 certification is only the first step towards the goal of sustainable building development. Architects and related professionals in the construction industry are encouraged to follow the ArchSD by taking similar steps to address and minimize work areas that will cause major environmental impact.

The Architectural Services Department (ArchSD) is one of the largest professional entities providing building design, procurement, maintenance and advisory services in Hong Kong. In 1997, the ArchSD completed 123 major projects, valued at around HK$8.8 billion. Being a multi-disciplinary organization with over 2 050 staff, implementing aspects of environmental protection measures, directly or indirectly through its consultants, is an important part of its daily work as it contributes towards creating a better living environment for ourselves and for our children.

Appendix
Green Manager Report Form
Green Measures Taken and Assessment of Effectiveness (1996)
Branch/Department: Architectural Services Department

Period	Paper consumption#		Envelope consumption (no.)	For Office at Architectural Property Branch Centre*	
	Photocopying, offsetting and A4 single-line (reams)	Computer 2 000 folds per box (box)		Electricity comsumption (kWh)	Waste paper collection (kg)
Reduction target for the year (%)	5%	1%	1%	2%	NA
January	1 600	22	3 000	142 040	605
February	1 500	20	3 100	115 020	607
March	1 400	30	3 150	143 180	598
Quarterly total (last year's quarter)	**4 500 (5 950)**	**72 (73)**	**9 250 (9 265)**	**400 240 (421 140)**	**1 810 (1 800)**
April	1 600	25	4 200	138 340	589
May	1 400	23	4 260	183 150	585
June	1 500	19	4 250	203 390	580
Quarterly total (last year's quarter)	**4 500 (6 520)**	**67 (194)**	**12 710 (13 000)**	**524 880 (605 100)**	**1 764 (2 080)**
July	1 600	16	6 300	221 950	587
August	1 500	14	12 000	222 360	584
September	1 400	13	12 450	211 140	591
Quarterly total (last year's quarter)	**4 500 (5 028)**	**43 (70)**	**30 750 (34 030)**	**655 450@ (634 380)**	**1 762 (1 999)**
October	1 800	17	3 200	205 430	599
November	2 000	18	3 000	171 810	582
December	1 700	18	3 470	140 760	583
Quarterly total (last year's quarter)	**5 500 (4 500)**	**53 (79)**	**9 670 (12 350)**	**518 000@ (474 060)**	**1 764 (1 877)**
Annual total (last year's total)	**19 000 (21 998)**	**235 (406)**	**62 380 (68 645)**	**2 098 570 (2 134 680)**	**7 100 (7 756)**

Paper consumption figures shown indicate quantities issued to various divisions of the ArchSD.
* Figures are not available for main office in QGO, a joint-user building.
@ Increase mainly due to conversion of unoccupied areas into two contractors' offices.

5

Use Of Technology To Assist Environmental Design: A Case Study Of Verbena Heights In Hong Kong

Kam-sing Wong

INTRODUCTION

The housing project *Verbena Heights* was first known by its address 'Area 19B of Tseung Kwan O New Town'. The design inception occurred in late 1992.

The developer was the Hong Kong Housing Society (HKHS), a quasi-governmental body providing subsidized housing on both a rental and sale basis. The gross site area was approximately 2.1 hectares of reclaimed land. The development brief entailed about 1 000 rental units and 2 000 flats-for-sale, plus a spectrum of amenity and community facilities (including a car park, a market-place, a nursery, a kindergarten, a children-and-youth centre, a rehabilitation centre for persons with disability, and a centre for the elderly).

The brief was similar to many other large-scale public housing projects in Hong Kong. However, the project was started with quite a *different* design agenda – the project team was committed to explore a proactive 'environmentally responsive' approach from the stage of design inception. The project team comprised not only the architect, but also the project manager and other professional and specialist consultants.

Verbena Heights was completed for occupation in two phases: the rental blocks in late 1996 and the for-sale blocks in mid-1997. In the light of its *alternative* design approach, the project had aroused interest from both academics and the public. The project design was not intended to be an end in itself, but to signify an effort made on the path towards sustainable housing design in the context of Hong Kong (Plate 1).

Plate 1 Verbena Heights (south view): An alternative housing form in contrast with the surrounding 'point blocks' prevalent in Hong Kong.

Feedback and evaluation on the project will hopefully be beneficial to the community at large. This chapter is an attempt to share some of the lessons learnt from the design and research process of *Verbena Heights*. The key points of this evaluation may be presented by the three Rs: research, responsiveness and review.

▌RESEARCH

Initially arising from general concern and a good intention, the substantive content of the environmental agenda has yet to be defined. What does *green* housing design mean?

A team of environmental consultants and specialists were appointed by the HKHS to conduct a comprehensive research on the environmental design and management aspects of the housing project. The principal environmental consultant was ERM (Hong Kong) Limited, which was supported by ECD Architects and Energy Consultants (from the UK) on solar- and energy-related issues, Vipac Engineers and Scientists Limited (from Australia) on wind- and energy-related aspects, and Arup Acoustics (from Hong Kong) on noise mitigation. The research covered the following key range of issues:

energy efficiency
water conservation
waste management
material conservation
occupants' health
environmental education

The research was generally carried out in two stages. The initial stage of research was intended to be of a generic nature, and the findings would have general relevance to all housing projects in Hong Kong. Based on the initial findings, the next stage of research was more focused in terms of scope and depth, being specific to Verbena Heights by taking into account its contextual and architectural particulars. The research findings set the scientific foundation that guided the progressive evolution of the architectural layout and design detail of the project.

The use of technology played a dual role in the project. On the one hand, consideration was given to the application of contemporary technologies that are potentially appropriate to residential buildings in Hong Kong. On the other hand, state-of-the-art technologies were adopted in the process of building design with a view to optimizing the performance of environmental responsiveness.

Technology (I)

The climate of Hong Kong is generally considered appropriate to the harnessing of solar energy for active use. Solar water heating and photovoltaic (PV) systems are options of proven technology presently available for selection. Undoubtedly, solar energy is a 'clean' and renewable source of energy for beneficial use.

However, systems and products need to be economically viable. A cost-benefit analysis is usually carried out to evaluate the merits of a particular system. The evaluation is typically presented in terms of the payback period, i.e. the number of years anticipated for a system to pay for itself (including the cost of initial installation and recurring maintenance of a system). Technological advancement is expected to progressively improve the system's efficiency and shorten the payback period.

In the case of Verbena Heights, the harnessing of solar energy for domestic water heating was considered sufficiently feasible for recommendation. The payback period was anticipated to be fewer than ten years. The choice of a PV system for supplemental power generation, which would take over ten years to pay for itself, was however eliminated.

High-rise residential buildings impose further physical limitations to the application of solar systems. In view of Hong Kong's solar path and the efficiency of solar collectors available nowadays, the roof of a high-rise building is usually the most appropriate location for installing a solar collection system. The façade-mount option will result in a substantially lower efficiency. Given the decreasing roof area ratio for high-rise residential buildings of increasing height, the lack of space for mounting solar panels is potentially a critical constraint to the extent of application. In addition, the increasing distance between the roof and residential units at lower floors leads to lower efficiency in terms of heat transmission.

For Verbena Heights, the final recommendation suggested providing the flats at the topmost five storeys with a solar water heating system, with a cantilevered roof structure to extend the surface area for solar collection. Despite technical and financial feasibility, the solar water heating system has not yet been provided as decided by the HKHS, but the option has been left open for future installation (probably when the technology becomes more affordable and efficient) (Plate 2).

Water is often another wasted resource in contemporary cities. The supply of sea water for toilet flushing in Hong Kong is a distinct and environmentally sensitive practice among opinions in world cities. However, some parts of the territory, including a number of new town development areas (e.g. Tseung Kwan O), are not yet provided with this type of sea-water infrastructure.

A 'low-flush water closet' was therefore envisaged as an alternative technology for use. According to the Building (Standards of Sanitary Fitments, Plumbing, Drainage Works and Latrines) Regulation, the flushing capacity for toilets is prescribed between 9 and 14 litres. Water closets of advanced hydrologic design can achieve the same or even better performance at 7.5 litres per flush (or even less). The alternative technology reduces not only household water use, but also the energy consumption in pumping and treatment. For Verbena Heights, given a promising result from the cost-benefit analysis, the HKHS approved the specification of low-flush water closets for use at all residential units.

Plate 2 Cantilevered roof: Extended roof surface area for solar collection while providing supplementary solar-shading and insulation.

Technology (II)

Energy use is central to addressing the environmental agenda, particularly relating to key global issues like climatic change due to greenhouse gas emissions, and regional/local issues like acid rain and air pollution. Reduction in the consumption of energy (especially fossil fuels) and emphasis on the harnessing of renewable energies are the key strategies.

With this premise, the consultant team of *Verbena Heights* conducted a preliminary survey to identify the patterns of energy consumption in typical households in Hong Kong. According to a life-cycle energy analysis of high-rise dwellings in Hong Kong, the recurring operational energy is the major sector of energy use. Among others, space conditioning represents the key component of domestic energy consumption in need of reduction. Like many vernacular dwellings, contemporary architectural design can be climate-responsive and capitalize on renewable natural assets, including the harnessing of ambient wind for cooling (Plates 3 and 4).

Plate 3 Solar-shading device: Study on alternative forms of external shading device.

Plate 4 Elevation detail: External shading screen wall, bay window with upper light shelves, and perforated metal panel in front of clothes-drying racks.

Solar shading devices placed at strategic locations are conducive to effective thermal control. The solar path is seasonal, but generally comprehensible. For a complex building layout, it is beneficial to use three-dimensional computer modelling to assess the solar effects, especially the implication of inter-block shading (Plate 5).

Plate 5 Inter-block shading study: Computer modelling for designers to understand the relationship between solar exposure and inter-block shading effect at different months and time.

The pattern of ambient wind is usually far less understood. Even though the record of prevalent wind data is becoming more readily available and reliable nowadays, it is still difficult for building designers to reasonably predict the wind effects at a particular site. The wind pattern is not only seasonal but also changes easily with the context. In Hong Kong, the mountainous topography and the densely built-up urban fabric further intensify the complexity of local wind patterns.

Besides computational means such as CFD analysis, wind-tunnel modelling is considered an appropriate technology for designers to acquire a deeper understanding of local wind behaviour. Through wind-tunnel modelling tests, the project team of *Verbena Heights* explored and adjusted the disposition and height of the residential blocks, with the intention of optimizing the effectiveness of natural ventilation for residential units in the summer months. Instead of the conventional cruciform point blocks at a uniform height, linear building plans and stepping tower heights were derived to catch the prevailing summer breeze from the seaside. Innovative 'breezeways', in the form of multi-storey balconies/openings dispersed throughout the high-rise building blocks, were explored to further enhance the wind permeability of building mass. Wind-tunnel modelling was adopted to assist in verifying the strategic juxtaposition of the breezeways (Plates 6 and 7).

Plate 6 Wind-tunnel modelling: Laboratory tests for designers to understand the complex relationship between built form and local wind environment, taking into account the surrounding context.

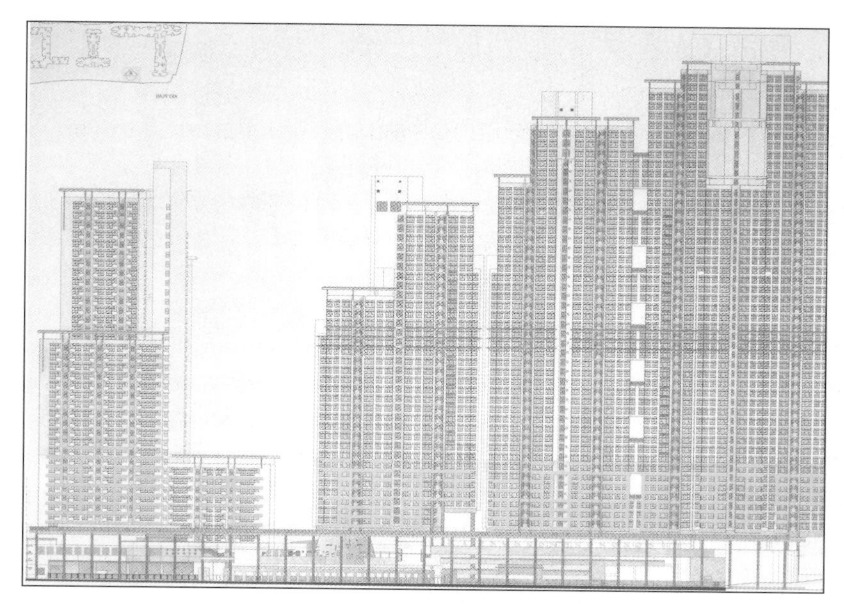

Plate 7 Verbena Heights (southwest elevation): Stepping tower heights and 'breezeways' in the form of openings through the building mass for effective wind-driven ventilation.

The wind effect on pedestrian levels adjacent to high-rise structures is also a concern for comfort and even safety. For Verbena Heights, wind-tunnel modelling tests were carried out to appraise the pedestrian-level wind climate at major circulation routes and outdoor sitting areas. At locations where the wind climate exceeded the recommended standards, amelioration measures were studied and then verified by wind-tunnel modelling (Plate 8).

Plate 8 Entrance portal with floating wind-break: An example of amelioration measures to manage the wind effect on pedestrian levels adjacent to high-rise structures.

▮ RESPONSIVENESS

Technology is a tool for designers to explore design options. The limit is the designers' innovation and 'value'.

For Verbena Heights, the project team had to confront an essentially economic-technocratic world-view, which is predominant in our society nowadays. However, an underlying thrust came from a commitment to exercising our responsibility and care to Planet Earth.

Presently, buildings use half of the annual energy consumption in the world, and the associated ecological impacts are phenomenal. The truth is that many things on which our future health and prosperity depend are in dire jeopardy: climatic stability, the resilience and productivity of natural systems, biological diversity, and the beauty of the natural world.

Building designers play a key role in environmental responsibility, and this should be the turning point for us to shift towards more responsive practices.

We can make a change.

▮ REVIEW

As admitted at the beginning of this chapter, *Verbena Heights* was not intended to be an end in itself, but to signify an effort made on the path towards sustainable housing design in the context of Hong Kong. Feedback and evaluation is considered an essential component of the process of research and development (Plate 9).

However, Hong Kong is characterized by its low level of investment on research aspects, even in comparison with many of its neighbouring Asian countries. It is time to review institutional investments, with a view to catching up towards sustainability.

Another suggestion is to review our education, especially for the building professions. Educational institutions should set a goal of ecological literacy for all of their students. Graduates should have the capacity to distinguish between development and growth, sufficient and efficient, optimum and maximum, use and 'abuse' of technology, and adopt a 'can do' attitude rather than a 'can't do'.

What should we do now?

Plate 9 High-density housing: Verbena Heights — the search for an environmentally responsive form in the high-density urban context and humid subtropical climate.

6

A Public Housing Experience

Ada Yin-suen Fung

INTRODUCTION

The Hong Kong Housing Authority has been called one of the largest homemakers in the world. Over 870 000 housing flats have been built, a further 270 000 flats will be completed in the next five years, and we have our ambitious housing production targets to achieve. The major development and construction processes include planning, design, demolition, site formation, piling and building. Development and environment go hand in hand. The concept and practice of environmental protection can be illustrated throughout these processes. Our core values – caring, customer-focused, committed – also apply to our environment. This chapter gives a brief overview of our care for the environment in respect of planning and design, waste reduction and the construction process right through to the living habitat.

▮ PLANNING AND DESIGN

Our concern for the environment starts right from the very beginning. Environmental assessment studies, site planning, standard design and project design all play an important role towards environmental protection.

Increasing The Supply Of Public Housing – Strategy And Challenge

The Chief Executive announced three major housing targets in his inaugural speech on 1 July 1997:

- to provide at least 85 000 flats annually in the ten-year period from 1997/98 to 2006/07;
- to reduce the average waiting time for public rental housing to three years over the same period; and
- to achieve a home ownership rate of 70% by 2006/07.

It is estimated that, of the annual production target of 85 000 flats, 50 000 will be for public housing under the ambit of the Housing Authority. There are three major actions to increase the supply of public housing:

- to seek additional allocation of new public housing sites, including sites in the strategic growth areas and supplementary housing sites;
- to maximize development intensities for existing and new public housing sites, like infill developments and an increase in development intensity;
- to shorten the production lead time, from site identification, planning and project approval to construction period.

These pose more challenges for the project teams. When they encounter sites that are heavily constrained by environmental factors, they need to be very skilful in striking a balance in order to optimize the development potential, meeting environmental requirements and business objectives. Indeed, project teams have been progressively practising these balancing acts throughout previous years.

Environmental Assessment Studies – An Early Focus

During different stages of a project, environmental assessment studies are required to confirm the environmental acceptability of the project. This is particularly important during the feasibility and planning stages, so that we can ensure early focus on environmental issues and allow timely action.

We have to make the best use of our land supply to achieve our housing production targets. In order to optimize the development potential and to meet environmental standards, we need to conduct environmental assessment studies for certain categories of sites, especially for sites involving land-use rezoning.

For public housing developments, environmental assessment studies are usually related to:

- traffic noise and air quality;
- industrial noise and air quality;
- sewerage discharge;
- landfill gas hazard and land contamination.

These studies help to identify the suitability and intensity of housing development, suitable mitigation measures, modification and upgrading works. Based on these recommendations, we then prepare the planning parameters, conceptual layout and client requirements. This is to ensure that the project design conforms to these environmental requirements, right through to project implementation and completion.

Designing To Meet Environmental Constraints

In 1991, the Housing Authority established an appropriate yardstick for the optimum development of public housing sites. As a result, a new concept based on 'development ratio' (DR) was introduced for use in lieu of the population density to guide the planning and design of its new housing projects. An appropriate DR within the range of 5 to 7 was selected, depending on the site constraints. DR is defined as follows:

$$DR = \frac{\text{gross floor space (domestic + commercial)}}{\text{net estate area}}$$

DR7 was normally applied to sites that are straightforward, and a lower DR would be applied to sites that are difficult, such as those with noise constraints. However, since 1994/95, the Housing Authority has been facing a pressing need to increase the housing production, and we have been striving hard to achieve higher DRs like DR7 even for sites that are heavily constrained by environmental factors, e.g. Tseung Kwan O Area 59 and Ma On Shan Area 77.

By 1997/98, we applied our strategies to further increase our housing production by increasing development intensities. We are now trying to achieve DRs that are higher than DR7. We try to achieve DR7.5 or DR8.

To overcome the problems of noise and air quality that we encounter in site planning, we have taken the following actions to achieve higher development intensities:

- Use non-noise-sensitive buildings, such as commercial centres and car parks, at the perimeter of the site to act as noise shields adjacent to major roads.
- Use single aspect buildings or reduced aspect buildings instead of standard domestic blocks.
- Use podium structures to form noise shields adjacent to major roads.
- Use blank gable-end walls to shield noise.
- Use non-noise-sensitive design to accommodate housing for senior citizens above car-park podia.
- Set back standard domestic buildings away from noise sources.
- Provide noise barriers where necessary.
- Site residential buildings away from atmosphere-polluting sources.
- Introduce optimal building disposition and orientation for residential blocks.

Designing For Energy Efficiency

Thermal efficiency is a major consideration in the design of windows in standard domestic blocks. Window areas at the façades of the Standard Harmony Block were revised in 1994 as a result of thermal efficiency studies. Overall Thermal Transfer Value (OTTV) has been adopted as thermal control in the design of building envelopes of commercial centres.

In respect of lighting installations, different light fittings of high efficacy together with the associated automatic controls are used to save energy:

- Specify minimum luminous outputs for fluorescent lamps and compact fluorescent lamps to make good use of electricity supply.
- Incorporate both automatic timer control and manual override control into the control of lighting in communal areas to save energy and to facilitate operational efficiency.
- Specify energy-efficient low bay floodlight fittings with SON lamps in the design of lighting systems for car-park buildings to save energy.

Other measures to save energy include:

- Specifying more energy-efficient lift drive systems.
- Adopting energy-saving features in air-conditioning installations such as

duty cycle, power demand monitoring, optimum start/stop time, temperature reset, etc.

- Specifying that water pumps shall be so selected that the operating points are located near the best efficiency points.

Designing For Noise Control In Domestic Buildings

Apart from mitigating external traffic noise, we also need to reduce the noise generated by services in buildings, like water pumps and air-conditioning plants, etc., so that a level of acoustic comfort to residents is maintained.

Low-noise motors for fresh and flush water pumps are being used to restrict the airborne noise to a specified level. To reduce the structural-borne noise, we adopt the following noise abatement measures to the ground-floor pump rooms of domestic blocks:

- Fully isolated slab for ground floor pump room.
- Pressure-reducing valves with acoustic enclosures at pipe inlets to fresh and flush water sump tanks.
- Modulating ball valves for fresh and flush water sump tanks.
- Pipe-mounting brackets with vibration isolators.
- Anti-turbulence pipes in water tanks.

Silencers are normally installed to attenuate the noise emitted from central air-conditioning units of commercial centres. External chiller plants are normally provided with acoustic enclosures to reduce the noise level. Emergency generators are now relocated to the roof of domestic blocks to minimize noise disturbance to residents.

Avoiding Ozone Depletion – A Global Responsibility

The Ozone Layer Protection Ordinance controls ozone-depleting substances and fulfils Hong Kong's obligations under the Montreal Protocol on Substances that Deplete the Ozone Layer. Accordingly, we have taken the following measures to avoid ozone depletion:

- Specify installation of environmentally-friendly refrigerant HFC 134a for air-conditioning to replace ozone-depleting refrigerants such as CFC-11, CFC-12 and HCFC-22.
- Use carbon dioxide as the fire-extinguishing agent for fire extinguishers, instead of halons which are ozone-depleting.
- Specify CFC-free thermal insulation materials for ductwork and pipework in air-conditioning installations to replace polystyrene insulation.

Use Of Fabric Reinforcement

In Hong Kong, steel reinforcement in the form of individual steel bars is a basic constituent of high-rise buildings under the traditional construction method.

Following the successful implementation of three pilot projects, fabric reinforcement has been specified for the construction of standard blocks since 1996. The tailor-made steel fabric reinforcement can optimize structural design, ensure quality workmanship, reduce labour, and reduce material wastage. As a result, wastes from construction sites can be reduced.

Fuel Ash

Following the success of pilot projects, the use of pulverized fuel ash in foundation concrete works for all domestic blocks became a mandatory requirement since late 1993. It marked a concrete step to mitigate the disposal problem of this environmentally unfriendly by-product of Hong Kong's power plants.

UPVC Windows

Following the pilot project that tried out the use of UPVC windows in Fanling Area 47B, UPVC windows have been specified with aluminium windows for use in some standard domestic blocks since 1997. UPVC windows have satisfactory environmental performance in the following aspects, albeit at a higher cost when compared to aluminium windows:

- Reduced noise transmission through the window frame.
- Reduced thermal conductivity through the window frame.
- Better resistance to aggressive atmospheric conditions, especially against moist, salient air and industrial fumes.
- Less energy consumption in production.

Automated Refuse Collection System

Two pilot estates have been provided with trial automated refuse collection systems: Shek Yam East Phase 1, and Wah Sum Estate in Fanling Area 47B. These will help to reduce the volume of refuse collected, as well as provide more hygienic estate conduits. A centralized refuse handling plant collects all the refuse by suction, compacts it and stores it in a container for the Regional

Services Department's special refuse truck to collect. The implementation review is being conducted to examine the cost and benefit of these installations.

The system at Wah Sum Estate is the first installation for residential use in Hong Kong (see Figure 1). A brief description of the system is as follows:

Design Concept

3 913 flats in six blocks; 11 000 occupants (approximate).
Estimated volume of waste: eight tonnes per day – one container.

Function

A pneumatic conveyance system for refuse collection consists of an integrated collection plant, piping, systems and discharge valves. Under each refuse chute there is a discharge valve in which the waste is stored between the emptying cycles. Separate pipe networks under the valves connect the chutes with the collection plant.

In the refuse collection plant the refuse and air are separated in a cyclone, and the refuse falls into a compactor that compresses the refuse into a container.

Piping System

The piping system consists of a steel underground refuse transport pipe of diameter 500mm, and 10mm wall thickness on straight pipes and 14mm in bends that connects the refuse chutes with the collection plant.

Discharge Valves

At the bottom of each chute there is a discharge valve over which the refuse is stored until its turn for emptying.

At the end of each branch pipe there is an air inlet valve for letting in transport air.

Operation

Twice a day: 8–10a.m., 7–10p.m.

The installation works automatically in accordance with a computer program. The computer monitors and registers the operation, and also generates alarm signals in case of problems, indicating what and where the problem is.

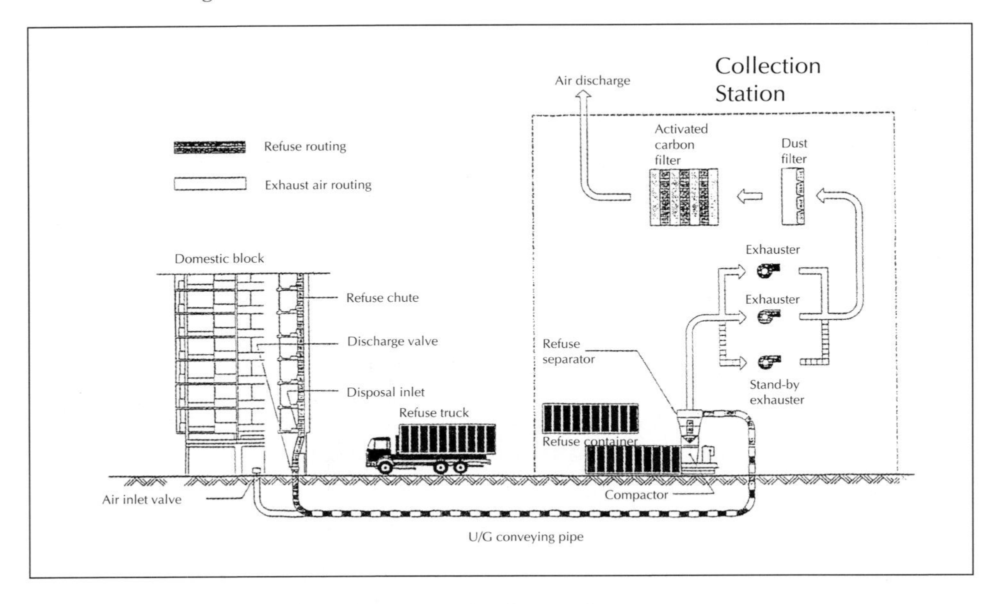

Figure 1 Operation Diagram of Auotmated Refuse Collection System

▌ WASTE REDUCTION AND CONSTRUCTION PROCESSES

In Hong Kong, construction waste accounts for about 50% of the total amount of solid waste. Our concern for the environment is exemplified in the handling of waste materials from demolition works and in asbestos management. Moreover, system construction on building sites reduces construction waste and minimizes the use of materials for temporary works.

Demolition And Landfills

To ensure the scarce landfill capacity is well utilized, all demolition contractors are required to sort and process all materials arising from demolition works at source and to remove all timber, steel, rubbish and other decomposable materials so that only suitable materials are dumped at landfills.

This measure will help to make landfills less hazardous to health and to the environment, and will reduce the need for more landfills.

Asbestos Management

Various types of asbestos-containing materials were used in buildings before asbestos was known to be harmful to health. As a result, the control of asbestos has become a concern in the demolition of buildings.

Apart from complying with the Air Pollution Control Ordinance, the Housing Authority has an established asbestos management policy, strategy and procedures associated with the management and abatement of asbestos-containing materials. Specification for asbestos removal works, responsibilities and procedures for the removal of asbestos-containing materials are clearly documented in demolitions contracts.

Moreover, sample documents for works procurement, including specifications and work methods, are also laid down for asbestos abatement works managed by the management and maintenance staff of the Housing Authority.

System Construction

To improve building efficiency, the department has adopted the use of standardized building designs for more than 20 years. The use of standard modules for different sizes of flats in the design of housing block types facilitates standardization and mechanized construction. This results in waste reduction on our building construction sites. Moreover, 'green' innovation in the choice and use of materials for temporary works further reduces waste and preserves the precious resources of the Earth. Standardization that leads to waste reduction features the following elements:

- Use precast façades and precast staircases to eliminate the need for traditional formwork during the construction process.
- Use factory-produced standard prefabricated components such as panel walls for internal partitions, door-sets, metal gates, cooking benches, sink benches and windows.
- Specify metal hoarding instead of timber hoarding.

With the use of repetitive modular units in domestic blocks, e.g. the Harmony Block series, contractors can also standardize their formwork and adopt different kinds of system construction techniques. These mechanized construction methods were pioneered in the 1980s, and more innovative systems have been developed progressively by contractors in the 1990s:

- Large panel steel wallforms have been developed to replace timber formwork for standard domestic blocks.
- Large panel steel 'table-forms' for slabs have also been developed.
- Semi-precast slabs for typical domestic units are used by some contractors

in conjunction with precast façades. They can act as permanent formwork, hence reducing the need for slab form altogether.

- Where the site condition is not suitable for large-panel steel formwork, smaller panels of aluminium formwork are used as an alternative to timber formwork for some projects.
- Coupled with the above construction methods, gondolas are used instead of bamboo scaffolding for the major façades of domestic buildings.
- Steel climbforms have been tried out by some contractors since 1997.
- Prefabricated proprietary site offices are being used by some contractors instead of constructing timber site offices.

These enable contractors to achieve operational efficiency and waste reduction by substantially reducing the use of natural resources of timber and bamboo as temporary works. Moreover, the use of metal is kept to a minimum where possible.

THE LIVING HABITAT

After the buildings are completed, it is the residents who carry on with the work to protect the environment. It is how they act and what they do with their habitat that matters. Promotional and educational activities on housing estates can foster their awareness in environmental protection.

In 1994, Friends of the Earth and the Housing Authority jointly organized an 'Adopt a Housing Estate Project', which was a new concept of community-based environmental education programme. Hau Tak Estate in Tseung Kwan O was the first estate selected for this scheme. Following the 'kick-off' ceremony, a series of activities had been held. These included Green Family Competition, Green Market Competition, Eco-forum, Green Seminar Workshop, Waste Minimization Campaign, etc., all aiming to inculcate the concept of 'reduce, reuse, recycle' in the minds of the residents.

Through better awareness of environmental issues and concern for their living environment, residents can also develop a better sense of belonging to protect their living habitat. Our estate management and maintenance teams will be working with residents to achieve the ultimate goal of protecting the environment.

CONCLUSION

The Housing Authority has pioneered a lot of initiatives to enhance environmental protection in the development and construction processes, extending right through to the living habitat. Following this path of environmentally friendly development for habitation, we shall take further steps to put in place an environmental management framework in the future. This will ensure that we can meet the challenges of achieving sustainable development for our city in the new millennium.

REFERENCES

Hong Kong Housing Authority. 1994. *Hong Kong Housing Authority Annual Report 1993/94*. Hong Kong: Government Printer.

——. 1997. *Hong Kong Housing Authority Annual Report 1996/97*. Hong Kong: Government Printer.

Howlett, B. 1997. *Hong Kong 1997: a review of 1996*. Hong Kong: Information Services Department.

Tang, W.K. 1996. A green challenge in building services. *Quest QA Newsletter* (Housing Department (Works Group), Hong Kong), January 1996.

Yuen, T.C. 1997. Increasing the supply of public housing to meet projected demand. Territorial Development Conference, October 1997, Hong Kong.

7

Passive Environmental Strategies For Architectural Design

Bernard Vincent Lim and Man-kit Leung

INTRODUCTION

This chapter examines the human comfort criteria in relation to the climatic as well as the urban context of Hong Kong, and investigates how these criteria can be met by applying various passive environmental strategies for architectural design in the region.

HUMAN COMFORT CRITERIA AND CLIMATIC CONTEXT

The concept of neutral temperature[1] can be used to ascertain the comfortable temperature range of three cities with different climatic contexts – Hong Kong, Shanghai and Kuala Lumpur. This is to compare the variations of comfortable temperature ranges in various regions in relation to the local climate, human activities and clothing. Clothing values of people taken in this analysis, though arbitrary by nature, are assumed to reflect different cultural backgrounds of different societies and their effect on neutral temperatures. Results are summarized in Figure 1.

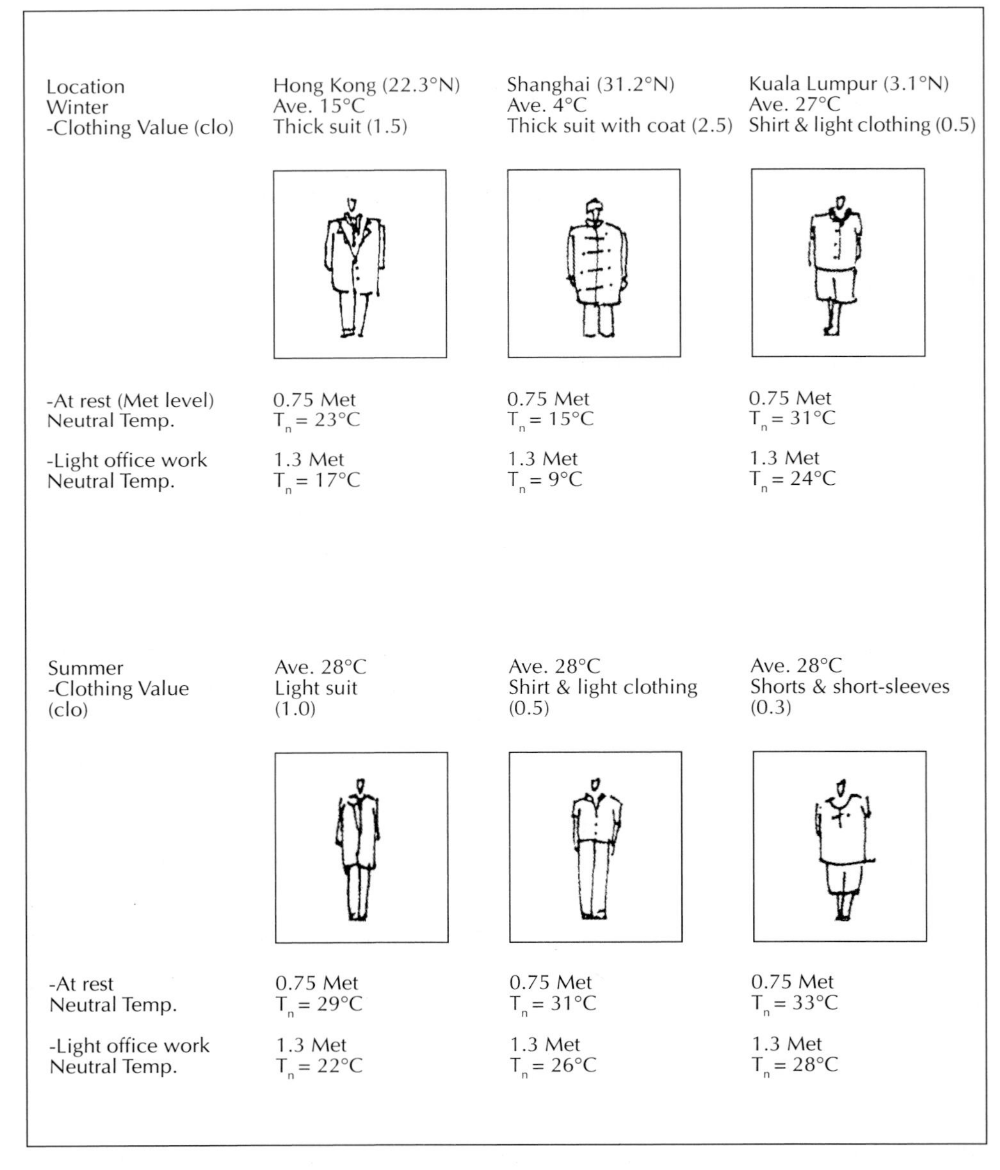

Figure 1 Comparison of neutral temperatures in Hong Kong, Shanghai and Kuala Lumpur.

The Climatic Context Of Hong Kong

Analysing the comfortable temperature ranges in our region, we can compare them with the actual ambient conditions throughout the year to find out the 'comfort zones' of Hong Kong in various months (Figure 2).

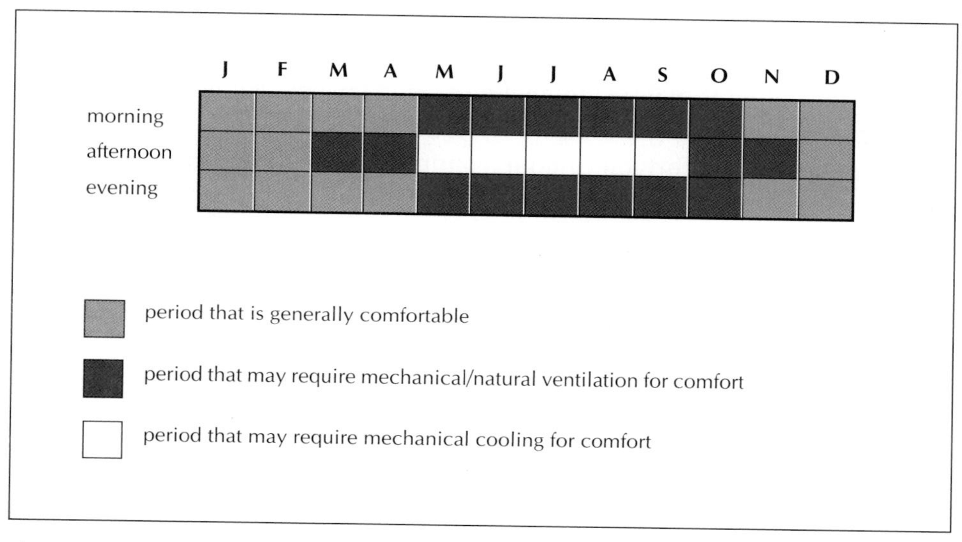

Figure 2 Comfort zones in various months of the year in Hong Kong.

We have employed outdoor temperature data for the computation of the above analysis. When indoor comfort is considered, we should be beware of the presence of internal and external heat gains. Appropriate strategies should be adopted to reduce these heat gains. These will be discussed in the next section.

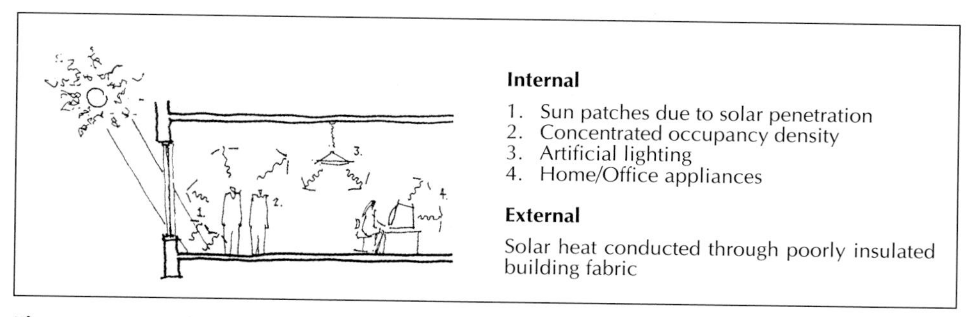

Figure 3 Examples of internal radiant heat sources that may significantly affect human comfort conditions.

Summary

Human comfort criteria differ in various climatic and clothing conditions as well as human metabolic rates.

The climate of Hong Kong is generally comfortable from November to April. Sufficient air movement by natural or mechanical ventilation is required from March to November to maintain human comfort. From May to September when the ambient temperatures in the afternoon are too high for passive cooling, mechanical cooling is required. A change of mode in environmental strategies, from passive to mechanical cooling, is desirable.

To maintain comfortable indoor conditions, we should minimize both internal and external heat gains to buildings.

▮ THE URBAN CONTEXT OF HONG KONG

The Urban Heat Island

In Hong Kong, skyscrapers, which strive to exploit the development potential of their land, are clustered in the urban core along the harbour. Deep 'urban canyons' are formed between these tall structures (Figure 4). Solar heat is trapped due to the canyon effect. No substantial wind is possible near the street level to take away the dust and smoke generated by vehicular traffic. This further escalates the elevation of air temperature inside the canyon and results in the urban heat island (Figure 5).

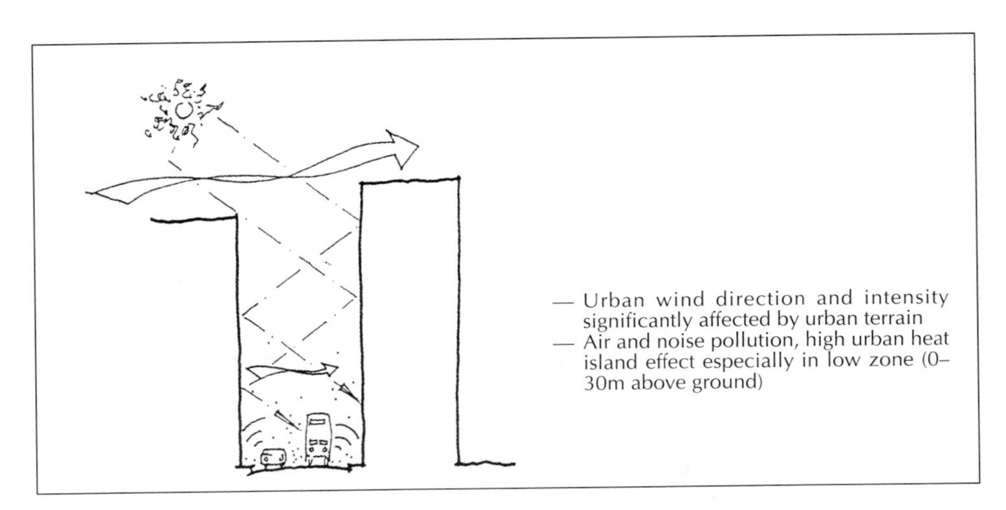

Figure 4 The Urban Canyon

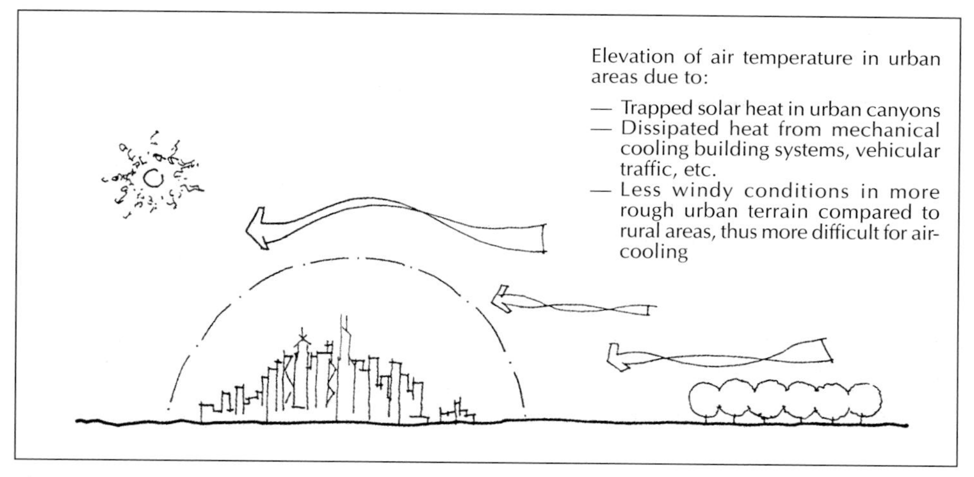

Figure 5 The Urban Heat Island Effect

The air temperature difference between urban and rural areas is known as the heat island intensity. Studies[2] reveal that its value is dependent on the proportional depth of the urban canyon and wind conditions. We can assess the urban heat island intensities in different areas of Hong Kong by using the chart illustrated in Figure 6.

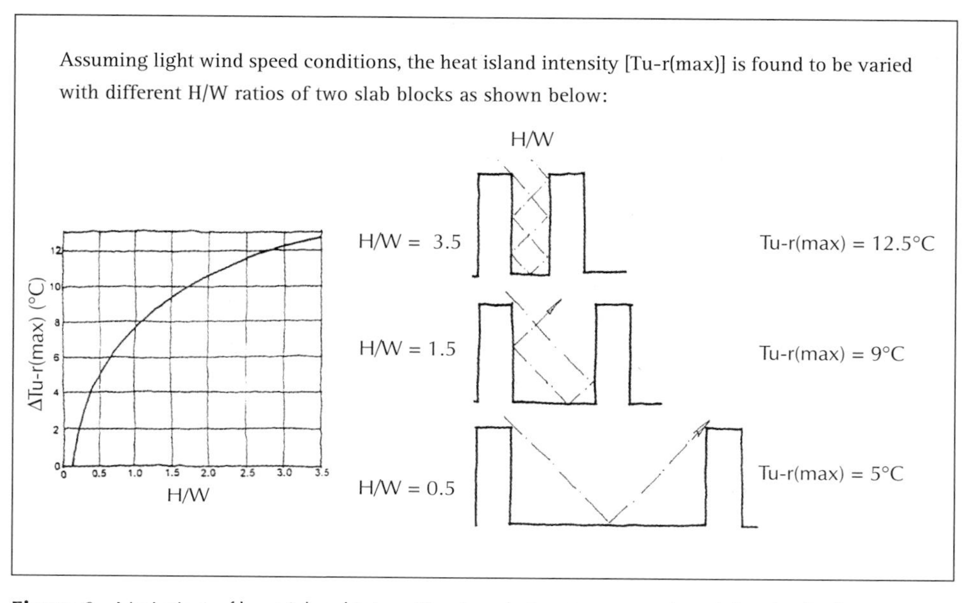

Figure 6 Variation of heat island intensities in relation to proportional depth of urban canyon (source: Oke, 1987).

As light wind speed is a prerequisite for high heat island intensity,[3] we can also ascertain the locations of possible heat islands by mapping the urban settlements and the principal topographically confined airsheds[4] as shown in Figure 7.

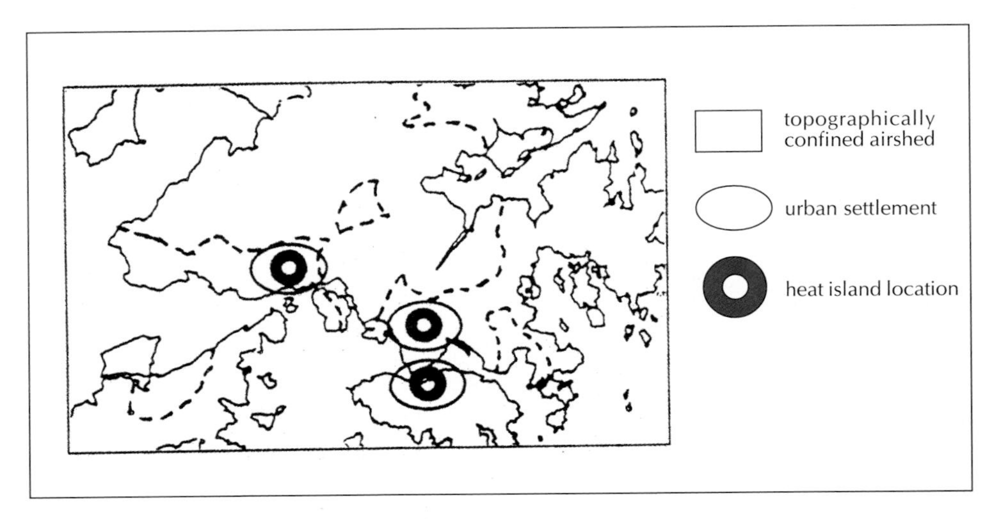

Figure 7 Mapping of Possible Heat Island Locations

Summary

The effect of urban heat island should be considered in the assessment of site context at the conceptual design stage. The rise in air temperatures in heat island areas should be added to the general meteorological figures to gain a full picture of the microclimatic conditions.

The locations and intensities of heat islands can be determined by studying surrounding buildings and topography.

▌ PASSIVE ENVIRONMENTAL STRATEGIES

Two broad approaches of environmental design strategies (Figure 8) can be identified as follows:

The SELECTIVE PERMEABLE approach aims to achieve environmental control by the selective use of natural agents, such as natural lighting and ventilation. It requires an integrated consideration of architectural space, built form, orientation and building envelope design to achieve optimum selective permeability in creating the desired environmental conditions.

The SEAL-OFF approach involves the use of efficient mechanical building services systems to control and achieve the desired internal living and working environment. The primary aim is to insulate and seal off the interior from the external environment.

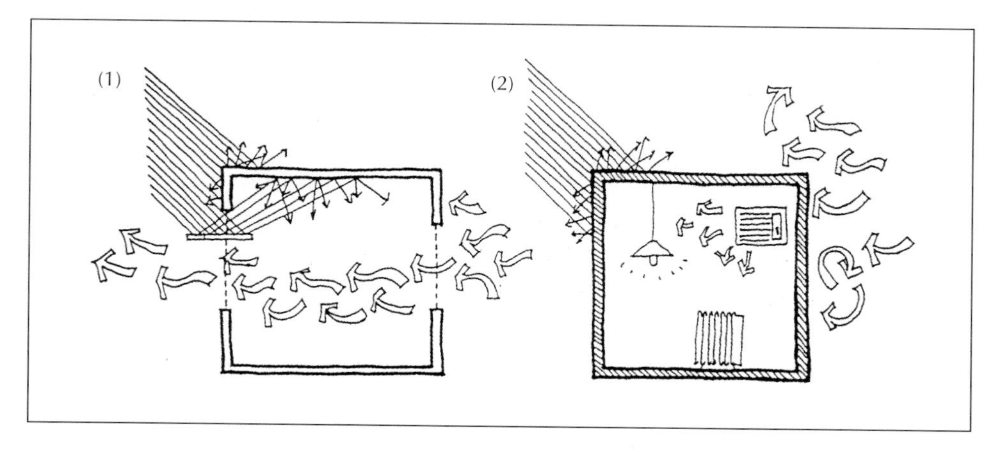

Figure 8 The selective permeable (1) and seal-off (2) approaches of environmental strategies.

As previously mentioned, the climate of Hong Kong is generally comfortable for most of the year. The use of the SELECTIVE PERMEABLE approach is desirable. When the ambient temperatures in the summer months become too high for passive cooling, mechanical cooling should be used. The approach then has to be shifted to the SEAL-OFF mode. In fact, in some of the most air-/noise-polluted areas in Hong Kong and in very humid conditions, the SEAL-OFF approach is more appropriate for occupants' health and normal functioning. This change of mode could be achieved by occupant control or building management systems.

In the next section, we shall investigate various passive environmental strategies that fall within the SELECTIVE PERMEABLE approach.

Built Form And Orientation

As discussed in earlier sections, minimization of solar heat gains and maximization of natural ventilation are desirable for the subtropical Hong Kong climate. First we shall examine the sun angle chart for Hong Kong as shown in Figure 9.

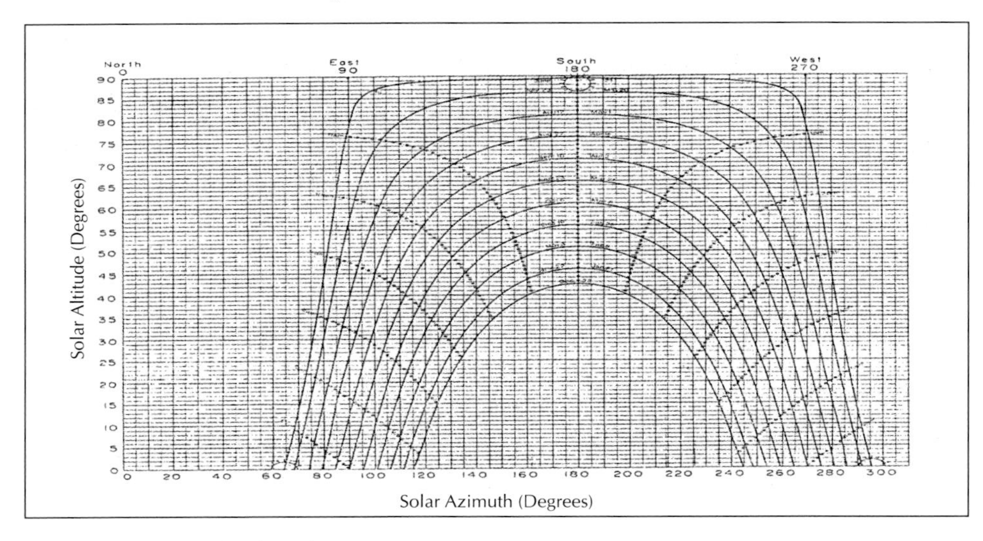

Figure 9 Sun Angle Chart for Hong Kong

Sun penetration angles for different orientations vary. The sun angle diagrams (summer) for the south- and east- /west-facing windows are illustrated in Figure 10. The high-angled summer sun in the south is less penetrating and is easier to shade from. The low-angled sun in the east/west may result in excessive heat gain, requiring relatively complicated and costly devices for sunshading.

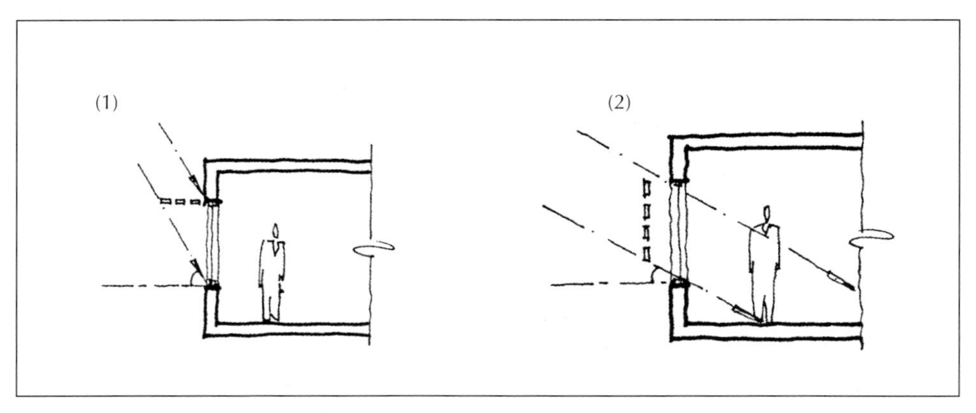

Figure 10 The sun angle diagrams (summer) for the south-facing (1) and east- /west-facing (2) windows.

It is suggested that a building length/width ratio of at least 2:1 is to be adopted whenever possible. With the long axis of a building orientated east-west, it enables a greater proportion of south-facing windows with sun penetration minimized in the summer and maximized in the winter. The narrow plan form shown in Figure 11 also facilitates the passage of air through the building and is desirable for naturally ventilated buildings.

One should also be aware of the afternoon heat transmitted through the building fabric on the western side of a building. Function facilities such as storerooms, escape stairways and E&M plant rooms may be planned at such locations to act as a buffer between the heated envelope and the living/working environment (Figure 11).

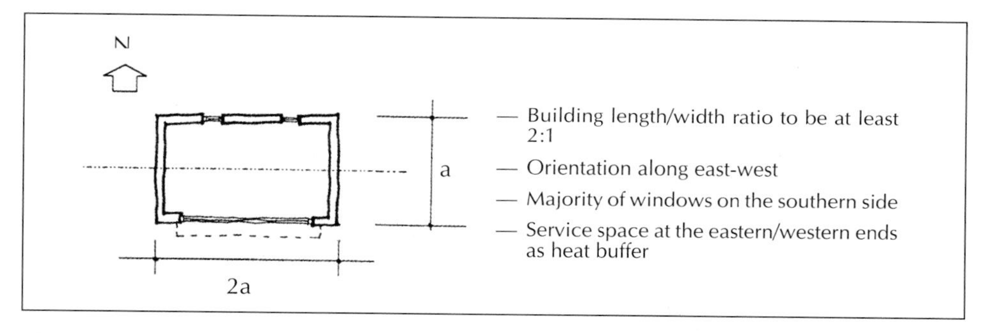

Figure 11 Desirable built form and orientation to minimize direct and indirect solar heat gains.

When several building blocks are planned together, they should be arranged so that they do not fall into one another's light or wind shadows. In Hong Kong, the prevailing wind is northeasterly for about 70% of the year. Building clusters could be arranged as shown in Figure 12 so that the wind can easily reach all façades of the buildings.

BUILDING ENVELOPE DESIGN

For Natural Ventilation

Ventilation openings, as large as practically possible taking into account other requirements such as sun control, security, privacy, weather protection and potential heat loss in winter, should be planned on both sides of occupied spaces to provide cross-ventilation. The following illustrates a number of key points useful in building envelope design for natural ventilation (Figure 13).

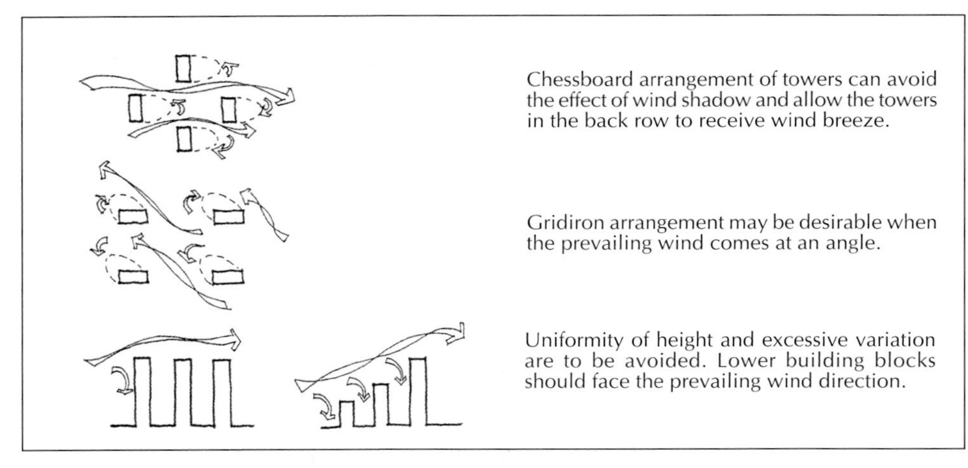

Figure 12 Planning of building clusters to maximize ventilation.

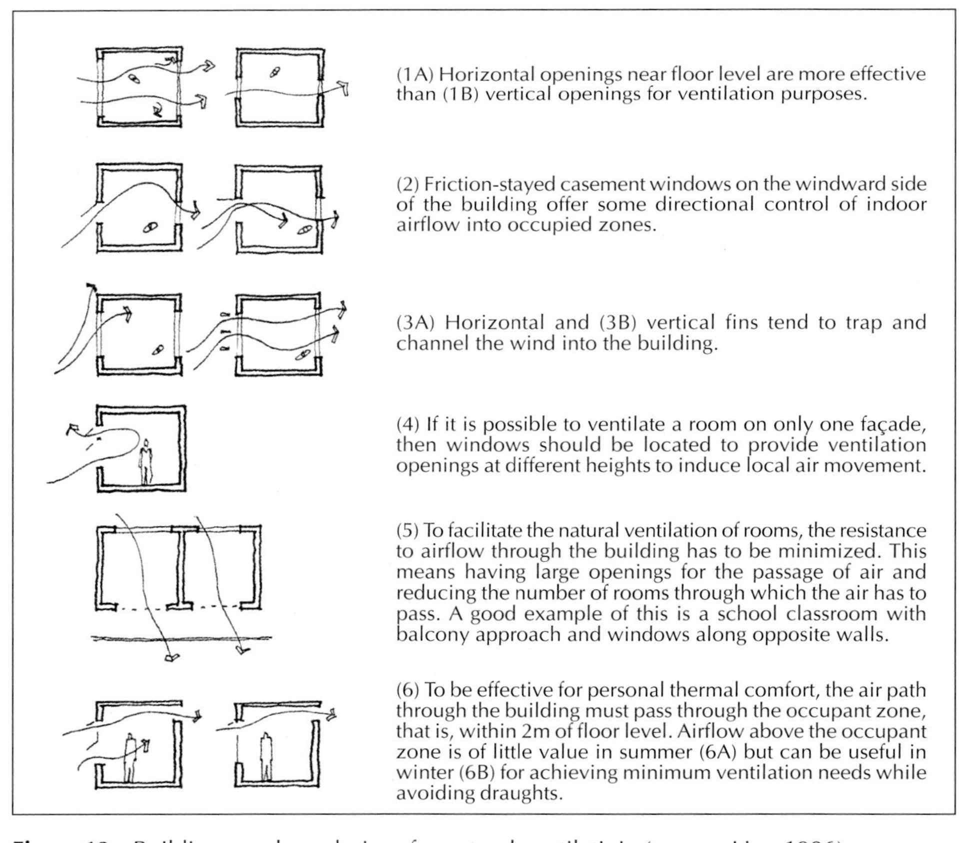

Figure 13 Building envelope designs for natural ventilatioin (source: Lim, 1996).

For Daylighting

Building envelope for daylighting in Hong Kong's climate should be designed in such a way that direct solar gain, which leads to glare problems, overheating and thermal discomfort, can be avoided. Diffused radiation should be utilized to provide the required illumination level for normal functioning. Common types of window and shading systems for daylighting are illustrated and analysed below (Figure 14).

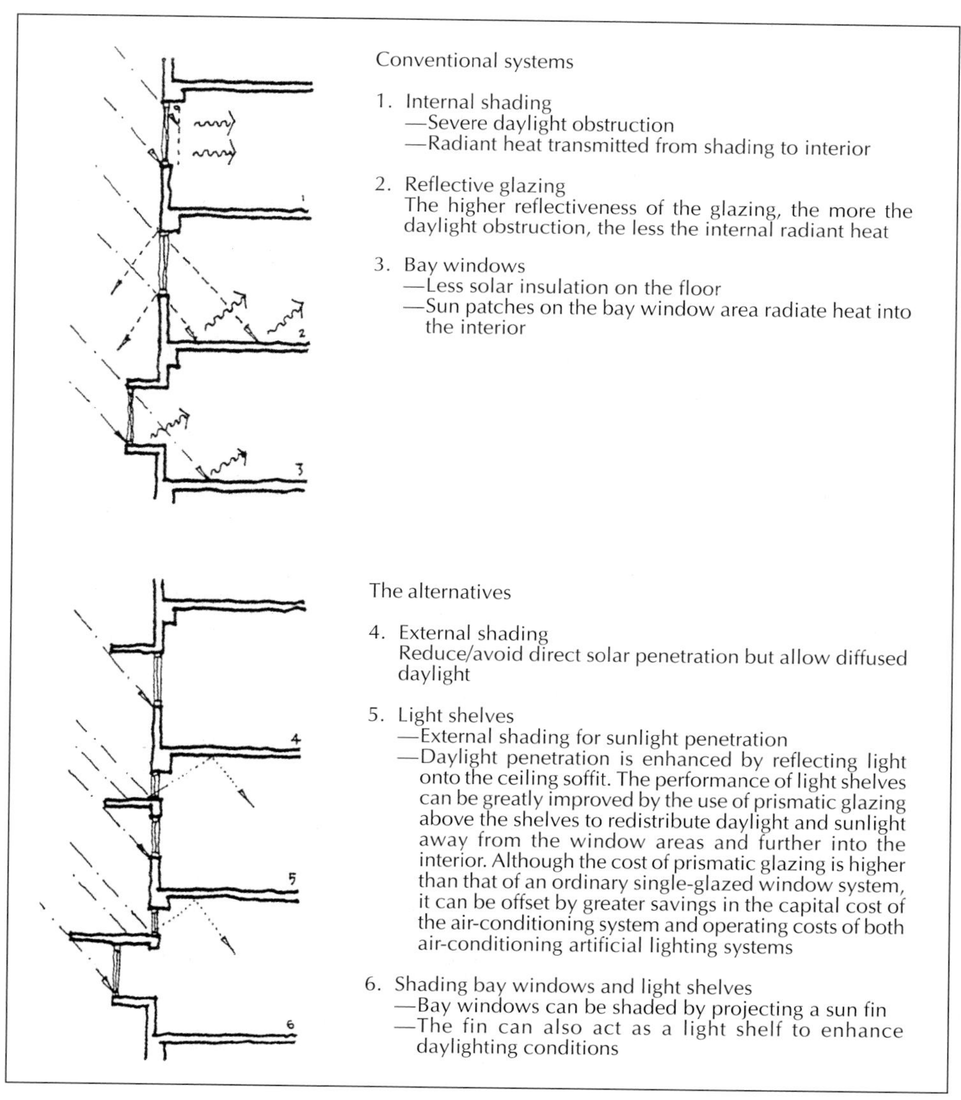

Figure 14 Conventional and alternative window and shading systems.

For Interception Of Solar Heat

Shielding of the building enclosure is desirable to intercept the solar heat gain transmitted through the building envelope sections. External skins can be designed on top of roofs or in front of the external walls (Figure 15). Ventilated air cavity between the external skins and the building enclosure will reduce the convective heat transfer from the outer skin to the inner ones. Applying radiative insulation, for instance, reflective foil, onto the inward-facing surfaces of the building skins can also reduce radiant heat transfer. One should be cautious regarding the potential maintenance problems of double-skin construction, such as pest problems. Such problems could be easily resolved if we allow maintenance access to the air cavity between the skins or even make functional use of the space.

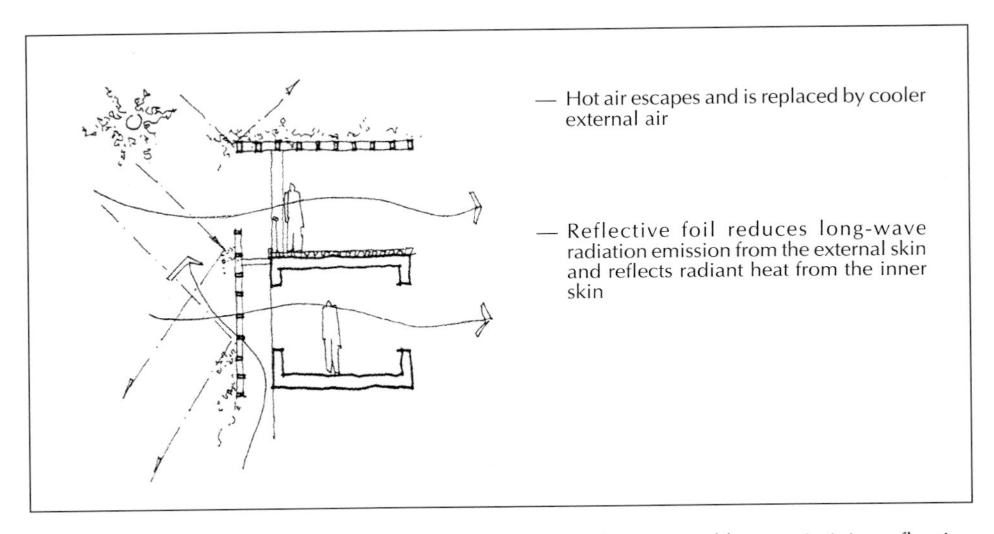

Figure 15 Double-skin wall/roof system with ventilated cavity and low-emissivity reflective foils.

▮ REFERENCES

Environmental Protection Department and Planning Department, Hong Kong. 1991. *Environmental guidelines for planning in Hong Kong.*

Humphreys, M.A. 1976. *Desirable temperatures in dwellings.* BRE Current Paper, United Kingdom.

Lin, X.D. 1996. *Planning for green architecture in hot and humid climate.* Taipei: Zhan shi shu ju, Min 85.

Oke, T.R. 1987. *Boundary layer climates.* London: Methuen.

NOTES

1. Humphreys (1976) discovered that our comfortable temperature range varies with the outdoor temperature (T_{amb}), our metabolic rate and clothing conditions. The temperature range is named as the neutral temperature (T_n) and is defined as:

$$T_n = 11.9+0.534*T_{amb}+/-2.5^{\circ}C$$

2. Oke, 1987.
3. Oke, 1987.
4. Environmental Protection Department and Planning Department, 1991.

Part C: Environmental Factors

This part is devoted to the environmental factors affecting the design of buildings. These include heat, light and noise.

The chapter 'Design for Building Environmental Performance', by Daniel Wai-tin Chan, John Burnett and Phillip Jones, discusses building services systems relating to environmental factors. It provides a brief on sick building syndromes, and healthy building is explored in a new perspective. Energy effectiveness – how energy is consumed in a building – is explained. Operation and maintenance are considered as important issues. The aims and context of an environmentally friendly building are also outlined.

In the chapter 'Noise and Design of Buildings in Hong Kong' by Stephen Siu-yu Lau and Dariusz Sadowski, the causes of Hong Kong's environmental noise problems are addressed. Architectural design and land use planning to minimize noise problems are discussed. Two case studies illustrate how noise problems are dealt with in Hong Kong.

In Wong Wah Sang's 'An Investigation on Overall Thermal Transfer Value', hypothetical designs based on varying dimensions, materials or detailing are compared for the thermal transfer value. The effect of adjacent buildings, which is not included in the current calculation of overall thermal transfer value, is discussed. The case studies show how different types of buildings react to the assessment of thermal transfer value.

8

Design For Building Environmental Performance

Daniel Wai-tin Chan, John Burnett and Phillip Jones

INTRODUCTION

Building Services Engineering At Stake

Almost every summer when we pass by the campus of The Hong Kong Polytechnic University, we are amused by a slogan on the wall of the large staircase stack which says 'Between Eleven and Three, Slip Under a Tree'. Indeed, there is no place to hide from potential exposure in the city of today. When we are enjoying the sunshine outdoors, we are warned that a high dose of ultra violet radiation (due to the depletion of the ozone layer) may induce skin cancer or eye cataracts. It is no safer back indoors, hiding within the building fabric that is the 'weather modifier'. We are then exposed to the potential ill effects of the sick building syndrome and building-related illness. In the first, building services engineers are accused of polluting the outdoor environment by venting chlorofluorocarbon (CFC), an ozone depletion agent, and excessive carbon dioxide (from excessive energy consumption), a global warming gas, into the atmosphere. Building services engineers are responsible for the design, installation, operation and maintenance of the services systems that contribute to global environmental problems, and which are often unable to sustain acceptable indoor environmental conditions. With the increasing awareness of environmental problems, both indoors and outdoors, building developers and end-users have a higher expectation of the performance of building services systems. This chapter reviews the conventional notion of the purposes of building services systems in a building, and discusses innovative and more systematic approaches to improving on their performance.

A Brief Moment In History

In the 1960s when Hong Kong started to develop its flourishing economy, air-conditioned offices became the expected standard. Energy was cheap and was not an issue to take into account seriously. Indoor air might be cool or even cold, but it was also fresh. There was a notion that air-conditioned space had to be cooled. Occupants put on jackets in over-cooled spaces. Air was fresh because we were generous to pump in sufficient outdoor air. The buildings were loose to air infiltration and the architects took little care to control. Architects and interior designers used lighting of high intensities to highlight the interior and features of the building.

The oil embargo in 1973 triggered a demand for energy conservation. Suddenly, the quantity of fresh air introduced into buildings was reduced to a barely acceptable level (2.5l/s/person) without considering it might jeopardize the indoor air quality. Buildings became tighter to reduce air infiltration. Lighting levels were lowered without the compensation of using natural lighting. The demand in the 1970s was for energy buildings.

The over-responsive cut-back in the quest for reducing energy use induced many complaints about the indoor environment as manifest in all kinds of sickness. These complaints formed a category of its own when the World Health Organisation (WHO) described it as the 'Sick Building Syndrome' (WHO, 1983). However, Hong Kong did not then take note of the occurrence. Instead, building automation technology and energy management systems found a market in Hong Kong. The general demand was for the so-called 'intelligent building' with the objective of better environmental control through distributed direct digital control systems. Unfortunately, without effective training, technical staff could not cope with the technology. Many of these 'intelligent' systems had subsequently become 'white elephants'.

Healthy building issues finally found significance in Hong Kong in the early 1990s. Unfortunately, professionals' idea of sustaining an acceptable indoor environment was to provide the maximum required outdoor air into indoor space at all times without due regard to optimization with the real need, which works against the requirement for energy conservation. The failure of many 'intelligent' systems deterred the fever for intelligent building. There was more concern about the air side system. For example, engineers started to review the actual benefits of putting in variable air volume systems due to the maintenance problems experienced. Fan coil systems seemed to regain favour due to their simplicity. By the mid-1990s, the underfloor plenum supply system started to emerge and has gained some momentum towards the end of the decade.

The energy crisis created design turmoil for building services engineers. Effective use of energy is now a classic requirement. Looking forward to the next century, the supply of drinking water will become an imminent problem. The United Nations, after a three-day conference in Paris in March 1998, recommended consideration of drinking water as a precious commodity. The objective is to appeal for the effective use of water resources. On the other hand, our atmosphere is continuously being treated as an effluent dumping reservoir, and there is a threat that in the near future, the atmosphere will become so polluted that 'cleaned' (conditioned) air will be supplied to our buildings as a utility, like electricity, water and gas. Hence the idea of 'green' building is gradually being put into practice.

In Hong Kong, the Real Estate Developers Association (REDA) initiated the development of a building environmental assessment method for office buildings. In 1997, two versions of Hong Kong Building Environment Assessment Methods (HK-BEAM) for 'new' office buildings and 'existing' office buildings were launched. They are tools to assess voluntary actions by building owners to reduce the environmental impact on the global and local environments, and to sustain an acceptable indoor environment. HK-BEAM for residential buildings is under development. The demand in the next century will be for environmentally friendly buildings equipped with green building services systems. The historical quest for building quality is summarized in Table 1.

Table 1 The historical quest for building quality

Decade	Issues	Building theme
1960s	Sense of coolness	Air-conditioning systems
1970s	Energy crisis	Energy-efficient buildings
1980s	Awareness of the sick building syndrome, and building-related illness, and application of computer-control technology in buildings	Intelligent buildings
1990s	Concern about the sick building syndrome, building-related illness and environmental impacts	Healthy buildings
Twenty-first century	Global effort for sustainable development	Environmentally friendly buildings

The following sections discuss the common notions of building services system design and examine the systematic approaches to improved performance.

DESIGN FOR HEALTHY BUILDINGS

Sick Building Syndrome And Building-related Illness

On 1 June 1988, Dr Sherwood Burge at a London seminar presented a speech in the Institute of Mechanical Engineers' main auditorium. He said that building services engineers had erred in concentrating on comfort conditions instead of on health issues in their design. Dr Burge, a consultant chest physician at Solihull Hospital, was initially sceptical about the sick building syndrome, but found the symptoms in sufferers well defined and very real. He opined that comfort was not the sole issue for the indoor environment, and sounded an alarm bell to building services engineers. Indeed, health-related problems were identified as early as in the 1960s (Black and Milroy, 1966). WHO (1983) named this category of problems 'sick building syndrome' (SBS).

The SBS is mainly manifested as sick symptom complexes. In layperson terms, European Concerted Action Report No. 4 (Molina et al., 1989) described them as follows:

- Nasal stuffiness
- Dryness and irritation of the mucous membrane of the eye
- Dryness and irritation of the throat
- Dryness and irritation of the skin, occasionally associated with a rash on exposed skin surfaces

- General manifestations such as headaches, generalized lethargy and tiredness leading to poor concentration

SBS is a vague term used to describe the complaints of occupants. Raw, Roys and Leaman (1990) proposed a working criterion for SBS diagnosis of having more than two work-related symptoms per person, experienced more than twice in a year. The work-related symptoms (Wilson and Hedge, 1987) are as follows:

- Lethargy
- Stuffy nose
- Dry throat
- Headache
- Itching eyes
- Dry eyes
- Runny nose
- Flu-like symptoms
- Difficulty in breathing
- Chest tightness

The American Society of Heating, Refrigerating and Air-conditioning Engineers Inc. (ASHRAE) gives a definition for SBS (ASHRAE, 1987) that is better understood by the building services engineer. If more than 20% of the occupants in a building complain about one or more of the symptoms, if these symptoms persist for more than two weeks, if the causes of the complaints are not readily recognizable, and if those persons affected recover from such symptoms on leaving the building, then SBS is said to exist in that building.

Accompanying the SBS are the building-related illnesses (BRIs). BRIs include legionnaires' disease, potential effects of radon gas and the presence of asbestos fibres, and other chronic effects due to overexposure to harmful pollutants such as formaldehyde, pesticides, etc.

Engineers At Stake: Legal Action On Sick Building Syndrome And Building-related Illness

Professional building services engineers can no longer claim work immunity from faults in their design, installation, operation or maintenance of systems. On 23 December 1993, five employees of the USA Federal Environmental Protection Agency were awarded a total of US$950 000 from the building's owners and managers for ailments they claimed were related to the indoor air quality at their place of work. The ailments cited were respiratory and neurological disorders, but there were no physical injuries.

Earlier in 1998, a former musician with the Hong Kong Philharmonic Orchestra was awarded HK$20 million in damages for his disablement caused by inhaling an overdose of pesticide during a rehearsal. What was astonishing was the requirement that the defendants pay for the ten years of legal fees that amounted to HK$200 million.

In 1989, it was reported that the British Broadcasting Corporation (BBC) was fined £3 600 with £3 196 costs over an outbreak of legionnaires' disease in April 1988 in which 3 people died and 79 were infected. The BBC pleaded guilty to all charges including negligence and failure to safeguard the health of its employees. The penalty was insignificant compared to the more than £1 million claimed by the victims.

Indeed, the number of indoor air quality (IAQ) lawsuits in the USA has risen sharply since the 1970s. In the late 1980s, the number of cases increased exponentially. Figure 1 is taken from Hansen (1993). The total amount of claims for damages, together with the legal costs, was in the order of US$1 billion.

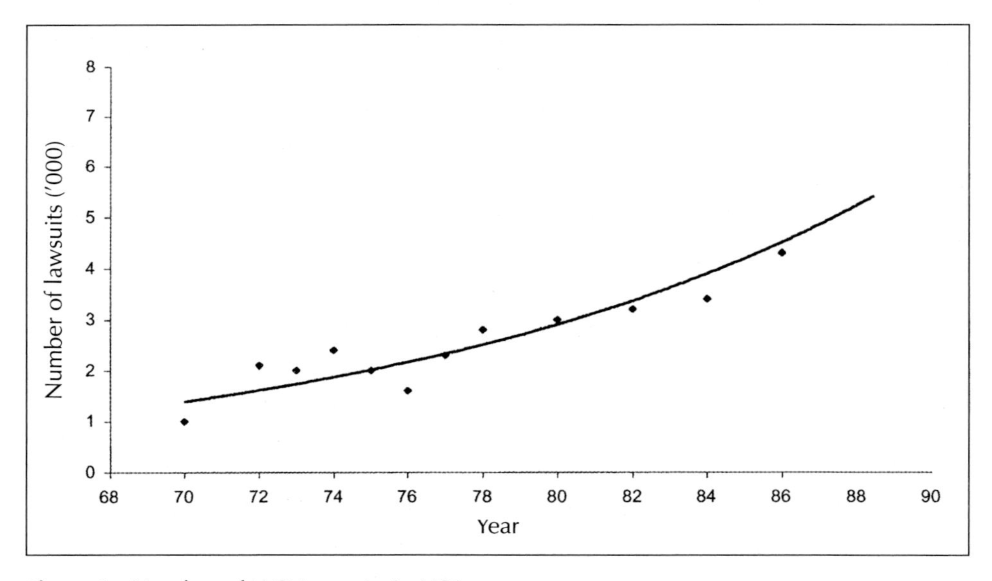

Figure 1 Number of IAQ Lawsuits in USA

Healthy Building From A New Perspective

In the Waterside Mall case, Jim Dinegar, Vice-president for Government and Industry Affairs, Building Owners and Manager Association (BOMA) International, remarked sarcastically, 'We're not trying to poison our tenants, but we feel frustrated in not knowing what we have to do.' It is not surprising because the causes of SBS are still not readily understood, and the risk of BRI is very difficult to determine.

A breakthrough may be sought from considering the SBS in a new perspective. In the UK, Jones et al. (1995), subsequent to a detailed study of eight buildings, hypothesized a new model for the SBS. The model proposes that the SBS is the outcome of the dynamic interaction between the occupants and their surrounding environment, and that symptoms occur when the response of the immune system is lowered below a certain threshold. The threshold is negatively affected by 'antagonists' and positively affected by 'alleviators'. Antagonists are anything that places a load upon the occupants' physical and psychological resources, while alleviators are anything that reduces or alleviates the level of load. Personal characteristics play an important role in this overall resource-load equation.

Like all other comfort qualifiers such as indoor air quality, thermal, lighting and acoustic conditions, the SBS is a response of the state of the mind to the physical, chemical, biological and psychological risk factors (antagonists) of the occupants' environment, whether they are in their offices or at home. In order to provide a healthy and acceptable indoor environment, the following points are noted:

- The term SBS is a misnomer. It is not the building that is sick. It is the interaction with the indoor built environment that gives rise to the sickness symptoms, not illness, in the building's occupants.
- The SBS is assessed by occupants themselves, not by experts. The SBS is a 'group' sensation. The 20% rate is an arbitrary criterion that rules out over-sensitive individuals. However, it does not mean that the minority should be ignored.
- The degree of 'sickness' or 'healthiness' can be indexed by the number of building-related (building and its services systems) symptoms per person over a period of time. In the study by Wilson and Hedge (1987), a mean number of symptoms per person was found to be 3.11 out of the 10 work-related symptoms. It must be noted that due to the non-standardized survey method, this index should be used only as a reference to the analysis of

the indoor environmental quality operation criteria and is not an absolute indicator of the extent of problems.

- The SBS is not only a consequence of the physical environment, but also a consequence of the company culture, a person's financial, social and family situations and his/her character. Therefore, the sustainability of a healthy working environment requires also close communication between the building management and the tenants.
- People who have a lower level of immunity for any reason are more susceptible to the SBS.

A Review Of The Current Design Practices In Office Premises

The most widely used design criteria set for indoor environmental quality are given in Table 2.

Table 2 Practices of design criteria for comfort in office premises

Indoor environment	Design parameters	Design criteria	Standard
Thermal comfort	air temperature	24°C	ASHRAE 55–1992 optimum comfort condition
	relative humidity	55%	
Air quality	outdoor air supply	10 litres/sec/person	ASHRAE 62–1989
	carbon dioxide concentration	1 000ppm	
	air distribution	as even as possible	by means of layout of air diffusers
Lighting	illumination intensity	500lux	CIE, CIBSE
Noise	noise criteria	35–40NC 55–60dBA	ASHRAE Handbook

However, genuine compliance to these standards by many design engineers is doubtful as these criteria are 'inherited', and the factors and assumptions behind these criteria are in most cases not thoroughly investigated. Most engineers tend to equate these numbers to the concept of 'building occupants should accept'.

In fact, building services engineers should take a 'total quality' approach instead of looking into the comfort parameters alone. The priority of built environmental quality is reconstructed and shown in Table 3.

Table 3 Priority of built environment quality

Category	Subcategory	Issues
Safety	structural	integrity
	fire	accidents arson
	BRI	legionnaires' disease radon and progeny asbestos other toxic materials
	security	theft robbery assault
Hygiene	air and water supply	cleanliness
	SBS	indoor environmental quality
	building materials	release of pollutants
	humidity and unnecessary accumulation of water or condensates	growth of bio-pollutants
	cleanliness	nursery beds of bio-pollutants dust mitres respirable suspended particles
Comfort (refer to Table 4)		
Transport	vertical	lift and escalators
	horizontal	travellators
	disabled	lifting mechanisms
Engery effectiveness	energy efficiency	system performance
	energy management systems	effective energy conversion energy audit and assessment monitoring and control
	building automation	sensors and instrumentation
	building envelope	overall thermal transfer value (OTTV)
Water	economy	conservation
	recycle grey water	for reuse in flushing or irrigation
Building intelligence	telecommunications	fax PABX computer network
	teleconferencing	audio-visual facilities interactive communication
	broadcasting	television cable television satellite television interactive television
Luxury	as required by client	e.g. gymnasium with showers for senior executives
Aesthetics	building fabric	fenestration
	interior decoration	services integration
	internal layout	comfort agent distribution people circulation meeting place for leisure

The achievement of an acceptable, comfortable indoor environmental condition is not simply a fulfilment of the criteria as listed in Table 2. The numbers quoted are only applied under very general conditions. The criteria chosen should in fact take into account the appropriate physical or physiological factors which depend on the individual occupants, the job nature and office environment. Table 4 lists the basic comfort factors in the choice of 'compatible' criteria. The appropriate model is also given for reference.

DESIGN FOR ENERGY EFFICIENCY

Target For Energy Efficiency

Energy conservation has become a key requirement in all building services system designs following the energy crisis in the 1970s. However, it pushes indoor environmental quality to the border of acceptability. The notion of energy efficiency also does not provide enlightenment to the building services engineer. The usual notion of energy efficiency is a measure of how 'little loss' a machine or process in a system can produce by converting the input energy into another form of energy or useful work. It does not give a clear description of how useful the energy is consumed in a building to produce an expected indoor environment. Energy consumed in a building takes effect to produce an acceptable level of the built environmental qualifiers in Table 3 and the comfort factors in Table 4. Engineers are prone to err when considering energy from the perspectives of conservation and efficiency only, without consideration of the human perception. Energy effectiveness emphasizes the proficient use of energy to produce a design intent and does not discourage energy use as the notion 'energy conservation' often suggests. Therefore, energy effectiveness is a total energy management concept and retains the drives for research into alternatives but cleaner forms of energy sources.

High energy efficiency increases the energy capacity of the energy source and minimizes the overhead of the machine and the system. The notion of efficiency in a machine is obvious. Sometime, a different type of efficiency is used if the output is not compatible with the input. The coefficient of performance (COP) of a chiller and efficacy of a lamp are examples.

Energy conservation is concerned with the reduction of unnecessary consumption and the recovery of 'waste' energy from systems. Reduction of unnecessary consumption is usually achieved by using an appropriate operation schedules and modifying the comfort consideration.

Table 4 Comfort factors

Comfort category	Physical or physiological factors
Thermal comfort (Fanger's/ASHRAE 55–1992 comfort and draft risks models)	Air temperature Air relative humidity Air velocity Air turbulence Radiant temperature and radiant anisotropy Work output Metabolic rate Clothing values (clothes insulation)
Air quality (ASHRAE 62–1989, Fanger's new Air Quality Comfort Equation)	Quantity (litre/sec/person, air change rate) Quality (air constituents) Cleanliness Distribution and circulation (ventilation effectivess) Odours Micro-organisms
Lighting (CIBSE, CSP index for comfort, satisfaction and performance)	Horizontal illuminance Cylindrical illuminance Brightness distribution in the visual field Glare Light directionality and shadows Colour rendering Work piece resolution
Sound and vibration (ASHRAE Handbook — Fundamental)	Noise and vibration level Frequency distribution Characteristic (tonal, flutter, etc.) Speech interference Impulsive or intrusive Spatial variation
Working space	Posture Repeated body movement Visual coordination Use of video display units Space for circulation inside the building Pattern of integration/ communication among individuals
Windows	View Natural light Solar effect Glare

Energy effectiveness is not as a straightforward concept as energy efficiency or energy conservation. In the context of building services engineering, energy effectiveness takes into account of the human perspective. Not only that the equipment and system should have a high efficiency, but energy should be conserved at all possible opportunities. In addition, the control criteria should be carefully chosen to be compatible to the activity, company culture, users' preference, environmental conditions, investment plan and etc. For example, the indoor air temperature set point, defined as the comfort temperature to a preferred percentage of satisfaction, can be a function of the immediate outdoor air temperature. This has the advantage of reducing the temperature shock, and may have a higher satisfaction rate. The control criteria depend on the maximum allowable percentage of dissatisfaction, which is a management decision based on the company culture and the investment plan. The maximum allowable percentage of dissatisfaction is between 10% and 30%, with 20% being the low margin for a healthy building, and 30% being the maximum tolerable choice with some degree of adaptation by the occupants. A 10% dissatisfaction rate is a target for Grade A buildings for which the owners are willing to pay more. The dissatisfaction percentage is unlikely to be below 10% due to diminishing returns. The concept of energy effectiveness is therefore a more pragmatic approach in avoiding energy saving at the expenses of satisfaction.

Knowing How Energy Is Consumed In A Building

In a total energy management design concept, it is important to know how energy is consumed in a building. Hong Kong is situated in a geographical zone classified as subtropical. Most of the year round, the weather is hot and humid. In Hong Kong, due to the city's air pollution and noise, the use of natural ventilation is limited to free cooling. Almost all offices are air-conditioned and use electricity. In a typical office building, the air-conditioning system constitutes around 60% of the building's electrical energy consumption. Lighting constitutes about 20%. The remaining 20% includes 'small power' appliances (12%), and lifts and miscellaneous items (7%). The breakdown for space cooling load for a typical office building is given in Figure 2. The main components of the air-conditioning load is shown in Figure 3 (HK-BEAM 1996).

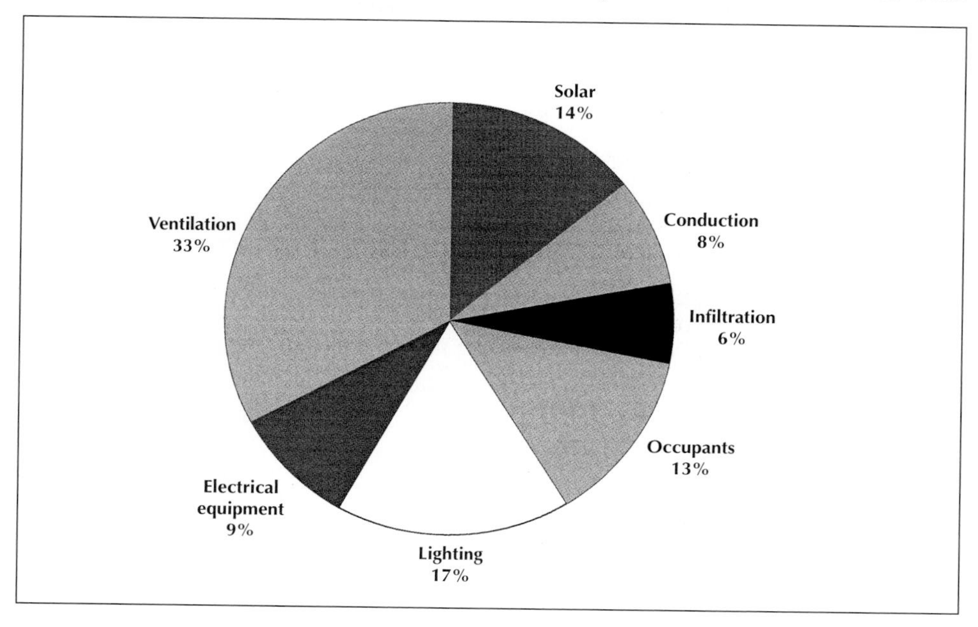

Figure 2 Pie Chart of Space Cooling Load

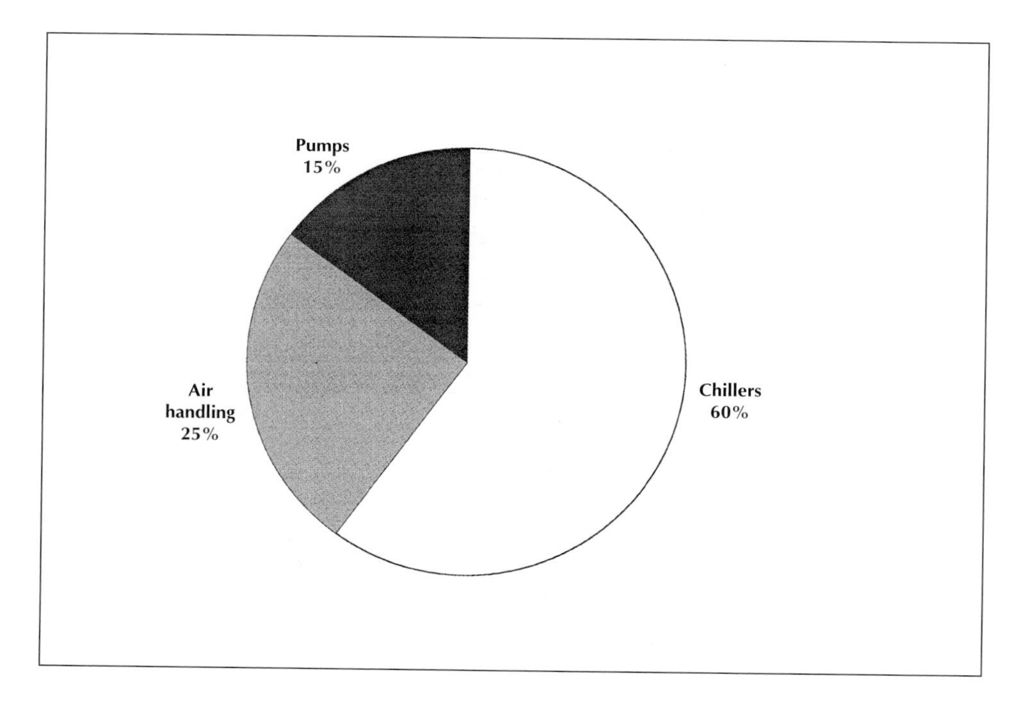

Figure 3 Pie Chart of Air-conditioning Load

Effective Energy Consumption Opportunities

'If we need it, we need it.' Energy has to be consumed if an acceptable indoor environment is to be sustained. Figures 2 and 3 throw some light on the strategy of designing for energy effectiveness. In Hong Kong, since buildings are cooled most time of the year, the following discussion is confined to cooling systems.

Opportunity From Selection Of Heat Rejection System

The air-conditioning system consumes the major part of the cake, it is therefore worthwhile to take a closer look. In a typical air-conditioning system, space heat is rejected either to the air or to the sea. The rejection processes can be configured as a series of heat exchanges in cascade (Figure 4). The system COP of different types of systems has an important bearing on the efficient use of energy. The factors affecting the choice are:

- Plant capacity
- System performance (machine efficiency and system COP)
- Life expectancy of equipment
- Availability of cooling agent (sea water, nullah water, underground water, etc., with acceptable quality)
- Space for intake cell (if sea water is to be cooled), plant accommodation, maintenance
- Access route (installation, maintenance, overhauling, replacement)
- Power supply, availability of high tension provision
- Cost: first, recurrent, maintenance, land lease, equivalent rental
- Ease of operation, control and maintenance
- Energy management system
- Project time

The list is not exhaustive. Some choices can be disqualified at the beginning once a project brief is released by the client and the project manager. Very often, a life cycle costing analysis is required to justify the final decision.

Opportunity From Selection Of Number Of Chillers

The number of chillers required in a system depends on:

- Maximum air-conditioning load
- The most energy-efficient operation mode in terms of reducing part load operation of any individual chiller
- Space available for accommodation, routine maintenance and overhauling maintenance

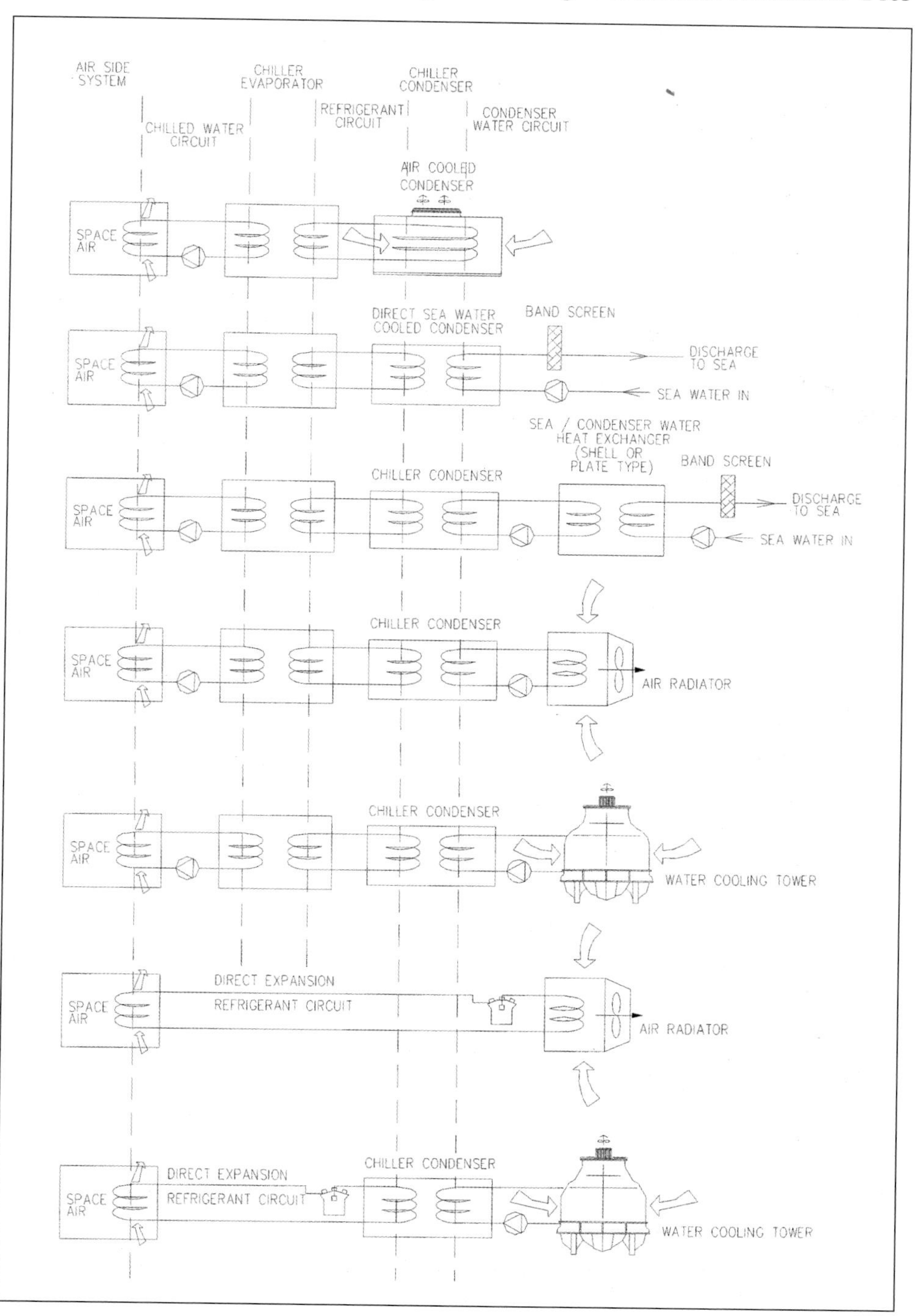

Figure 4 Different Types of Heat Rejection Method

- Available capacity of individual chillers
- Initial cost for the total number of chillers installed
- Standby percentage of capacity
- Backup percentage of capacity
- Noise
- Type of refrigerant

Given the available space and availability of the required individual chiller sizes, the determining factors are the standby and backup strategies and energy consideration, and how limited the initial cost budget is. The standby and backup strategies are important when the plant is running at the rated load with all the available chillers. Usually, all chillers in the chiller group have the same rated capacity. Occasionally, a chiller of smaller size will be installed to serve small air-conditioning loading requirement during off-peak hours or for condenser heat reclaim purpose. For a chiller group, assuming the risk of chiller breakdown is limited to one, the backup percentage of capacity is defined as the percentage of total loading, in addition to the installed total capacity, so that full-load operation can be maintained at all times with a maximum of one chiller breakdown. In many situations, full-load operation is required for only a small percentage of time in the annual cycle. The standby percentage strategy can be used for guaranteed usage of the system. Without an extra chiller, the standby percentage is defined as the percentage of capacity of one chiller. So, for most of the year when there is more than one chiller in idle, there is always a 'standby'.

The number of chillers must be such that chillers are running close to their rated capacity most of the time. Figure 5 shows the daily load profile generated from a model year in a model building (Yik et al., 1998). Figure 6 is the frequency distribution and cumulative frequency function of the bin cooling load. The maximum cooling load is 2 884 refrigeration tons. The obvious choices are six- and four-chiller groups of 500 and 700 tons for each chiller respectively. The solid and dotted hatch lines indicate the percentage of time in an annual cycle during which a certain number of chillers are operating at their rated capacity. Table 6 summarizes the operating conditions. With the selection of 500-ton chillers, two of the chillers will be idle most of the time, while choosing 700-ton chillers would mean one chiller idling 90% of the time. For the 500-ton chiller choice, the chillers operate at less than 80% of the rated capacity for 37% of the operating hours. For the 700-ton chiller choice, the chillers operate at less than 80% of the rated capacity for 52% of the operating hours. Selecting smaller-capacity chillers always has the advantage of the chillers operating at a higher percentage of the rated capacity and hence is more energy-efficient. However, coupled with other factors including operation and maintenance management, 700-ton chillers may be a better choice.

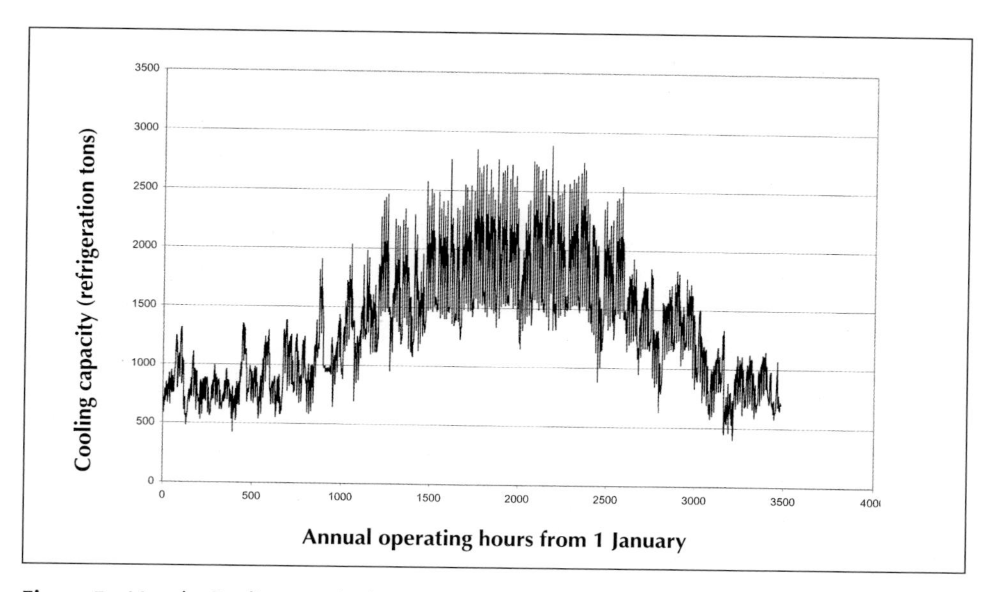

Figure 5 Hourly Cooling Load of a Model Building in a Model Year

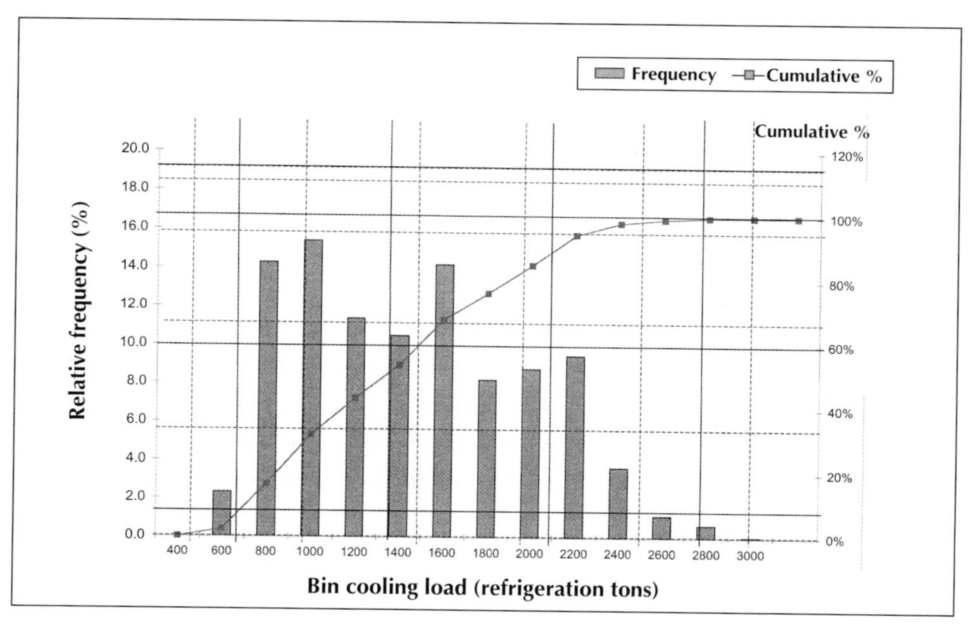

Figure 6 Determination of Number of Chillers

Table 6 Two chiller group options

Number of chillers	500-ton chillers (up to six)		700-ton chillers (up to four)	
	cumulative %	% of operation time	cumulative %	% of operation time
1	4	96	10	100
2	32	64	54	90
3	61	39	90	46
4	85	15	99.8	10
5	99	1		
6	99.5	0		

Opportunity From Reducing Heat Transfer Through Walls (Conduction) And Solar Radiation

The heat transfer through walls depends on weather conditions, building envelope construction and internal operation criteria. Given a set of internal operation criteria, energy-efficient designs to limit heat gain from outside depend very much on the building envelope construction. The government of the Hong Kong Special Administration Region issued a Code of Practice for Overall Thermal Transfer Value (OTTV) in Buildings to control heat gain through the building fabric. The recommended OTTV for a building tower should not exceed 35W/m^2, and for a podium, the limit is 80W/m^2. The Code provides a methodology for the calculation of maximum allowable OTTV.

In fact, the energy transfer behaviour of the building envelope should be viewed as an integrated concept incorporating weather conditions, building orientation, aspect ratio, shading, building shape and profile, window-to-wall ratio, glass shading coefficient, opaque wall thermal mass, internal operation parameters, and types of building services systems. Chan (1996) provides a comprehensive development of the OTTV calculation and the sensitivities of all related factors. In particular, Chan points out that:

- Window-to-wall ratio and the glass shading coefficient have much larger impact on the envelope's heat gain than does any other parameter.
- A heavier building fabric can reduce the conductive heat gain by more than 10%. The limit is 600kg/m. Beyond this, the energy reduction credit has a diminishing return.
- Increasing U-values from 0.5 to 3.79W/m^2/°K will reduce the conductive heat gain by 15.5%. Again, there is a diminishing return for further increasing the U-value of the building fabric material.

- The opaque portion of the envelope absorbs less solar radiation. The practical range of 0.2 to 0.88 can produce a difference of conductive heat gain by 17.7%.

Opportunity From Limiting Infiltration Of Outdoor Air

Air infiltration includes the mechanical infiltration of fresh air and the 'uncontrollable' infiltration through building components such as doors, windows and air openings. Fresh air is supplied indoors to maintain a healthy and comfortable air quality. ASHRAE Standard 62-1989, BS5720-1979, and CIBSE Guide A, European Guideline No. 11 specify clearly the amount of outdoor air required. Reduction of the quantity of outdoor air is permissible when the number of occupants varies at different times of the day. Demand control by the concentration of carbon dioxide is a common practice to save energy for cooling fresh air. Any attempt to use a variable outdoor air quantity system should not render the accumulation of indoor pollutants to the level of concern as set out by relevant international standards, such as those from National Institution of Safety and Health (NIOSH) and American Conference of Governmental Industrial Hygienists (ACGIH).

Opportunity From Daylighting And Lighting Design

The use of daylight is affected by the design of windows. Since the beginning of 1970, buildings in Hong Kong have changed to adopt a centre service core type of design, instead of having a perimeter balcony and central lightwell. This change is not surprising as buildings get taller and building services systems become more sophisticated. The centre service core renders the vertical distribution of services and the lift system design easier and more convenient. The use of daylight is therefore limited to the perimeter zone, which is generally taken to be 2m to 4.5m from the perimeter wall. A large window-to-wall ratio would, of course, make the use of daylight easier.

Based on a general 500lux horizontal lighting level and with a typical electrical lighting system with dimmer control, Chan (1996) estimated an energy saving of 8.3% from the peak space cooling load, as the greater part of the energy feeding the lighting system would ultimately turn into heat. The amount of lighting energy reduction by daylighting depends on the ratio of daylighting utilization zone to the total floor area. It can be as much as 27% of the design lighting energy, which in turn can reduce the annual chiller load by 7.7% (Chan, 1996). However, due to the increase in the use of desktop computers, daylighting may be deterred by interior shading used to reduce glare and the discomfort arising from asymmetric radiant temperatures induced by direct sunlight and irradiance. Nonetheless, daylighting is encouraged. The associated problems can be solved by a proper design of external shades such as overhangs and side fins. The advancement in glazing not only reduces the transmission of the longer wavelength heat waves, but also permits a higher transmission of visible light waves.

The energy-saving opportunities in lighting design are tremendous due to the lighting technology upgrades in the past few years. The new technology available and the general recommendation can be found in USEPA (1996). For Hong Kong buildings, the installed lighting load is typically 20-30W/m^2. With modern low-energy lighting systems installed, loads of down to 12W/m^2 can be achieved without detracting from lighting performance.

In general, it is sufficient to take note of the following issues:

- Use high-efficacy lamps, with durable electronic ballast.
- The choice of lamps and system design should be integrated well with the ceiling structure, and preferably with wall and ceiling surfaces at a higher reflectance.
- Further savings can be achieved by improved provisions for manual switching, dimming, automatic lighting control or occupancy sensors.
- Frequent cleaning improves the maintenance factor of the lamps and maintains efficacy.

Opportunity From Air-side Systems And Control

The most common types of air-side systems used in Hong Kong's air-conditioning buildings are the Variable Air Volume System (VAV), Constant Air Volume System (CAV) and Fan Coil System (FCS), or a hybrid with various combination of the three.

Air distribution is mainly based on perfect mixing, with air supply from and air return to the plant room at the ceiling level. One common problem with the perfect mixing system is that there is a significant bypassing of supply air back into the mechanical plant room through adjacent return air routes. Sometimes, the bypass factor can be as much as 0.5. That is, half of the supply air is bypassed back to the system without circulating in the breathing zone. Energy is wasted and the intended indoor environment is not maintained. Ideally, any individual occupant should be free from pollutants generated by another person occupying the same space. With a mixing distribution, all occupants share the pollutants generated by each other. A displacement-type air distribution system is able to improve the situation. Air is supplied through low-level diffusers, preferably at low exit velocity. When the air gets into contact with the occupants' hot air plumes caused by body heat, air rises upward, which then returns into the plant room via the ceiling plenum. Since air is distributed more effectively to meet occupants' needs, energy is saved as well.

Very often, the supply air is not able to achieve the intended effect, because the air temperature sensors are not located at the right positions. Almost in all cases, the locations of the sensors controlling the capacity of the system are chosen as a matter of convenience during installation, rather than at locations representing the conditions experienced by the occupants. In other cases, the air temperature is set too low by the maintenance engineers to reduce complaints from occupants about being too hot. In a large-scale in-office study conducted by the authors, from 1 198 measured work stations in over 30 air-conditioned offices in Hong Kong, over 60% of the locations had thermal environmental conditions outside the ASHRAE 55-1992 thermal comfort zone on the low side (Figure 7).

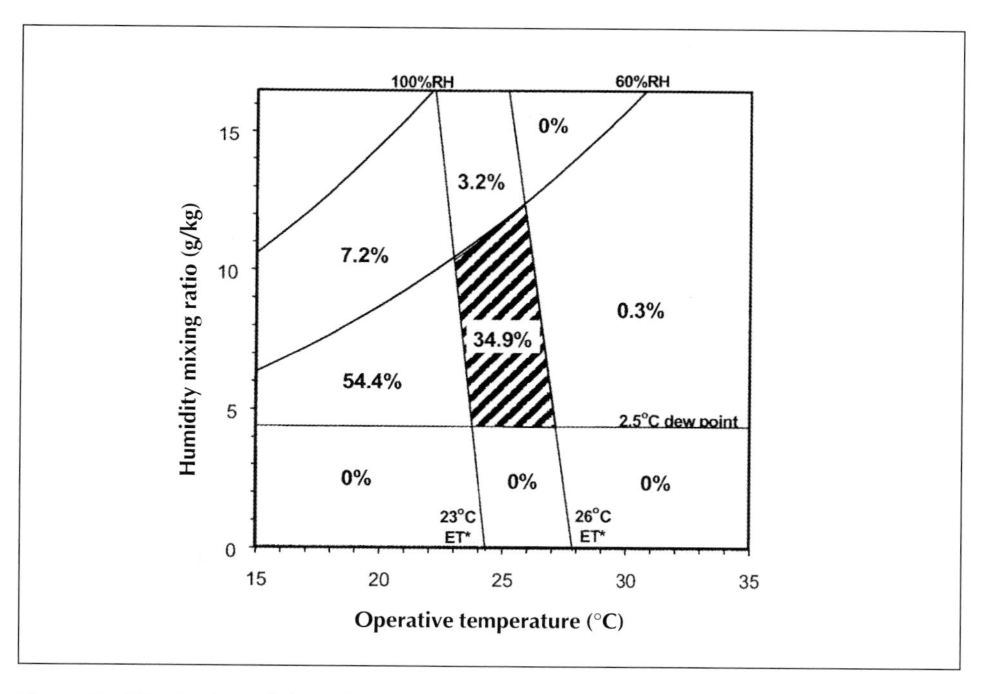

Figure 7 Distribution of the indoor climatic measurements on the ANSI/ASHARE Standard 55–1992 summer comfort chart defined in terms of New Effective Temperature (ET*), dewpoint temperature and relative humidity.

Much of the energy is used to cool down clothing. In many designs, the occupied space is pressurized by the mechanically drawn in outdoor air quantity used to dilute the indoor pollutant levels. It normally amounts to about 10% to 15% of the total supply air. A small portion of this cooled air helps to ventilate the toilets, added by the toilet exhaust fan. A larger portion of it finds its way leaking out of the building through windows, openings, doors and entrances. Supply air temperature reset based on an algorithm of comfort temperature and outdoor air temperature can help to reduce over-cooling of the indoor air.

Cooling energy from exhaust air can be recovered by heat wheels, heat exchanges or desiccant beds. Enthalpy control in air economizers coupled with demand control based on carbon dioxide concentration, with regards to avoiding accumulation of unacceptable levels of pollutants, is also encouraged.

Air circulation and distribution is carried out by a combination of the following fans:

- supply air fan (SAF) for thermal comfort purpose;
- return air fan (RAF) for reuse of cooled air;
- exhaust air fan (EAF) for removal of indoor air pollutants;
- 100% supply air capacity outdoor air fan (FCF) for free cooling and indoor air quality;
- minimum outdoor air fan (OAF) for acceptable indoor air quality;
- fire-rated emergency exhaust fan (EMF) for smoke management.

Various operating modes can be effected with the scheduling of designated fans with the necessary balance and control dampers, diffusers and registers, filters, ductwork system, economizers and required sensors at the right locations. The operating modes will produce an effective air distribution, energy recovery, use of natural resources, and safety management. Limited by space and cost, most of the systems comprise of only an air handling unit that serves the purposes of SAF, RAF and sometimes OAF. A higher-quality system has RAF and EAF to better balance the resistance in the supply air and return air sides. The provision of FCF is limited by the space requirement for pipe ducts. EMFs are mostly limited to large-volume spaces such as exhibition halls, atria and public areas. In Hong Kong, only the ventilation systems in the underground stations of the Mass Transit Railway system have such a sophisticated combination of all these fans and an effective operational model schedule to cater for most situations including normal, congested, energy-saving and emergency situations.

DESIGN FOR OPERATION AND MAINTENANCE

Mr Hubert Poon, former Assistant Director (O&M) of the Electrical and Mechanical Services Department of the government, once commented, 'The design engineers design the building services systems, supervise the installation and hand over to the client, wash their hands and leave the operation and maintenance (O&M) staff to suffer the rest of the 50 years.' Building services design engineers are too keen to apply available technology without due regard to the required provisions for O&M. O&M staff have little chance to participate in the design stage. On the other hand, the know-how in industrial plant O&M is not adequately applied in buildings. The following sections highlight some important issues for consideration.

Design For Operation And Maintenance

This is something easier said than done because design for O&M is more an awareness than a technical issue. It is important for the design team to have a vision of the life cycle of the building, rather than considering it simply a new building to hand over.

All members in the project team should have:

- a clear policy that the design is to meet the O&M requirements during the life cycle of the building;
- a position statement to elaborate the essential issues to meet such requirements;
- an action plan to execute all possible procedures for meeting the O&M requirements.

During the design stage, there are, nevertheless, grey areas where it is not clear who has the responsibility. It is then advisable for the project manager to:

- assign such responsibility to designated person;
- have a schedule to constantly receive progress reports to avoid confusion and shirking of responsibility.

The project team should implement the integration of services with the building envelope and structure through coordination meetings. It should also:

- include O&M staff (or someone with O&M experience) in the project team;
- properly integrate the horizontal and vertical distribution of services;
- carefully review space-proofing for equipment, access requirements for

transport, operation, maintenance, overhauling and testing of equipment and devices.

For all the systems and materials to be used, the project team should perform risk assessments on possible threats from:

- fire, robbery, theft, structural failure;
- BRI;
- smoke and odour (air) movements;
- noise and vibration propagation;
- CFC and other toxic gas releases.

It should provide, in particular, access to cleaning and servicing of air ducts, refuse chutes, high hygiene risk areas such as toilets, refuse collection and compaction rooms, grease tanks for restaurants, manholes and sump pump rooms.

O&M engineers often face the problem of not being able to evaluate the performance of the installed equipment and system, because of difficulties in the process of performance assessment brought about by the design. Such difficulties can be classified into two types:

- Inadequate or unsuitable *in situ* instrumentation: either there is not enough instrumentation for performance assessment, or there is a lack of proper calibration, or the instruments are installed at unsuitable locations creating a big error.
- Measurements requiring sophisticated procedures: some measurements, such as assessment of thermal comfort or ventilation effectiveness, require a sophisticated setting up of measurement systems. During commissioning and testing, an evaluation of the performance with higher accuracy is required. At other times, the system performance can be monitored by much easier and cheaper alternatives with a slight sacrifice in accuracy.

In Hong Kong, the focus of commissioning and testing is not strong. For better O&M management in the occupancy period, it is important to include the following in the project contract before the hand-over of the building to the client:

- a comprehensive operation and control strategies according to the design intent;
- the systems to be tested and commissioned according to these strategies under simulated occupancy conditions, which should be as practically reasonable as possible;

- fine-tuning of the control systems when the building is occupied;
- a spare parts-and-maintenance contract co-reviewed with the O&M team.

A set of complete documentation includes at least the design intent, design criteria, design conditions, tender document, contract document, variation orders, site constraints, as-fitted drawings, material suppliers and equipment manuals, O&M procedures, recommended routine maintenance, troubleshooting, lists of suppliers of original equipment, material and spare parts, defect list, and testing and commissioning reports. Parsloe (1994) gives an extensive list of responsibility for allocation to members in the project team and suggestions for proper documentation.

Design For Flexibility

After World War II, all affected countries were competing to revive their economies. Both the social and financial environments have undergone a tremendous change since then. Building technology is innovating the design of the building envelope and structure. The advent of the Internet greatly affects business strategies and hence company structures. Company structures at present tend to have flexible organization and grouping to meet the sales and services requirements. This has a large impact on office layout and office furniture. In order to adapt to the changing indoor functional requirements, all designs call for flexibility to cater for any requirements in layout changes, variation in occupancy density or indoor activities. Sometimes, flexibility is called for to cater for different phases of development of a building complex or possible modifications of investment plans.

The following are a few points for consideration when designing for flexibility:

- Extent: no matter how much a design takes flexibility into account, it is not possible to meet sudden changes of a functional requirement of a building. There must be an agreed limit of adaptation and a clear expectation of the kind of changes and schedule in the future.
- Constraint: a high degree of flexibility requires a higher cost in terms of spare provision for services, system capacity, space and labour. A ceiling on the budget will create constraints on future adaptation.
- Zoning: flexibility can be enhanced by integrating the building services provisions with the building structure and layout by careful zoning. Very often, the architects and interior designers will repeat a certain pattern of design in line with the structural orientation or alignment of the building. Changes can then be effected in accordance with the zone layout.

- Plug and service: flexibility can be effected through a careful planning of the horizontal and vertical services' distribution headers and plenums in either a spinal or circumferential pattern. Header and plenum outlets can be designed at strategic locations for easy plugging-in and can provide service to sub-systems when sudden changes occur.
- Market availability: the degree of flexibility is discounted if it relies solely on proprietary design of the systems, especially for the adapting fittings. Some manufacturers provide equipment, devices and systems with a higher level of flexibility in services systems.

Design For Refurbishment

Before a detailed design for refurbishment of a building is carried out, the project team should make the objectives of the work clear to all members involved. Normally a refurbishment is required in the following situations:

- adaptation to a new demand as required due to change of function, activities;
- replacement of degrading and risky systems, building materials or structures, especially after disastrous damage to the internal content;
- change to meet new expectations due to change of company culture or organization, which is very often aimed at promoting the image of the company or after re-engineering;
- enhancement of system capacity to meet additional loading requirement conforming to business expansion or use of new technology;
- improvement of system performance to compensate for inherent deficiency of the original system;
- facilitate the change or enhancement of investment plans to meet new market potential or market variations;
- compliance to new laws, standards, codes of practices, or to satisfy pressure groups on more stringent requirements on safety, health and environmental issues;
- increase the revenue of the building by extending the operational life span of systems, structures and the building;
- replacement of deteriorated systems or reinforcement of deteriorated parts of the building;
- keep in pace with social stance and cultural changes;
- recycle, regenerate and reclaim of waste material, waste energy and grey water, and reduce abusive use of material, energy and water;

- reinstate the building, its system and internal environment during economy downturns or the company may otherwise move to a new office;
- establish a new corporate image, etc.

When the objectives of the refurbishment are clear to all members in the project team, action plans can be taken to review or assess:

- the use of new technology;
- the building's external and internal environments;
- refurbish for flexibility;
- new laws, codes and standards;
- previous design intents and criteria, and compare them with current intents and criteria;
- the available capacity of utilities connected to the building or premises;
- risk analysis;
- trend analysis of the existing building and system performances;
- financial implications;
- current status;
- recurrent faults and complaints;
- careful scheduling and phasing of the work;
- performances of potential contractors, suppliers and spare parts management;
- the provision of temporary facilities;
- public relations to minimize disturbance to occupants;
- relocation schedules for existing internal activities, etc.

DESIGN FOR ENVIRONMENTAL FRIENDLY BUILDINGS

Aims

Like energy conservation, which has been considered as a classic requirement in building services design since the energy crisis in 1970s, reducing the environmental impact of a building in a site has become a mandatory requirement in all building projects. The Building Research Establishment in the UK produced two reports for the environmental performance assessment of new (Prior, 1993) and existing (Prior, 1993) offices respectively. The Department of Building Services Engineering of The Hong Kong Polytechnic University produced two corresponding versions for Hong Kong buildings (HK-BEAM 1/1996, 2/1996). These two latter documents provide the tools for assessing the 'environmental friendliness' of new and existing offices. The aims of the assessment are:

- to reduce the long-term impacts that buildings have on the environment;
- to raise awareness of local environmental issues and of the large contributions that buildings make to global warming, acid rain and depletion of the ozone layer;
- to promote and encourage the design for energy-efficient buildings, and the use of energy-efficient and building services systems and equipment;
- to reduce the unsustainable use of increasingly scarce resources such as water, timber and natural materials;
- to improve the quality of the indoor environment and hence the health and well-being of occupants;
- to provide recognition for buildings where the environmental impact has been reduced;
- to set targets and standards that are independently assessed and so help to reduce false claims or distortions;
- to enable developers, operators and users to respond to a demand for buildings that have less impact on the environment, and to help stimulate such a market.

Context Of An Environmentally Friendly Building

How a building is judged to be 'environmentally friendly' is best illustrated by the building environmental assessment methods as stipulated in the two HK-BEAM documents. The assessment examined the following categories of environmental issues:

- global issues and the use of resources;
- local issues;
- indoor issues.

Global issues cover the environmental impact on habitats when preparing the site, the use of resources, and the damages to the global environment caused by the effluents and wastes produced as a result of the activities inside the building. The global issues cover the following areas:

- the building owner's/operator's overall environmental policy, environmental purchasing policy and energy-efficiency programme;
- carbon dioxide emissions due to electrical energy consumption;
- ozone depletion due to CFCs, HCFCs and halons;
- storage for recycling of materials.

Local issues cover the environmental impact on the local environment in the vicinity of the building. These issues include:

- electricity maximum demand;
- water conservation;
- external noise;
- legionnaires' disease arising from wet cooling towers;
- transport and pedestrian access;
- vehicular access for servicing and for waste disposal.

Indoor issues are in line with the concept of design for healthy buildings. Such issues cover:

- building maintenance;
- O&M of building services systems;
- metering and monitoring equipment;
- biological contamination;
- indoor air quality;
- mineral fibres;
- radon;
- lighting;
- indoor noise. .

CONCLUSION

Building services engineering is an engineering practice in the design, implementation, operation and maintenance of all the electrical and mechanical systems in a building to sustain a safe, healthy, comfortable and environmentally-friendly outdoor and indoor environment in a professional and ethical manner within the financial and time limits. Hence, building services engineering differs from other primary engineering disciplines in that it needs a strong sense of coordination and integration with other professions like architecture, structural engineering, quantity surveying, interior design and environmental engineering. A building services engineer should also be a good site engineer and project manager. Unfortunately, building services engineers have long been accused of being handbook engineers or catalogue engineers. The problem is that the developing Asian economy is dominated by the building industry. Fast-track programmes and quick-return policies lower building quality. Apart from standard engineering problems, there is ample pressure from politics generated by people involved. Building services engineers have to stand firm on their position to carry out their duty. They should keep in pace with technological advancements and be sensitive to the trend of quality demand, especially when the building market is at a downturn.

This chapter highlights the major design directives for young engineers to consider. Further information on each of the quality qualifiers can be obtained from the listed references. The two HK-BEAM documents contain a comprehensive reference list for building services engineers.

REFERENCES

ASHRAE. 1987. *Indoor air quality position paper.* New York: American Society of Heating, Refrigerating and Air-Conditioning Engineers, Inc.

ASHRAE 62-1989. *Ventilation for acceptable indoor air quality.* New York: American Society of Heating, Refrigerating and Air-Conditioning Engineers, Inc.

Black, F.W. and E.A. Milroy. 1966. *Experience of air-conditioning in offices. Journal of Institute of Heating and Ventilating Engineers* 34: 188–96.

British Standard BS5720-(1979). *Code of practice for mechanical ventilation and air conditioning in buildings.* British Standard Institution.

Chan, D.W.T. 1990. Don't panic or be misled. *Journal of Hong Kong Institution of Engineers* (September 1990).

Chan, K.T. 1996. A study on the feasibility of using the overall thermal transfer value (OTTV) on assessing the use of building energy in Hong Kong. Ph.D. thesis, Hong Kong Polytechnic University.

Chartered Institution of Building Services Engineers. Environmental criteria for design. In *CIBSE guide A – design data.* London: Chartered Institution of Building Services Engineers.

Hansen, W. 1993. Conservation of capacity. *Journal of Facility Management* (April 1993): 6–11.

HK-BEAM 1/1996. *Hong Kong building environmental assessment method: an environmental assessment for new air-conditioned office premises, version 1/96.* A joint project by Centre of Environmental Technology Ltd. and Hong Kong Polytechnic University, sponsored by Real Estate Developers Association of Hong Kong.

HK-BEAM 2/1996. *Hong Kong building environmental assessment method: an environmental assessment for existing air-conditioned office premises, version 2/1996.* A joint project by Centre of Environmental Technology Ltd. and Hong Kong Polytechnic University, sponsored by Real Estate Developers Association of Hong Kong.

Jones, S.P., P.E. O'Sullivan et al. 1995. *New guidelines for the design of healthy office environments.* LINK Initiative, Welsh School of Architecture and the Barlett School of Architecture.

Molina, C., C.A.C. Pickering, O. Valbjørn and de Bortoli. 1989. *European concerted action on indoor air quality and its impact on man, report no. 4, sick building syndrome – a practical guide.* Commission of the European Communities.

Parsloe, C.J. 1994. *The allocation of design responsibilities for building services engineers.* Building Services Research and Information Association, Technical Note 8/94.

Prior, J.J., ed. 1993. *BREEAM/new offices, version 1/93: an environmental assessment for new office designs* (second edition). Garston, Watford: Building Research Establishment.

——. 1994. *BREEAM/existing offices, version 4/93: an environmental assessment for existing office buildings.* Garston, Watford: Building Research Establishment.

Raw, G.J., M.S. Roys and A. Leaman. 1990. *Further findings from the office environment survey: productivity.* Proceedings of the 5th International Conference on Indoor Air Quality and Climate, Toronto, Canada, Vol. 1: 231–6.

United States Environmental Protection Agency. 1995. *Lighting upgrade manual, EPA's green lights program.*

Wilson, S. and A. Hedge. 1987. *The office environment survey: a study of building sickness.* London: Building Use Studies.

World Health Organisation. 1983. *Indoor air pollutants: exposure and health effects.* EURO reports and studies no. 78. Copenhagen: WHO Regional Office for Europe.

Yik, F.W.H., J. Burnett, P. Jones and W.L. Lee. 1998. Energy performance criteria in the Hong Kong Building Environmental Assessment Method. *Energy and Buildings* 27: 207–19.

9

Noise And Design Of Buildings In Hong Kong

Stephen Siu-yu Lau and Dariusz Sadowski

INTRODUCTION

Noise is becoming an increasingly important factor in the way it affects the lives of thousands of people. As a result, it is also becoming more permanently imprinted through the physical manifestations in the building fabric of whole cities and individual buildings alike. This creates a unique situation where human activities and their impact on the environment are producing an acutely negative effect on those using the environment in the first place. This chapter deals with the reactions to one type of such negative effect: noise.

While it is recognized that the production of solutions, especially after-measures, is not the ultimate answer in the long term, it is beyond the scope of this chapter to delve fully into the machinations and mechanisms of our society and economy, and the underlying issues of this situation. Nevertheless it is precisely those social issues which in the end will provide the ultimate solution, while technical and professional advances will provide only temporary relief at their best.

▮ THE CAUSES OF HONG KONG'S ENVIRONMENTAL NOISE PROBLEMS

Hong Kong is well known to the world in two respects: its economic achievements over the last few decades, and its high population density. These two phenomena were not realized without an impact on the environment and the associated quality of living standards, which are rarely taken into account in a country's economic figures. While the total land area of Hong Kong is 1 075km^2, only a small part – approximately 9% – is developed. This is largely due to the mountainous terrain, which is generally not economically viable for construction. The resulting density in the metropolitan areas reaches over 40 000 persons per km^2 – among the highest in the world. The high population density and the apparent need to accommodate often conflicting land uses in a tight area result in frequent cases of unacceptably high noise levels for the residents. In fact, Hong Kong has been dubbed the noisiest city in the world.

Lack of flat land on which to build is part of the reason for such high density. Another is the constant influx of immigrants, both legal and illegal, from mainland China. This influx is a continual trend, which began in the late 1940s and was marked by several significant, major immigration waves in the 1950s and 1960s. In 1990, the number of reported illegal immigrants from China arrested was as high as 76 persons per day while the number of legal immigrants from the Mainland was 28 000 persons per year.

The combination of a heavy influx of immigrants with general population growth created an overwhelming demand for jobs, housing and social services. To provide adequate housing for the growing population, the government has set up the Housing Society and the Housing Authority, among other departments, to be responsible for the design and building of new homes. The 1980s prompted the government's decision to develop suburban areas due to the unprecedented overcrowding and construction coupled with economic growth within existing metropolitan areas. A vast demand for housing, given Hong Kong's limited buildable land, led to the emergence and spread of apartment-type residential developments, the average height of which sprang from 10 storeys in the 1960s to 60 storeys in the late 1990s.

Keeping in pace with economic growth, large sums of money were spent on the expansion of supporting infrastructure, facilities, and buildings for various kinds of developments. Superimposed over existing roads within the city were highway networks and mass transit railway systems (Plate 1). The introduction of different kinds of franchised public transport facilitated the smooth running of a fast-paced city, which took shape almost overnight.

Spontaneous planning, or the lack thereof, to accommodate needs as they arose, was soon changing the city into a chaotic world of general confusion and conflicts of interest. For example, it is not uncommon to find multi-storey factories facing high-rise apartment buildings. Similarly, some multifunctional complexes may maintain facilities such as medical and day-care centres located on different floors from fresh meat and vegetable markets.

Plate 1 Lack of planning resulted in elevated roadways next to residential buildings.

Today, most parts of Hong Kong's physical environment are a result of mixed land-use development. For most areas in Hong Kong, the priority was placed on housing and economic concerns rather than the environment. This was still the case in the 1970s when the government planned and constructed elevated roadways often found, unsurprisingly, to be as close as one metre away from the windows of residential buildings on either side of the road. The extent of damage from environmental pollution suffered by the residents as a result of inadequate planning called for major mitigation measures.

It was not until the 1980s that the government had the economic resources to do something about the massive pollution problems.[1] In January 1981, the Environmental Protection Agency was established, which at the time was small and ineffectual. By April 1986, it was upgraded to become a more powerful department, with major fiscal and administrative input coming from the central government body. In the following year, the Environmental Protection Department (EPD) was formally instituted to look after noise, air, water and solid waste – the four major environmental threats to the city of Hong Kong. Its major tasks covered legislation, formulation of policies, educational programmes, law enforcement and prosecution, and control of construction, industrial and community noise.[2]

Through these times of rapid change, it was often the initiatives of government bodies that attempted to ensure the well-being of the residents. As such, similar efforts made by architects were less often hindered by the interests of their clients. With the endorsement of the legislative powers and the goodwill of the architects and other professionals, the public can now expect a quieter and cleaner environment.

▮ NOISE AND ITS IMPACT ON PEOPLE

Noise is defined as unwanted sound and it can affect people in a number of adverse ways. High-level noise can, for example, cause temporary or even permanent loss of hearing. Besides, noise often interferes with communication. On a deeper level, exposure to unwelcomed noise causes annoyance that not only distracts attention resulting in decreased productivity, but also contributes to the cultivation of a generally hostile environment. Furthermore, noise can disrupt necessary sleep, and most noise-related complaints are actually related to this aspect.

According to the EPD, an estimated one million people in Hong Kong live with traffic noise levels considered to be unacceptably high[3] – well over the standard of 70dB(A) set in the Hong Kong Planning Standards and Guidelines (HKPSG). There are a number of sources of environmental noise, all of which are a result of human activities. The major sources include traffic, industrial, construction and community noise. Very few, if any, sources of noise are natural.

In the urban areas of Hong Kong, dwellings, offices and schools share their neighbourhoods with industrial or commercial buildings that generate often unacceptably high levels of noise. The primary source of environmental noise nevertheless is surface transport, which is as severe on Hong Kong Island as it is in the rapidly urbanizing New Territories. In order to provide sufficient transport facilities to meet growing demands, the capacity of existing railways was expanded, leading to an increase in the corresponding noise pollution. Furthermore, as convenient transport makes land valuable, the number of residential buildings and estates being built along these railway lines continues to boom, consequently subjecting more people to railway noise.

Aircraft Noise

In addition to the 'usual' sources of noise, eastern Kowloon and the northeastern part of Hong Kong Island were exposed to noise produced by aircraft during landing and take-off. The problem was further exacerbated by the economic development in Kowloon, compounded by a high population concentration, and a growing number of residential buildings, factories, schools, hospitals and other urban facilities. Ironically, it was the very proximity to the airport that had helped spur such development.

With the relocation of the airport facilities to Chek Lap Kok, the noise problem in Kowloon has been greatly reduced. However, the move has only transferred the problem to the new town of Tung Chung, which must inevitably bear certain degrees of aircraft noise. However, the intensity is of a much lesser extent due to appropriate and early planning and design measures, which have minimized, but not totally prevented, aircraft noise.

Construction Noise

The fast-paced life in Hong Kong is linked directly to the nature and intensity of local business and economic climate, which has been fuelling the demand for ongoing construction activities. The resulting noise, which continues nearly non-stop adjacent to high-density commercial and residential areas, is an unwanted part of reality for most living here.

Due to the temporary or at least non-stationary nature of construction noise, there is no real basis for any mitigation measures, other than temporary ones, such as barriers and enclosures, once the noise is produced. This leaves little room for effective architectural input, which in general renders long-term benefits, but would be economically impractical in these situations. As such, the role of the government is very important here, as without legislation, there would be little incentive for contractors to employ quieter (and often

Plate 2 Noise from construction activities is a standard part of life in Hong Kong (source: *Environment Hong Kong 1999*).

more expensive) construction methods. In fact there have been cases where specially-designed noise enclosures for particularly acute situations were adopted, with the help of acoustic specialists.

With legislative enforcement though, the effects of construction noise can be ameliorated at least to some extent through persuasive prevention measures. Percussive piling is often the major component in the noisiness of construction sites. The EPD is planning to further intensify its fight against problems associated with percussive piling by proposing to phase out its use completely, thereby forcing contractors to look for quieter alternatives such as auger piles. Naturally, such measures are not very popular with contractors because shifting to a new construction technique would mean that new investment has to be borne by the contractors. This situation clearly illustrates that a sense of responsibility must be shared among all parties in order for design or legislative measures to succeed.

Noise From Mechanical Equipment

In Hong Kong, it is not uncommon to find large commercial complexes, complete with shops, cinemas and offices, located in the vicinity of residential buildings. As a central air-conditioning plant services the commercial complex, the noise produced by associated machinery can be a major menace to the residential neighbourhood. Input from architects at the early design stages would have a significant effect in reducing noise levels affecting neighbouring residences. Fortunately, through the intervention of legislative powers, this is not only possible but in most cases achieved.

In this case, a noise assessment report will be required for submission to the EPD during the design stage of the commercial complex to evaluate the noise impact. The assessment is done within the framework of the Noise Control Ordinance, which sets the maximum permissible background noise levels in accordance with the land-use situations in Hong Kong. Noise consultants are called in to prepare such reports and to recommend noise mitigation measures to reduce the noise to within legal limits. The remedies typically involve specifications of noise attenuators for both air-intake and exhaust of the air-conditioning plant.[4]

Plate 3 Proximity of flatted factories to residential buildings is aggravating the noise problem in Hong Kong (source: *Environment Hong Kong 1998*).

Naturally, the steps taken recently by the government to curb noise pollution by controlling the mechanical noise level produced by new buildings, thereby reducing its disturbance to the neighbourhood, will signify quieter environments for many. Unfortunately, they will be of little significance to the thousands living in some of the older areas. The problem is especially acute in areas such as Tsuen Wan or Wong Nai Chung, where buildings with industrial uses, also called flatted factories, are located next to residential buildings. The noise produced by the machinery and the air-conditioning plants from such buildings can be a serious problem affecting many residents in those areas. Usually very little can be done in retrospect.

Plate 4 The problem of noise in a mixed-use environment in Hong Kong — a case of multiple dwellings sitting on top of a commercial podium (source: *Environment Hong Kong 1998*).

Noise From Surface Transport

One of the most widespread and serious factors contributing to Hong Kong's noise pollution is vehicular traffic noise. To serve the transport needs of both new developments and the existing localities that are growing in population size, massive road networks are constructed to cater for the increasing number of vehicles. These new roads invariably take large volumes of traffic in close proximity to housing estates and other types of residential developments and educational facilities.

The double-sided nature of Hong Kong's increased economic prosperity has meant not only more private cars with increased pressure on existing road networks, but also raised expectations for higher standards of living with a cleaner and quieter environment. These concerns are slowly being reflected in the response from the relevant authorities. The construction of major road projects for example, has been required for the past ten years to undergo environmental impact assessments (EIAs).[5] Likely noise pollution hazards can thus be recognized at an earlier stage, and this allows noise mitigation measures to be provided if necessary and feasible.

The saying that 'prevention is better than cure' certainly holds true here as well. With proper planning and design at the early stage, a lot of the problems can be minimized or even avoided. In older or already-established areas, the situation may be more difficult, as it is too late for prevention and hence remedial measures have to be applied.

▌ADDRESSING THE PROBLEM IN HONG KONG

A problem cannot be addressed unless it is first recognized as such. Thus the responsibility of voicing the problem falls on those affected by pollution in order to prompt any government action. In real terms, because few, if any, complaints are brought to light, the problem continues in many areas of Hong Kong, so in such cases only after-measures – with the assistance of the EPD – can be applied. There are, however, several other directions from which the issue can be addressed. The initiative can come from either legislative or governmental bodies, developers or those responsible for the production of the built environment, or architects during the design stage. The ability to make an input and its consequent effectiveness depend largely on the power and responsibility structures within the society at any given time.

Naturally, early consideration during the planning and design stages provides the best and most efficient opportunities to keep noise pollution in check. Architects and urban designers play a key role in this respect. Not only have they been trained to confront such issues effectively, they also have a social and environmental responsibility to do so. Their efforts, in the particular case of Hong Kong, are unfortunately often muted. Due to the nature of the Hong Kong Special Administrative Region (SAR) economy and the structure of the construction industry, the architects' efforts to avoid or to lessen the effects of noise on the occupants of their buildings are often lost on their clients' agenda, which is mainly profitability.

Nevertheless, some architects' more innovative approaches to environmental control are being increasingly accepted, as a result of both the society's greater awareness and expectations, and the ensuing reaction from governmental bodies to act on those expectations. Only through the awareness and cooperation of all those elements within our society can sustainable environments, natural and artificial, become an achievable reality.

▮ NOISE CONTROL PRINCIPLES

There are three basic approaches to controlling noise: at the source, in the pathway, and at the receiving point (see Figure 1). While efficiency in terms of the total amount of noise being mitigated is at its most in the first case, the chosen and most appropriate approach will depend on each individual situation.

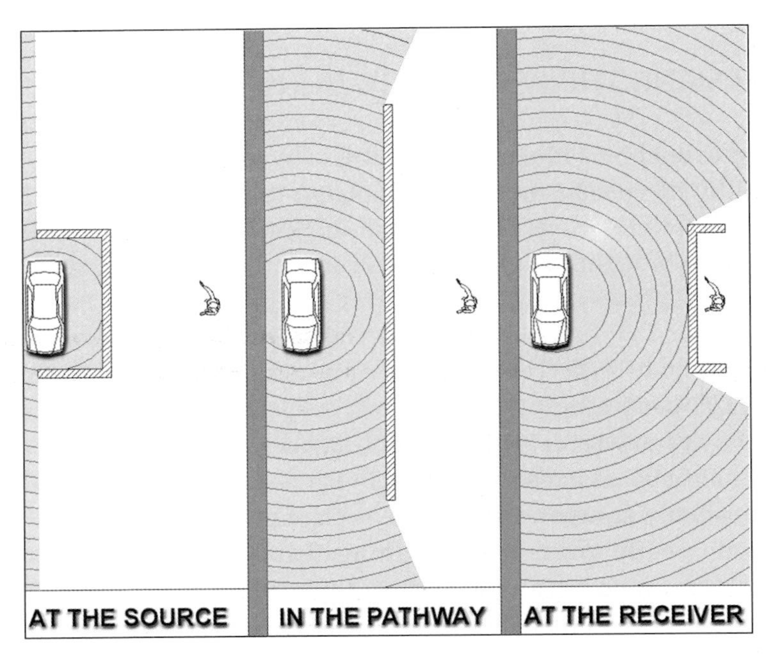

Figure 1 Three approaches to noise control (adapted by J.K. Liu from the unpublished report 'Environmental-friendly noise barriers for the roads of Hong Kong', by P.S.S. Lau, D. Sadowski and S.S.Y. Lau, 1998).

Controlling noise at its source is most sensible option, as it prevents any current or possible future problems, but it is often not the most practical. The problem of delegating responsibility arises, especially where multiple factors are involved as in road traffic-induced noise. However, in some cases, such as noise incurred by railways, the MTRC sets an example by taking the responsibility to ameliorate the problem at its source in the early stages of train and track manufacture process. In fact, it is taking this further by implementing noise control measures both at the source and in the pathway for greater effectiveness.

The second and often the most practical way to prevent sound waves from reaching the receiver is to place a physical obstacle, termed as a noise screening structure, between the two. The effect of noise screening is achieved when this structure obstructs the line of sight between the noise source and the receiving point (i.e. target of protection), and as a result places the receiver in the acoustical shadow zone illustrated in Figure 2.

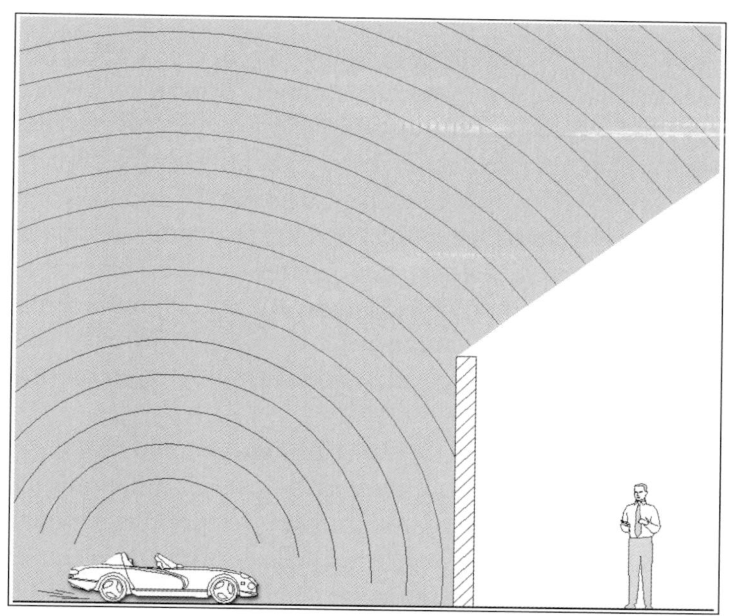

Figure 2 Shadow zone — mitigation is achieved through placing a noise-sensitive receiver within a 'noise shadow zone' produced by a screening structure (such as a noise barrier) (adapted by J.K. Liu from the unpublished report 'Environmental-friendly noise barriers for the roads of Hong Kong', by P.S.S. Lau, D. Sadowski and S.S.Y. Lau, 1998).

Mitigation of noise in the pathway is probably one of the most common approaches in Hong Kong. To optimize the noise screening effect, the distance between the source and the screening structure is reduced. In this way, the shadow zone is increased and consequently the total amount of area protected is maximized. As growing vehicular traffic noise in existing urban areas is the most common cause of noise pollution, use of screening structures is often the most practical form of mitigation.

The third approach is to control noise at the receiving point. In many cases, this method is effective in protecting only the desired target, which, if small enough, may be a preferable approach. Over time, however, it may prove to be the most expensive solution, because if new development takes place near the source of noise, additional measures will have to be implemented according to need. The conditions and past development pattern of Hong Kong suggest that this would be the most likely case in Hong Kong.

ARCHITECTURAL DESIGN

Balancing out the different factors affecting development is naturally indispensable, so that the best possible solution can be arrived at for each particular case. As expected, architects, whose training covers all aspects of building design that enables them to weigh all factors objectively and appropriately, are vital in the balancing process. If and when noise is a significant element affecting the building and its inhabitants, it should naturally be given proper attention. Hence, both views and features can be taken advantage of at the same time, for example, by not having a front facing a noisy road. Similarly, a less ideal but nevertheless ameliorative solution can also be found, if necessary, through alternative means without sacrificing other determinant design issues.

The screening of noise under these circumstances could be done effectively through the use of noise barriers, quieter road surfaces, other noise mitigation methods, or a combination thereof. Acousticians or specialists in related fields can then be consulted to provide more detailed specifications according to the solution most appropriate for the particular situation. The vision and leadership of architects in the design team with the resulting cooperation and support of all parties involved will thus produce a building responsive to all factors, environmental and otherwise, in an objective manner.

Currently, noise control measures are often considered only after the problem has already been felt. The ability to recognize problems in advance is key to properly addressing potential problems arising from various sources and will result in the employment of preventive steps being taken. Although preventive measures may result in additional initial costs to the development and may be limited by practical issues such as site conditions and structure of the building, they will almost always lead to inevitably quieter and more liveable environment that may ultimately prove more cost-effective than employing after-measures later. With experience and increasing efforts, advances continue to be made in the architectural building design stage, from the choice of site, layout planning or structure design to the smallest details such as a door seals or small pipe brackets carrying vibration waves. These are of particular importance in dense, high-rise urban developments and redevelopments. The approach of planning against environmental noise includes land use separation through layout, self-protective building design and additional noise barriers.

Since around 1991, there has been a gradual increase of lease-imposed conditions being applied by the government in respect of new urban and suburban planning redevelopment carried out by private developers, whereby the latter must undergo the EIA process for approval by the Environment Authority. With the help of legislative powers, architects were finally able to produce designs according to their training through theoretical and practical study during university years. Sceptics who doubted the relevance of architectural education in the hard and fast world outside must now reconsider their argument. The opportunity has finally come along, at least in the realm of noise as it affects buildings and their occupants, to make a difference. For the first time, architects, noise consultants and planners work in conjunction to consider the impact of external noise on potential inhabitants in the light of by-law requirements. The analytical process involves prediction, calculation and evaluation of the nature and magnitude of noise nuisance, followed by proposed mitigation measures.

Active research and close consultation between architects and the EPD have actually resulted in modifications to the standardized plans of high-rise residential buildings that made them more effective in curbing noise pollution. In the last two decades, building codes were revised to facilitate a maximum plot ratio of eight times that of the land area with a corresponding maximum site coverage of 66.6% to 33.3% for a single street high-rise residential building.[6] The long-awaited legislative backing which necessitates such modifications has finally enabled architects to produce alternatives to standardized plans – and so far this has been well received. Given such statutory guidelines, architects and developers promptly developed a plan type that was considered to be the most efficient to attain the maximum permissible floor area, view, ventilation and daylight to comply with all building code aspects and to maximize financial returns on the initial investment.

A typical floor plan (Figure 3) would, for example, consist of up to eight residential units serviced by a central service core of lifts, stairs and the necessary

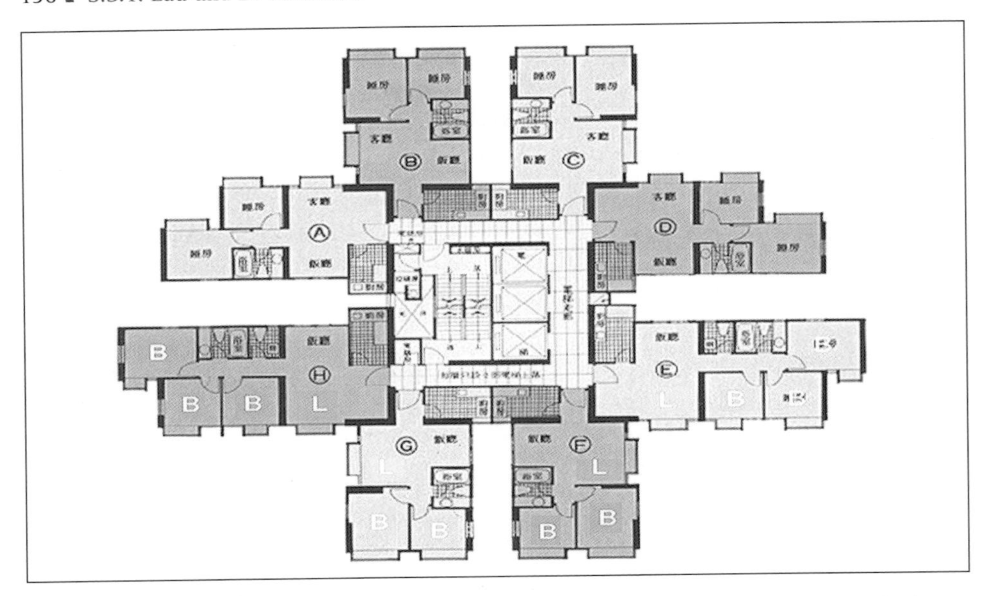

Figure 3 Typical cruciform-type floor plan commonly applied in Hong Kong to high-rise apartment buildings (source: *Application of screening structures to abate noise from surface transportation*, Environmental Protection Department, Noise Policy Group, Hong Kong).

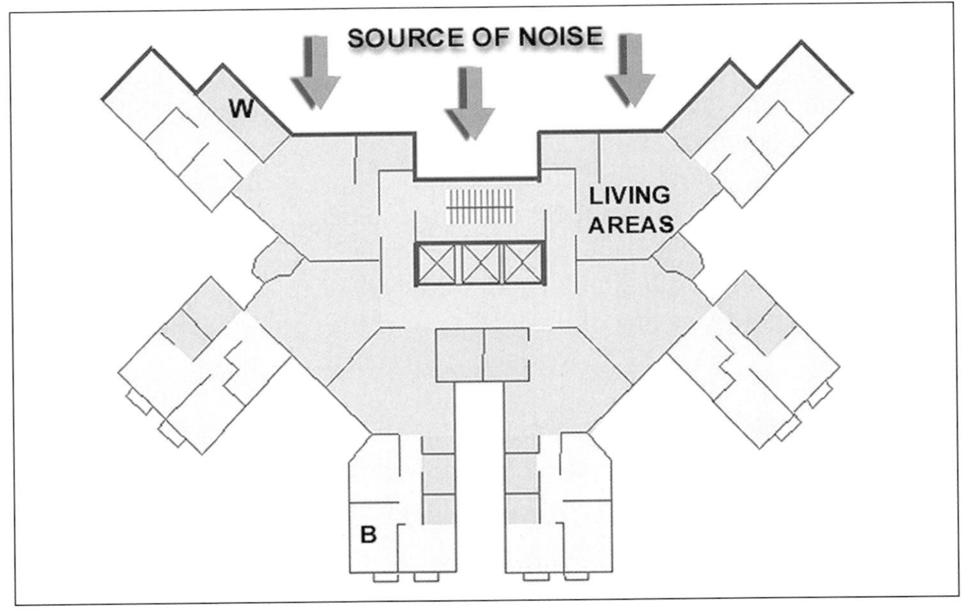

Figure 4 Alternative floor plan for situations where noise is a major design factor (source: *Application of screening structures to abate noise from surface transportation*, Environmental Protection Department, Noise Policy Group, Hong Kong).

rises, ducts and chutes. A new floor plan made possible through the cooperation with the EPD resembles a butterfly with up to six or eight units and a central service core similar to the original plan's (Figure 4). The difference between the two plans comes from the relocation of less noise-sensitive elements such as kitchens, washrooms and stairs, in such a way that they confront the direction of the invading noise. In this way, they form a noise protective screen to rooms which are noise-sensitive, such as bedrooms and living rooms. This particular active cooperation between architects and the noise control authority is extremely important at the present time when both the government and private sectors are implementing rapid infrastructural and housing construction programmes to meet Hong Kong's growing needs in the twenty-first century.[7]

Self-protecting Building

When buildings are located close to highways or railway lines, for example, they often need to adopt self-protecting characteristics. As mentioned above, this can be done through an appropriate layout of units to provide a buffer zone for the living areas, thus effectively insulating them from external noise sources. The 'weak' elements of a building, like doors and windows, can be protected or screened from direct exposure to noise pollution through the use of balconies and courtyards, or projected elements which, solely or in combination with other uses, provide protection. An example of where this principle has been applied successfully is Verbena Heights (discussed in more detail under the case study section on page 204).

Structure-borne noise can be a source of particular nuisance to neighbouring buildings which require particularly quiet environments, especially in the case of schools or hospitals and other health facilities. To reduce structure-borne noise, major enclosing elements of the building should be treated acoustically, while in exceptional cases, the building may be fitted out with double-glazed windows or even acoustic doors.

Other measures become particularly important when it is not desirable to rearrange the layout of the housing units because such action would result in, say, loss of view for the living areas. Rearrangement of unit layouts may indeed satisfy requirements of the EPD in obtaining lower noise levels affecting noise-sensitive areas of the housing units. However, if it means the neglect of other considerations and a consequent lowering of the overall dwelling quality, then alternative or additional forms of mitigation measures are then called for. Among them would be additional insulation, noise barriers, use of non-noise-sensitive buildings for screening purposes, or enclosures of the noise source.

▮ BUILDING INSULATION

When, due to site and other constraints, allowable noise level standards specified in the HKPSG cannot be met, it may be necessary to fall back on building insulation as an additional noise mitigation measure. Architects have here a number of choices at their disposal that can be implemented with notable results. More commonly, though, effective building insulation usually means only installing better-quality window glazing and assuming that the windows will be kept shut all year round. Given the hot and humid nature of Hong Kong's summers, this inevitably leads to a need to provide central air-conditioning. Unfortunately, this is not the best solution as the resulting higher usage of electricity will eventually contribute to other environmental problems like air pollution for example, resulting from higher demands placed on the generation of electricity.

The enactment of legislation is often required to prompt necessary steps to be taken. The recent involvement by the EPD is a welcome and encouraging step in this respect, not only for new developments but also for existing ones. To protect residents affected by traffic noise after opening of new roads near existing residential buildings, the EPD provides liaison with other departments to minimize the negative impact. Commonly, the EPD's involvement results in ensuring an adequate buffer distance between the new road and the buildings or in provision of mitigation measures such as roadside barriers or enclosures, wherever practicable. However, on some occasions it is necessary to adopt building insulation. This usually involves upgrading of the electrical supply, improvement of windows and installation of air-conditioning units, with each case considered on its own merits.

One source of noise that traditionally has not been addressed with warranted importance is the low-impact and airborne sound generated within the housing units, which nevertheless affects neighbours by transmission through walls and floor slabs. It is also one area that the legislation does not yet cover, but is becoming an increasingly apparent one as living standards and expectations for quieter environments and greater privacy rise. It also serves as an example where the necessary cooperation and expertise for the provision of countermeasures by architects is available, but due to a lack of necessary incentive and legislation, most residential developments still fail to address this issue.

Comparison with regulations in the UK, for example, indicates that there is room for the introduction of similar steps here in Hong Kong, since both technical and financial means are available for that purpose. The regulations in the UK for dividing walls and floors put forth specifications for insulation with respect to noise. Already in Hong Kong, guidelines such as the imposition of the required EIA in April 1998 are already addressing to a large degree the current concerns of society. In order to address these concerns in a more comprehensive manner, much can be learnt from effective models abroad.

▮ LAND USE PLANNING

Land use planning to reduce noise begins with the siting of different buildings in relation to noise sources within or outside the buildings. Depending on the nature and mix of industry in a particular geographical location, various zoning principles can be applied to divide the different uses so as not to allow the incursion of noise sources on sensitive receivers. While originally this was believed to represent progress in the quality of living standards, it quickly became apparent that such simplistic measures could not be applied to complex structures like cities. There are many examples of 'zoned' cities around the world, which are devoid of any spontaneous living qualities and end up with inefficient transport options, which in turn contribute to even more environmental problems.

Given that, there are many opportunities where some form of land use planning or zoning can be applied sensitively, especially where the separation from, say, heavy industry is required. Traffic between these zones is diverted into a peripheral route while boundaries between the zones can be crossed only by pedestrians and by public transport. The potential benefit is reduced impact of industrial and transport noise on residential and other noise-sensitive uses. Currently, the possibilities of land use planning for Hong Kong lie mainly in the development of new towns, which, when applied sensitively, can produce not only quieter but much more pleasant living environments. Nevertheless, years of experience and expert knowledge are required to achieve the fine balance between 'protected' but sterile environments and a lively neighbourhood.

On a slightly smaller scale and in cases where it can be applied, site planning is based on the different uses of various buildings or their parts, the potential of noise pollution from external sources, as well as their noise impact on neighbouring buildings. Noise surveying and advanced prediction methods are needed to evaluate the growth and extent of the environmental impact. Traffic noise plays an important part in the prediction of noise pollution in urban areas, where the noise impact on buildings on both sides of the road should be assessed during the planning stages of the traffic network system.

If, after such site planning considerations, noise levels still exceed the desirable ones, other control methods, such as noise barriers, should be considered. Another option is to identify noise-sensitive buildings, like housing units or schools, and locate them further away from the highway. Instead, noise-tolerant uses like storehouses, car parks or commercial facilities could be located to take the brunt of the noise, and in effect act as a screen for other uses.

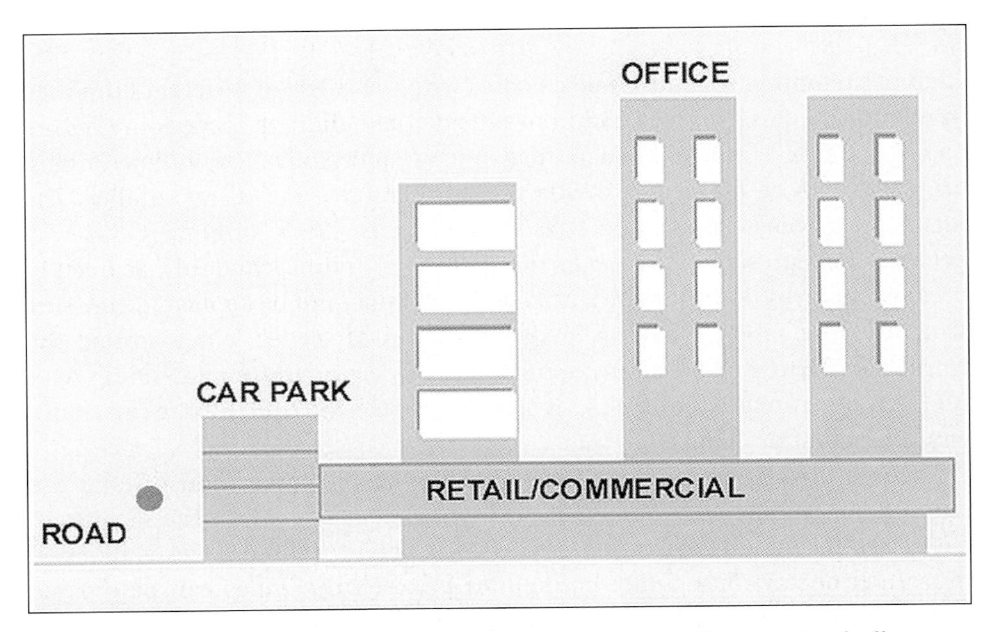

Figure 5 Site planning where noise-tolerant structures provide a noise buffer zone (adapted by J.K. Liu from the unpublished report 'Environmental-friendly noise barriers for the roads of Hong Kong', by P.S.S. Lau, D. Sadowski and S.S.Y. Lau, 1998).

▮ NOISE BARRIER

Addressing the noise pollution problem at the building design stage is a very effective way to guard against known and predictable effects of noise pollution. This is achieved through an alternative layout of the units by placing noise-sensitive parts of the building or development away from the noise. Occasionally, this can present problems for other functions that dwellings are supposed to provide for their residents. A common example is where, in an effort to meet Noise Standards and Guidelines for private and public housing, the overall planning of these properties suffers as a consequence. Overzealous adherence to noise control legislation can result in views from living areas being ignored, or kitchens, bathrooms and corridors forced to face noise-polluting highways. Noise barriers become an indispensable tool in such situations by reducing the impact of noise on all parts of a building, with minimum compromise to aspect and planning considerations.

A noise barrier is a structure designed to reduce the sound pressure level in the noise-sensitive zone that is exposed to noise emission (Plate 5). The barrier is often located on the side of a road to block the direct propagation of traffic noise, and its effectiveness depends mostly on its length and height. There are, however, other variables. For example, the barriers may be categorized according to the material used (solid or transparent barriers), or acoustic characteristics (e.g. absorbent barrier or acoustic panel). Natural land may also be formed and planned to act as a screen for sensitive buildings, even though there would be few instances possible for this in Hong Kong, simply because of the scarcity of land. Other variations relating to shape and extent of protection are also possible. A curved barrier would, for example, provide further protection to higher-floor units without increasing the barrier's height.

Plate 5 Noise barriers are among the most common mitigation measures adopted in Hong Kong.

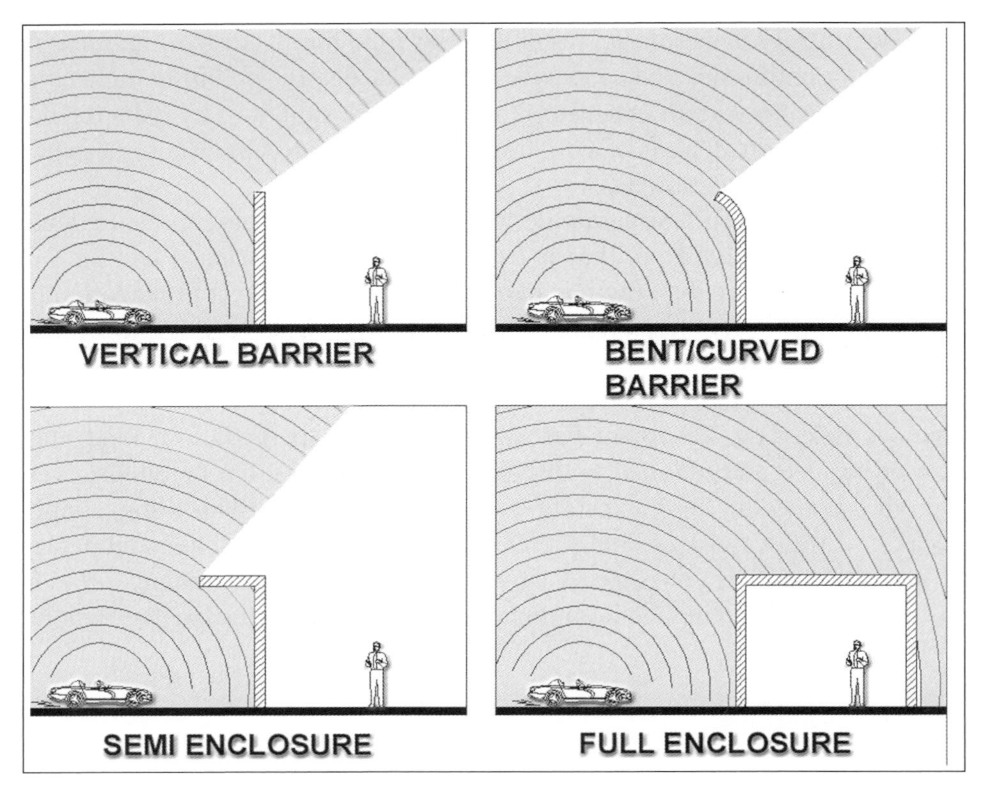

Figure 6 Types of noise barriers (adapted by J.K. Liu from the unpublished report 'Environmental-friendly noise barriers for the roads of Hong Kong', by P.S.S. Lau, D. Sadowski and S.S.Y. Lau, 1998).

As sound attenuates with distance, the most acute traffic noise nuisance occurs at lower-floor levels, and thus it is the lower floors that require the most attention. A study conducted by the University of Hong Kong[8] showed that the difference in noise levels originating from the same source can amount to 10dB(A) between the lower and higher floors. Therefore, the type of barrier used must conform closely to its purpose and particular situation.

Input from architects in the design and integration of noise barriers in existing or new urban environments is increasingly more common, as a result of both a greater professional responsibility and an increased demand for such structures to form a cohesive part of those environments. The unique education that architects receive allows them to apply their expertise in situations where such interventions have aesthetic and visual consequences. Noise barriers are becoming an increasing part of our constructed environment and thus should be properly integrated with it.

Plate 6 KCRC staff quarters: Noise barriers are a part of the built environment and they warrant the same aesthetic attention as more conventional building parts (source: *Application of screening structures to abate noise from surface transportation,* Environmental Protection Department, Noise Policy Group, Hong Kong).

ROAD RESURFACING

Road resurfacing is a method of reducing noise produced by the interaction of tyres with the road surface. Rough surfaces are particularly prone to producing increased amounts of noise, with brushed concrete, commonly used in Hong Kong, falling into this category. To redress the impact from the resulting traffic noise, the EPD has developed a number of criteria, which would enable certain road sections affecting noise-sensitive developments to qualify for resurfacing.

To make resurfacing of brushed concrete effective the road should be flat. Similarly, frequent use by heavy vehicles would also render resurfacing much less effective, as the engine noise would also drown out other types of noise.[9] The Island Eastern Corridor is a successful local example where resurfacing was carried out and resulted in a noise level reduction of up to 5dB(A), measured during a trial period from 1987 to 1989. This promising result led the EPD to initiate the 'Quiet Road Surface Programme' to include other applicable road sections in the SAR for resurfacing with quieter, open-textured material.

Since then, a number of further road sections have been similarly treated, with the result of a total of 8.2km of roads resurfaced up to 1995, bringing noise relief to some 11 000 dwellings. Furthermore, there are certain situations in which resurfacing provides other benefits. One such situation is where the road runs between high-rise buildings, and the resulting noise cannot be treated easily by conventional ways. As the noise is produced at the surface, it is first reflected upwards, and then back and forth between the buildings, resulting in what is called a 'canyon effect'. The best way of minimizing the impact of noise in such a situation is through resurfacing, as it deals directly with the noise at its source.

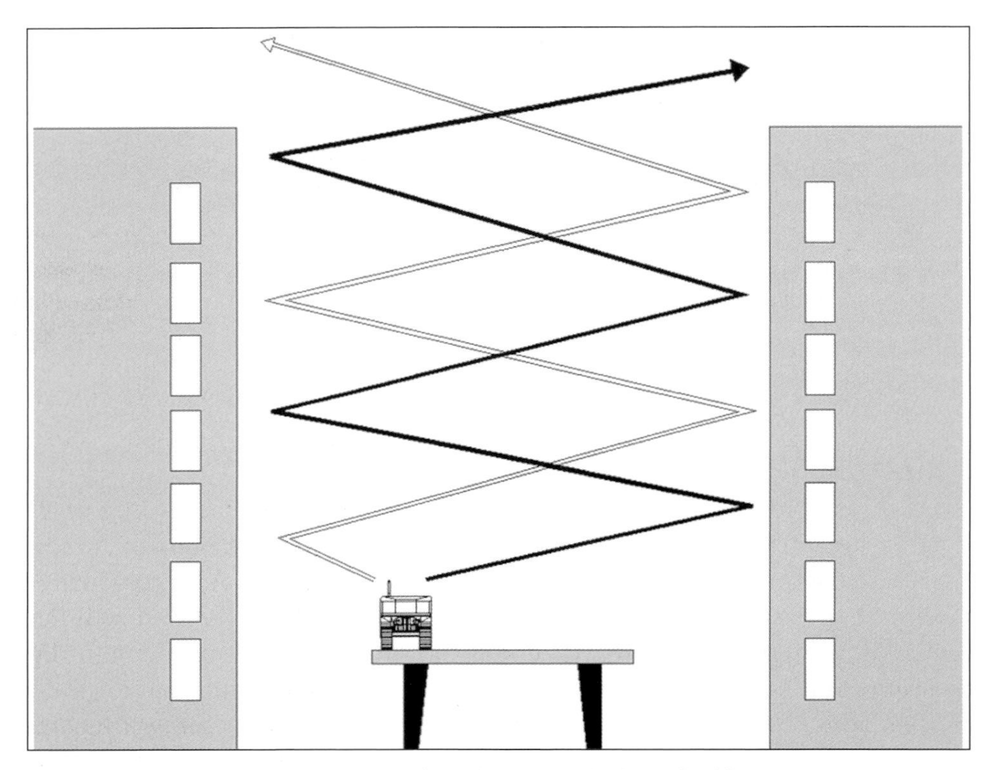

Figure 7 Noise problem gets worse with roads between high-rise buildings, creating a 'canyon' effect (adapted by J.K. Liu from the unpublished report 'Environmental-friendly noise barriers for the roads of Hong Kong', by P.S.S. Lau, D. Sadowski and S.S.Y. Lau, 1998).

▮ TWO CASE STUDIES

Tsing Yi Station

The design and planning of the new Tsing Yi Station can be helpful in illustrating some useful techniques in noise abatement which are available to designers in Hong Kong and which can be applied equally to other sites.

Tsing Yi Station is one of several stations located along the Lantau and Airport Railway, which connects the new Chek Lap Kok airport with Kowloon and Central. The development comprises a six-level podium that accommodates a railway station, commercial shops and food parlours, together with a residential development built on top of the podium to house about 10 000 people. Heavily trafficked Tsing Tsuen and Tsing King roads bind the site and pose serious threats to an acceptably noise-free environment for residential flats. The Lantau and Airport Railway also runs through and bisects the site, making rail noise a serious additional concern.

The initial response to this design problem was to provide a layout of twelve standard cruciform towers on both sides of the centre line of the podium. About 32% of the flats in this layout would be exposed to noise levels exceeding the limit of 70dB(A) L_{10} (1hr), with the highest levels reaching as much as 80dB(A). Therefore, three further mitigation methods were employed to deal with this problem. Increasing the separation distance between the towers and roads was one method, as doubling the distance between the source and the receiver would render a 3dB(A) reduction. The second mitigation measure was to increase the screening effect, while the third factor in reducing noise levels was achieved through site layout, building orientation and unit layout within the towers.

As the setback distance from both main roads had already been maximized, further setback could be achieved only by redesigning the unit layout of towers. Noise impact is normally more difficult to abate in standard cruciform towers, as they have openable windows on all sides. The towers were rotated, and the unit layout revised accordingly, which reduced the angle of exposure to the two main roads. After evaluating further aspects, modified cruciform tower designs with windows facing away from roads were adopted. Furthermore, three of the twelve towers were redrawn to contain seven flats per floor instead of the standard eight, which created a higher proportion of protected flats conforming to the noise level standards.

Covers on roads, it was determined, were to give the most effective screening effect. Since the residential towers sit on a podium, which already served as a type of noise screen, mounting cantilevered barriers on the podium edge extended the shadow zone and thus further reduced the noise levels potentially affecting the residents.

Verbena Heights In Tseung Kwan O

One of the innovative developments recently constructed in Hong Kong that deals with environmental issues with an active approach is Verbena Heights, designed by one of Hong Kong's architects. It is mainly a residential development, with the total number of residents estimated at 7 650. The site is located in Tseung Kwan O, in the southeast New Territories. Two roads to the southwest and northwest form the boundaries for the development. An MTR station is proposed to the northwest of the site, and access for residents to the station will be provided via a footbridge, which will also link it with another proposed development to the northwest of the site.

The most acute source of noise affecting the site is vehicular traffic. The three main external noise sources are the two roads at the side boundary and the Tseung Kwan O Tunnel Road. In the HKPSG 1990, the EPD recommends a peak hour maximum noise level of 70dB(A) L_{10} at the façade of a residential building, and a lower limit of 65dB(A) is required for schools and kindergartens that rely on openable windows. In this development, two ways of reducing the traffic noise impact have been considered. One is to implement a noise barrier wall at street level, while the other is to apply externally-mounted acoustic screens to protect the windows directly.

Using calculations, it has been shown that a large podium-mounted noise barrier can provide reductions in noise for the lower flats. However, a podium-mounted barrier cannot as effectively screen noise impact to the higher floors.

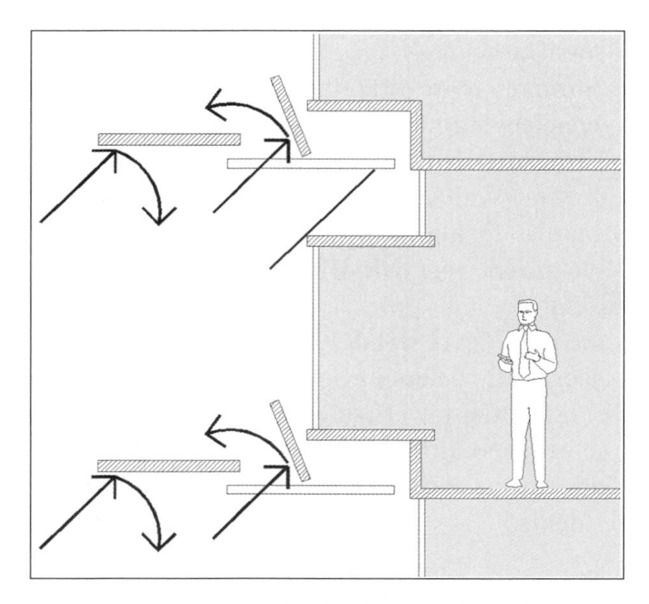

Figure 8 Effectiveness of street-level noise barrier in protecting residents at Verbena Heights (adapted by J.K. Liu from the unpublished report 'Environmental-friendly noise barriers for the roads of Hong Kong', by P.S.S. Lau, D. Sadowski and S.S.Y. Lau, 1998).

The effectiveness and feasibility of the externally-mounted acoustic screens were investigated, the use of which could be beneficial to lowering noise levels and reducing thermal gain at the same time. To achieve this, the screens are mounted outside the windows of the affected rooms, and their specially shaped profile provides reduction of noise levels, with the dual benefit of reducing the penetration of direct sunlight.

Hopefully, Verbena Heights marks the beginning of a new era for Hong Kong, where architects are given the opportunity to take on a more active role in addressing the ever pressing issues of environmental control in the environments they create. If this is to result in more sustainable housing with a degree of sophistication and responsibility to produce living spaces adequate for the twenty-first century, then more architects need to become aware of their responsibility, while the public should be made aware of their right to quieter, cleaner and more comfortable environments.[10]

THE ROLE OF EDUCATION

The cooperation of the whole of society is needed if attempts to create a better environment are to succeed. Education's role is to act as an alternative to strict policing against adverse activities. This is achieved through increasing society's awareness of the consequences of its seemingly 'little' actions, like driving a car or tuning on an air-conditioner, and of its rights to a healthier, cleaner and sustainable living environment. The first step is for those who know to inform others. Architects play an important role in this respect as they have to deal with other factors, apart from the environment, which gives them an opportunity to arrive at a well-balanced solution.

The government's role in this aspect is also very important, not only because of its general responsibility, but also due to its ability to reach the necessary breadth of audience with the available resources to do so. Education is, in fact, one of the major five steps undertaken by the government to combat noise pollution. Education will generate increased environmental awareness among the public. The University of Hong Kong is the first to organize a Master of Science programme in environmental management, which is geared towards professionals in the field and those from other disciplines, while the EPD regularly organizes exhibitions and talks by various specialists. The main objective of such educational activities is to cultivate a better understanding of the delicate nature of our living environment, and thus bring a heightened concern for protecting it for the enjoyment and use of future generations.

▮ FUTURE DIRECTIONS

Wisdom dictates that future actions are based on lessons learnt from the past. Society today is aware of past experiences and is able to contribute to continual advancements in technology. In turn, the technology redefines the way in which society interacts with its environment. This fuels the challenge of, and the need to channel, new developments in a responsible, beneficial manner to society now and in the future.

While advances in technology enable the development of increasingly sophisticated responses to everyday predicaments, an underlying current of environmental concern has promoted the natural and 'low-tech' approach to tackling our problems since the 1960s. The low-tech solution refers to increasing public awareness of existing and potential problems in society and its environment. Not only does awareness facilitate the identification of these situations, but it also encourages responsibility both in solving and preventing future occurrences.

The key to the low-tech approach is education, which in the case of environmental awareness includes information regarding the limitations of energy and other resources, and the resultant damage caused by overuse. The goals of this approach are hence threefold: to reduce general energy consumption, to output fewer corresponding by-products, and to advocate public responsibility for the preservation of a comfortable, liveable environment.

A second approach to defining the direction of future actions is termed as the 'high-tech' solution, which relies directly on technology as the key factor in addressing environmental concerns. This includes the designing and use of mechanisms that are environmentally friendly in terms of solving existing problems and preventing potential problems. In the case of noise pollution, a home of the future may opt to use what has been termed as active noise control.

While UK carmaker Lotus Engineering has integrated active noise control to deaden the noise of moving cars, and NASA has applied it to counteract the noise of jet engines, active noise panels have already proved to be a success in dealing with noisy cars, aircraft and ventilation shafts. They are similar in construction to double glazing, with the outer pane of glass acting as a microphone and the inner pane behaving like a loudspeaker. The principle is: the outer pane of glass detects the noise coming on to it, meanwhile the second pane of glass is vibrating in the opposite direction. As the noise starts to pass across the air gap, it is met by a noise coming in the opposite direction that in effect cancels the first.

Some specialists in 'intelligent buildings' predict the employment of active noise control in the home as the next logical step. Building engineers are currently experimenting on this. Active noise panels could be placed on the internal walls of a house to act as a layer of intelligent skin that responds to noise and controls it in a desired way. A sound level reduction inside a room of up to 50% has been shown in experiments.

As both low-tech and high-tech solutions share a common goal, the ultimate solution will likely require a combination of the two. At present, the urgency of environmental concerns has only recently gained priority internationally. Despite the promising progress in understanding and addressing these concerns, the realization of that goal may yet lie far in the future.

▌RELATED RESEARCH

To date, a majority of research on noise barriers has been conducted by noise barrier manufacturers. While the relatively small size of Hong Kong limits the extent on how much research can be done locally, several local distributors of noise barriers can tap into the resources of their parent companies overseas. In some cases, governmental organizations also conduct tests to certify the quality and specifications of products. Additionally, an Italian company, the Autostrade, has conducted studies on several types of noise barrier like the biowall, and the concrete or wood absorptive transparent wall.[11]

There are many aspects of noise pollution that are not immediately apparent but can still be investigated through appropriate studies. One such consideration is the varying types of noise generated by different road situations. For example, at speeds higher than 50km/h, the tyre-road noise is higher than the emission noise of the vehicles. This would necessitate the provision of a different type of barrier, designed for mitigation of noise at different frequencies. Other studies have concentrated on barriers by type, like the New Jersey low-frequency absorptive barriers. The first prototypes of the New Jersey phono-absorbent barriers with a selective configuration to absorb low frequencies have, for example, given excellent performance between 100Hz and 500Hz.[12]

Other lines of research has also led to a number of very promising developments in the area of computer simulation models, where a number of acoustical factors vital in the development of noise mitigation methods are explored in conjunction with often-ignored external factors.

▮ THE ARCHITECT, NOISE AND SOCIETY

Protecting the occupants of artificial environments from possible hazards, such as noise, is an important component of the overall health and well-being of the population. The liveability of our macro-environments often depends on the care, skills and determination of architects to contribute to making them a pleasant or at least a liveable place. The very definition of home or dwelling carries with it the notion of protection from outside elements. While in the past those elements meant excessive wind, rain or heat, today's greater challenge facing architects are adverse external factors created by humans. This is increasingly true in densely populated areas, such as Hong Kong, where the pressures from some activities invariably affect others.

It is the responsibility of architects to become not only aware of, but also well-versed in, the newest technological developments and latest possible solutions so as to provide the society with the best possible environments to live in. As professionals whose work literally shapes the community, it is also the moral responsibility of architects to create a secure environment that is resistant to external hazards. For the most part, this means producing designs attractive to the general public while subtly integrating mechanisms that protect occupants from noise pollution and other elements. In the process, this also means educating the architect's clients, who are rarely the future occupants of the buildings, of the benefits of these designs not only for the users but also in financial terms for themselves. As times are changing and society is becoming more aware of ecological and environmental issues, the architect's role is becoming increasingly important in terms of being part of the solution rather then the cause of such problems.

Ironically, environmental problems are compounded by notions of democracy and capitalism under which the 'have-nots' suffer consequences induced by activities by those who 'have'. A prominent example is the use of private vehicles, the number one problem for noise and other types of pollution. In general, the innovations and regulations adopted by architects and legislature can provide only a surface solution to the overall problem that affect everyone. Despite ongoing debates regarding economic costs and corresponding social benefits, the ultimate solution lies mainly in the public's attitude and its willingness to forego individual habits and luxuries for the benefit of its surroundings.

As we approach the twenty-first century, such issues are bound to only grow in importance and are unlikely to simply go away, no matter how long society in general may want to look the other way. Unless they are addressed promptly, they are going to only become bigger and more acute, further affecting our lives.

▌CONCLUSION

Despite the progress in environmental noise abatement techniques and the great efforts made to control environmental noise in past decades, the achievements have not been sufficient to render great improvement. Many of the problems result from a lack of early consideration of noise aspects in the planning stages at various scales of development. Careful planning and cooperation of architects and acousticians can hopefully solve these problems in the future.

Planning against transport noise in a busy metropolis like Hong Kong is definitely not an easy task. However, if we focus on potential noise problems at the very outset of planning, we have a high probability of meeting the noise exposure standards as set out in the HKPSG. More importantly, however, we can go beyond satisfying certain numerical objectives blindly, by realistically looking at all the variants and allowing the residents of Hong Kong to enjoy environments that are pleasant to live in with all the different aspects considered.

At the same time, it is an encouraging sign that the government is also taking a much more active stand on environmentally responsible changes, as reflected in the recent enactment of new regulations and issuance of related Practice Notes for Authorized Persons (Architects) by the Building Authority. A number of such regulations have been gazetted in recent years, covering many issues related to environmental sustainability. Their presence is an encouraging sign, but generally speaking, their effects are still of modest proportions.

The currently widespread development of residential buildings based on the cruciform plan is essentially a product of the Building (Planning) Regulation. Aided with regulatory changes, a shift towards a more proactive design approach on the part of architects, who can be the necessary catalyst, will be the critical element in changing our built environment.

However, controllable artificial factors such as noise pollution are social problems and, as such, responsibility should be borne by the whole of society. The sources of environmental problems lie within the activities of humankind, therefore, the solution lies in the modification of these activities. For example, those responsible for producing noise should also assume responsibility for reducing its effects on the community affected. A good example of a company taking responsibility for its noise-inducing activities is the Mass Transit Railway Corporation, whereby active effort and initiatives are coming from the company itself in trying to reduce the noise from the railways it operates.

The example of Hong Kong may be a lesson for other cities throughout the world that are faced with the problem of environmental deterioration due to noise pollution caused by rapid population growth and economic development. Although the pollution issue has yet to be resolved, the first invaluable step is to increase awareness at all levels so as to prompt a cooperative effort in discovering a viable solution. In Hong Kong, the implementation of this first step has already persuaded many of its people to take active measures to address environmental problems. So far, cooperation from the Hong Kong government has greatly facilitated the ongoing efforts by architects and other concerned citizens to reduce pollution. With the continual commitment and firm resolve of both the people and the government, the SAR may someday achieve a noise-free environment.

▮ ACKNOWLEDGEMENTS

Part of the research work in this chapter owns credit to the University of Hong Kong's Committee on Research and Conference Grants, 1997 and 1998.

▮ REFERENCES

Environmental Protection Department, Hong Kong. 1990, 1994, 1995, 1996. *Environment Hong Kong.*

——. 1996. *Hong Kong – The environmental challenge, 1986–1996.* Hong Kong: Government Printer.

——, Noise Policy Group. 1993. *Application of screening structures to abate noise from surface transportation.* Hong Kong: Government Printer.

Ginn, K.B. 1978. *Architectural acoustics.* Naerum, Denmark: Bruel and Kjaer.

Land Development Policy Committee, Hong Kong. 1990. *Environmental guidelines for planning in Hong Kong.*

Lau, Stephen and Ho-yin Lee. A review on environmental noise protection in Hong Kong. Unpublished paper presented at the 4th International Congress of Sound and Vibration, 24–27 June 1996, St. Petersburg, Russia.

Lau, S.Y.S. and G.R. Jiang. 1993. Environmental noise impact on architectural design. Thirty-fourth International Congress Asian and North African Studies, held at the University of Hong Kong, 22–28 August 1993.

Ng, A. and K.S. Wong. 1997. Sustainable housing in Hong Kong: Verbena Heights (TKO Area 19B) and beyond. *HKIA Journal*, Environmental Issues 9 (2nd quarter 1997).

Organisation for Economic Co-operation and Development (OECO), Paris. 1995. *Roadside noise abatement.*

Wong, Kam and Yeung. 1993. A review of noise mitigation works for new public roads in Hong Kong. Unpublished report, Environmental Protection Department, Hong Kong government.

▮ NOTES

1. *Environment Hong Kong*, 1996, p. 11.
2. *Environment Hong Kong*, 1990, pp. 84–91.
3. *Hong Kong – The environmental challenge, 1986–1996*, p. 19. The noise standard is given as L_{10}(1 hour) > 70dB(A), where L^{10}(1 hour) is the noise level exceeded for 10% of the one-hour period – generally used to describe traffic noise during the hours of peak traffic flow.
4. Lau and Lee, 1996.
5. A review of noise mitigation works for new public roads in Hong Kong (Wong, Kam and Yeung).
6. Building Planning Regulations, First Schedule, 1984, Hong Kong government.
7. Lau and Lee, 1996, p. 4.
8. The study 'Field measurement of environmental noise and calculation of reverberation time' was conducted as part of a HKU Master of Architecture project in December 1996.
9. Environmental Protection Department, 1996, p. 65.
10. Ng and Wong, 1997, p. 56.
11. Organisation for Economic Co-operation and Development, p. 164.
12. Organisation for Economic Co-operation and Development, p. 165.

10

An Investigation On Overall Thermal Transfer Value

Wong Wah Sang

INTRODUCTION

Control on the overall thermal transfer value (OTTV) is enforced by one of the Building Regulations for Energy Efficiency. The aim is to establish a comprehensive building energy code to control the total energy consumption in a building. OTTV measures the energy consumption of a building envelope which depends on the materials, glazing type and size, provision of shading, colour, as well as the orientation of the walls. The building enclosure controls the heat and light emitted into the interior, which affects the extent of use of artificial lighting and mechanical ventilation.

According to the Code of Practice for OTTV in Buildings 1995, the OTTV of external walls should be calculated using the following formula:

$$\text{OTTV} = [(A_w \text{ x U x a x } TD_{EQw}) + (Af_w \text{ x SC x ESM x SF})] \,/\, A_{ow} \qquad (1)$$

where

A_w	= Area of opaque wall, m^2
U	= Thermal transmittance of opaque wall, $W/m^2 {}^\circ C$
a	= Absorptivity of the opaque wall
TD_{EQw}	= Equivalent temperature difference for wall, °C
Af_w	= Area of fenestration in wall, m^2
SC	= Shading coefficient of fenestration in wall
ESM	= External shading multiplier
SF	= Solar factor for the vertical surface, W/m^2
A_{ow}	= Gross area of external walls, i.e. $A_w + Af_w$, m^2

The thermal transmittance (U) of the opaque wall or roof should be derived by the following formula:

$$U = 1 / (Ri + X_1/k_1 + X_2/k_2 + \ldots + X_n/k_n + Ra + Ro) \qquad (2)$$

where

Ri	= Surface film resistance of internal surface of the wall or roof, $m^2 {}^\circ C/W$
X	= Thickness of building material of the wall or roof or part thereof, m
k	= Thermal conductivity of the building material, $W/m^\circ C$
Ro	= Surface film resistance of external surface of the wall or roof, $m^2 {}^\circ C/W$
Ra	= Air space resistance, $m^2 {}^\circ C/W$

▌ HYPOTHETICAL DESIGN TO COMPARE OTTV DETAILS

Different Dimensions

To investigate the OTTV of typical wall sections, the wall sections A1, A2 and A3 of different dimensions as shown in Figure 1 are considered.

Conditions: a. South facing
b. White mosaic tile on concrete panel and concrete beam
c. Density of wall construction > $570 kg/m^2$
d. Tinted glass: shading coefficient = 0.7

By using equation 2, U (600mm concrete beam)= 1.51
U (100mm concrete panel)= 2.32

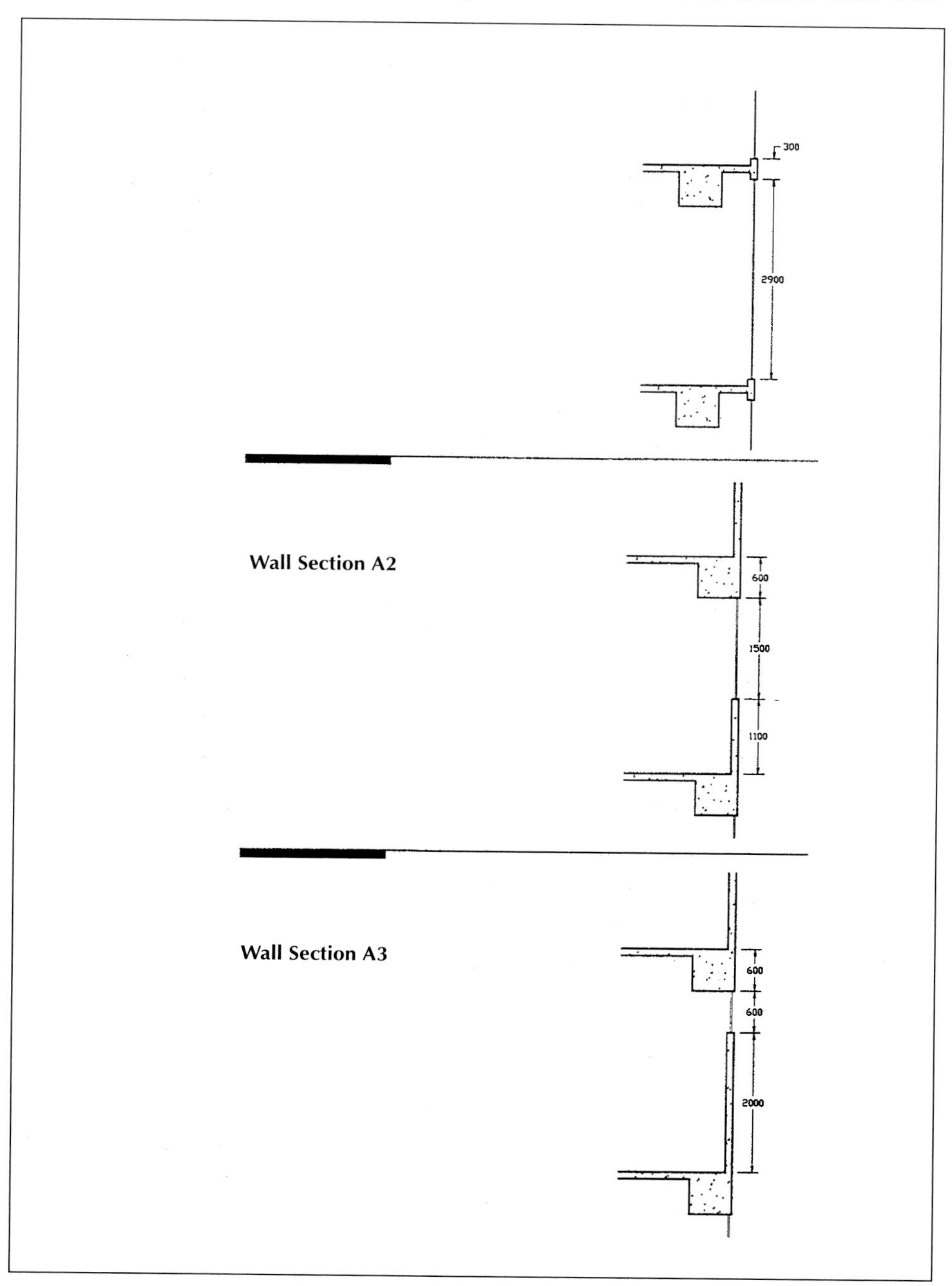
300
2900
Wall Section A2
600
1500
1100
Wall Section A3
600
600
2000

Figure 1

Wall Section A1 OTTV = [(2900 x 0.7 x 1 x 191)
+ (300 x 2.32 x 0.58 x 1.4)]
/ 3200
= 121.34W/m^2

Wall Section A2 OTTV = [(1500 x 0.7 x 1 x 191)
+ (1100 x 2.32 x 0.58 x 1.4)
+ (600 x 1.51 x 0.58 x 1.4)]
/ 3200
= 63.55W/m^2

Wall Section A3 OTTV = [(600 x 0.7 x 1 x 191)
+ (2000 x 2.32 x 0.58 x 1.4)
+ (600 x 1.51 x 0.58 x 1.4)]
/ 3200
= 26.47W/m^2

Observation

The above calculation shows that the OTTV is mainly governed by the ratio of opaque wall area to fenestration area, especially in cases where glass with a high shading coefficient (SC) is used. The SC can range from 1 for clear glass to 0.35 or even lower for various types of coated glass. Extensive use of glass will lead to high OTTV (poor thermal performance).

The above calculation shows that the OTTV is roughly inversely proportional to the area of fenestration in the wall. However, OTTV is subject to many variables such as the orientation of the façade being considered and the SC of the glass.

Different Materials

To investigate the effect of different types of glass on OTTV, reflective glass (SC of 0.4) is used instead of tinted glass (SC 0.7).

Conditions: a. South facing
b. White mosaic tile on concrete panel and concrete beam
c. Density of wall construction > 570kg/m^2
d. Reflective glass: SC = 0.4

Wall Section A1 OTTV = [(2900 x 0.4 x 1 x 191)
+ (300 x 2.32 x 0.58 x 1.4)]
/ 3200
= 69.41W/m^2

Wall Section A2

$$\text{OTTV} = [(1500 \times 0.4 \times 1 \times 191) + (1100 \times 2.32 \times 0.58 \times 1.4) + (600 \times 1.51 \times 0.58 \times 1.4)] / 3200 = 36.69\text{W/m}^2$$

Wall Section A3

$$\text{OTTV} = [(600 \times 0.4 \times 1 \times 191) + (2000 \times 2.32 \times 0.58 \times 1.4) + (600 \times 1.51 \times 0.58 \times 1.4)] / 3200 = 15.73\text{W/m}^2$$

Observation

Compared with the case when tinted glass (SC 0.7) was employed, reflective glass (SC 0.4) obviously provides a better thermal performance.

From the calculation above, for wall sections of the same dimension, the OTTV is reduced from 121.34W/m^2, 63.55W/m^2 and 26.47W/m^2 to 69.41W/m^2, 36.69W/m^2 and 15.73W/m^2 respectively when reflective glass is used instead of tinted glass.

This shows that the type of glass being employed is a crucial factor for OTTV.

Different Detailing

To investigate the effect of overhang size on the OTTV, wall sections of 2900mm height glass area with different overhang dimensions (W) as shown in Figure 2 are considered.

Conditions:
a. South facing
b. White mosaic tile on concrete panel and concrete beam
c. Density of wall construction > 570kg/m^2
d. Tinted glass: SC = 0.7

Wall Section B1 (W = 300⇨ESM = 0.926)

$$\text{OTTV} = [(2900 \times 0.7 \times 0.926 \times 191) + (300 \times 2.32 \times 0.58 \times 1.4)] / 3200 = 112.38\text{W/m}^2$$

Wall Section B2 (W = 550⇨ESM = 0.856)

$$\text{OTTV} = [(2900 \times 0.7 \times 0.856 \times 191) + (300 \times 2.32 \times 0.58 \times 1.4)]$$

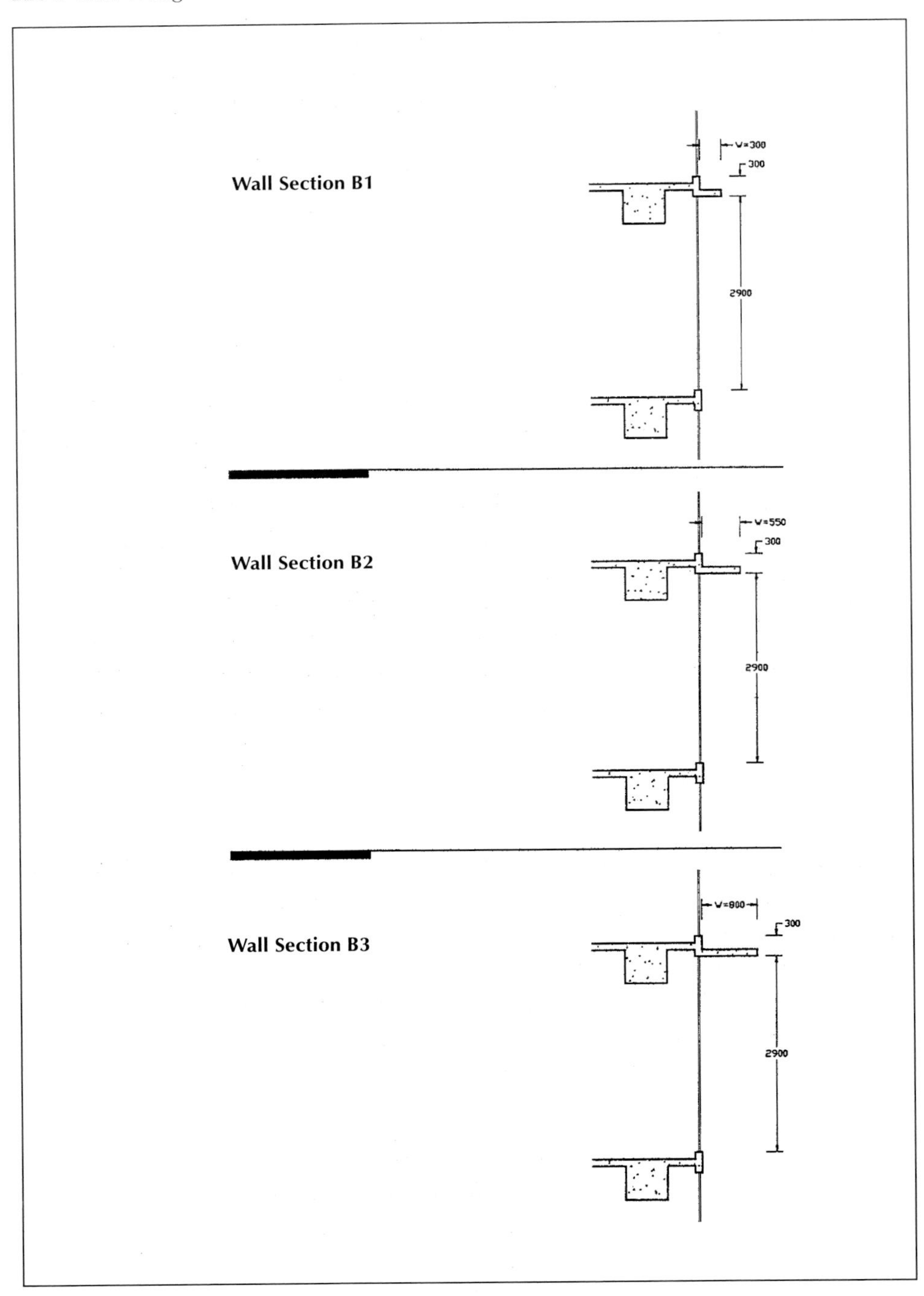
Wall Section B1
W=300
300
2900
Wall Section B2
W=550
300
2900
Wall Section B3
W=800
300
2900

Figure 2

/ 3200
= $103.89 W/m^2$

Wall Section B3 (W = 800⇨ESM = 0.79)

OTTV = [(2900 x 0.7 x 0.79 x 191)
+ (300 x 2.32 x 0.58 x 1.4)]
/ 3200
= $95.9 W/m^2$

Observation

Although the overhang contributes in reducing the OTTV, the difference is not very significant. According to the above calculation, the OTTV is reduced by only 14.67%, from $112.38 W/m^2$ to $95.9 W/m^2$ when the overhang length is increased from 300mm to 800mm.

The OTTV is partially governed by the length of overhang in terms of the External Shading Multiplier (ESM). The ESM is a ratio of the length of overhang to the length of the fenestration measured vertically. This means that the length of fenestration is also a determining factor of OTTV.

In this example, because the size of fenestration is large, the increase in the overhang size from 300mm to 800mm is not very significant in comparison with the 2900mm fenestration. Therefore, the OTTV does not change very much.

Different Overhang Sizes

To investigate the effect of overhang size on the OTTV, wall sections of 1 500mm height glass area with different overhang dimensions (W) as shown in Figure 3 are considered.

Conditions: a. South facing
b. White mosaic tile on concrete panel and concrete beam
c. Density of wall construction > $570 kg/m^2$
d. Tinted glass: SC = 0.7

Wall Section C1 (W = 300ρESM = 0.926)

OTTV = [(1500 x 0.7 x 0.856 x 191)
+ (1100 x 2.32 x 0.58 x 1.4)
+ (600 x 1.51 x 0.58 x 1.4)]
/ 3200
= $54.52 W/m^2$

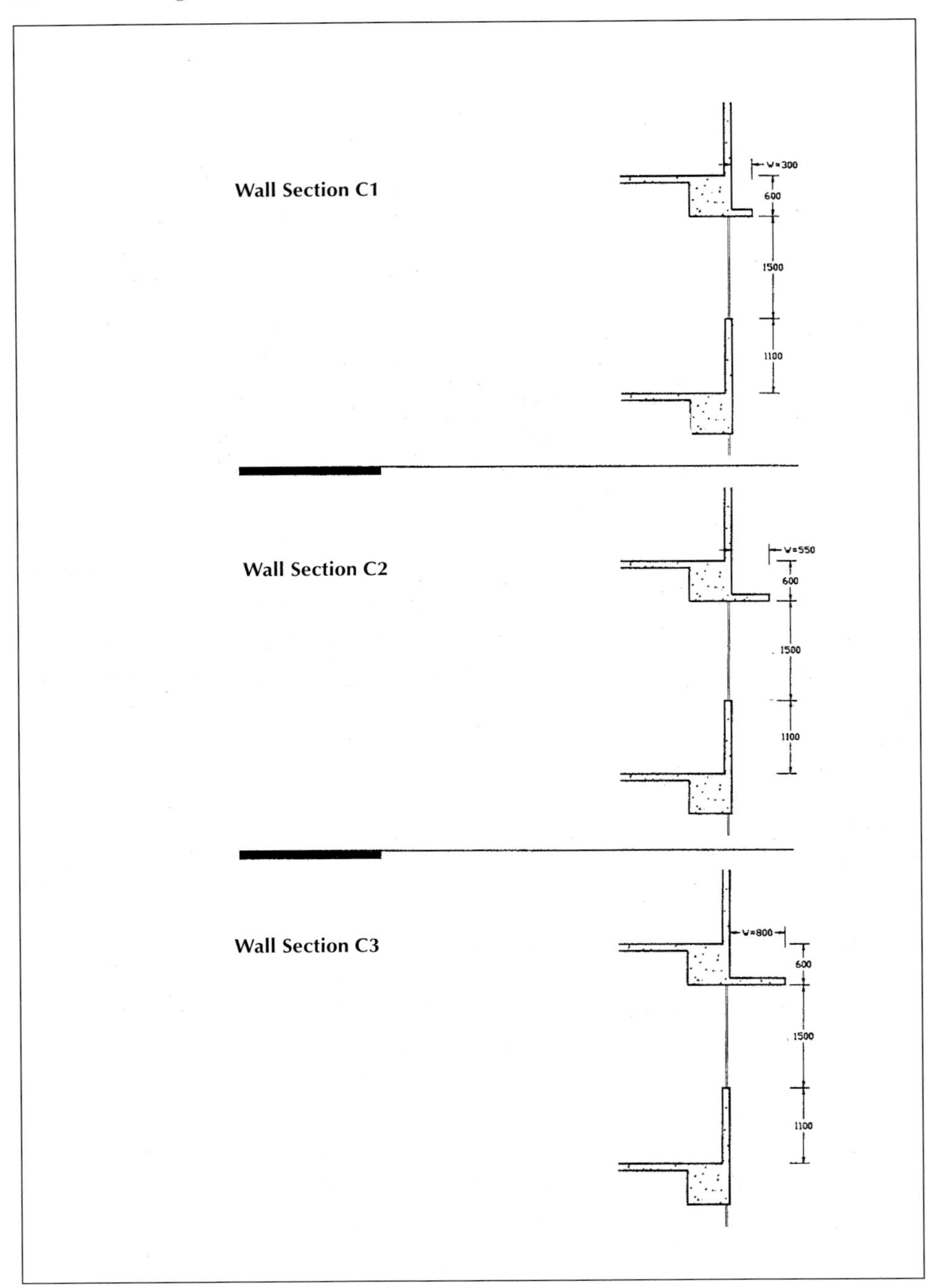
Wall Section C1
V=300
600
1500
1100
Wall Section C2
V=550
600
1500
1100
Wall Section C3
V=800
600
1500
1100

Figure 3

Wall Section C2 (W = 550ρESM = 0.856)

$$
\begin{aligned}
\text{OTTV} &= [(1500 \times 0.7 \times 0.729 \times 191) \\
&+ (1100 \times 2.32 \times 0.58 \times 1.4) \\
&+ (600 \times 1.51 \times 0.58 \times 1.4)] \\
&/ \ 3200 \\
&= 46.57 \text{W/m}^2
\end{aligned}
$$

Wall Section C3 (W = 800ρESM = 0.79)

$$
\begin{aligned}
\text{OTTV} &= [(1500 \times 0.7 \times 0.645 \times 191) \\
&+ (1100 \times 2.32 \times 0.58 \times 1.4) \\
&+ (600 \times 1.51 \times 0.58 \times 1.4)] \\
&/ \ 3200 \\
&= 41.3 \text{W/m}^2
\end{aligned}
$$

Observation

The OTTV is reduced by 24.25%, from 54.52W/m^2 to 41.3W/m^2, when the overhang size changes from 300mm to 800mm. As shown in the previous example, the OTTV is partially governed by the length of the overhang together with the length of fenestration measured vertically. As the length of fenestration is reduced to 1500mm in this example (2 900mm in the previous example), the contribution from the increase in overhang size to the OTTV is more significant.

Conclusion

The building skin is a composite structure of different materials put together. OTTV is one measure of its performance, which depends on the materials, glazing type and size, provision of shading and the orientation of the wall.

Among these elements, from the above cases, the OTTV is mainly governed by the area of fenestration, especially in the case where glass with a high SC is used. The SC of fenestration is the ratio of solar heat gain through a particular type of glass under a specific set of conditions to the solar heat gain through double-strength sheet clear glass under the same conditions. Glass with a high SC means that the heat gain through this type of glass is high. Hence, most of the heat gain by a building is caused by the fenestration area. The SC can range from 0.35 (coated glass) to 1 (clear glass), so the heat gain through coated glass is only 35% of that of clear glass. The SC is a determining factor in the OTTV of a building, especially when the fenestration area makes up a significant portion of the total façade area.

A shading device undoubtedly benefits the OTTV, but this is only so when the length of the overhang is compatible to the vertical length of the fenestration area. Because the OTTV depends on the ESM which is a ratio of the length of the overhang to the size of fenestration, an overhang considerably reduces OTTV only if a longer overhang and smaller fenestration are used.

EFFECT OF ADJACENT BUILDINGS

Hypothetical Design

In most Hong Kong cases, buildings are situated so closely that only a street lies between buildings of 20 to 30 storeys. In some extreme cases, buildings adjoin each other to form clusters of single elevation buildings due to a maximum use of frontage. In such cases, it is possible that only one out of the four façades of such a building is exposed to the street.

Unfortunately, in OTTV calculation, a building is assumed to be self-standing. In order to investigate the effect of adjacent buildings on the OTTV, 21 June (夏至) and 22 December (冬至) were chosen as examples to represent typical days in the summer and winter respectively to see the variation of shaded area on the elevation of the building under investigation. (For details, see Figure 5.)

In this example, the building being investigated is surrounded by eight other buildings. There are 14m-wide streets between each building, and the dimensions of all buildings are 30m x 30m plan and 8 400m high. Shadows that vary with the sunpath are cast on the central building by the other surrounding buildings. The percentage of the shaded area on the façade was recorded from 7:00a.m. to 5:00p.m. at one-hour intervals.

Figure 4

Summer (21 June)

	7:00	8:00	9:00	10:00	11:00	12:00	13:00	14:00	15:00	16:00	17:00
East											
South											
West											
North											

Winter (22 December)

	7:00	8:00	9:00	10:00	11:00	12:00	13:00	14:00	15:00	16:00	17:00
East											
South											
West											
North											

under direct sunlight shaded by other buildings not under direct sunlight

Figure 5

Actual Case In Central (Kam Sang Building)

In this section, Kam Sang Building, 255–257, Des Voeux Road, Central, is treated as an example of a typical commercial building in Hong Kong. It is studied for the effect of the shadow cast on its external walls by adjacent buildings. The dates 21 June and 22 December were chosen as examples representing typical days in the summer and winter respectively.

Plate 1

Kam Sang Building is a commercial building of 24 storeys, located at a corner site at the junction of Hillier Street and Des Voeux Road. At the back of the building are other commercial blocks, of about 25 storeys, which are

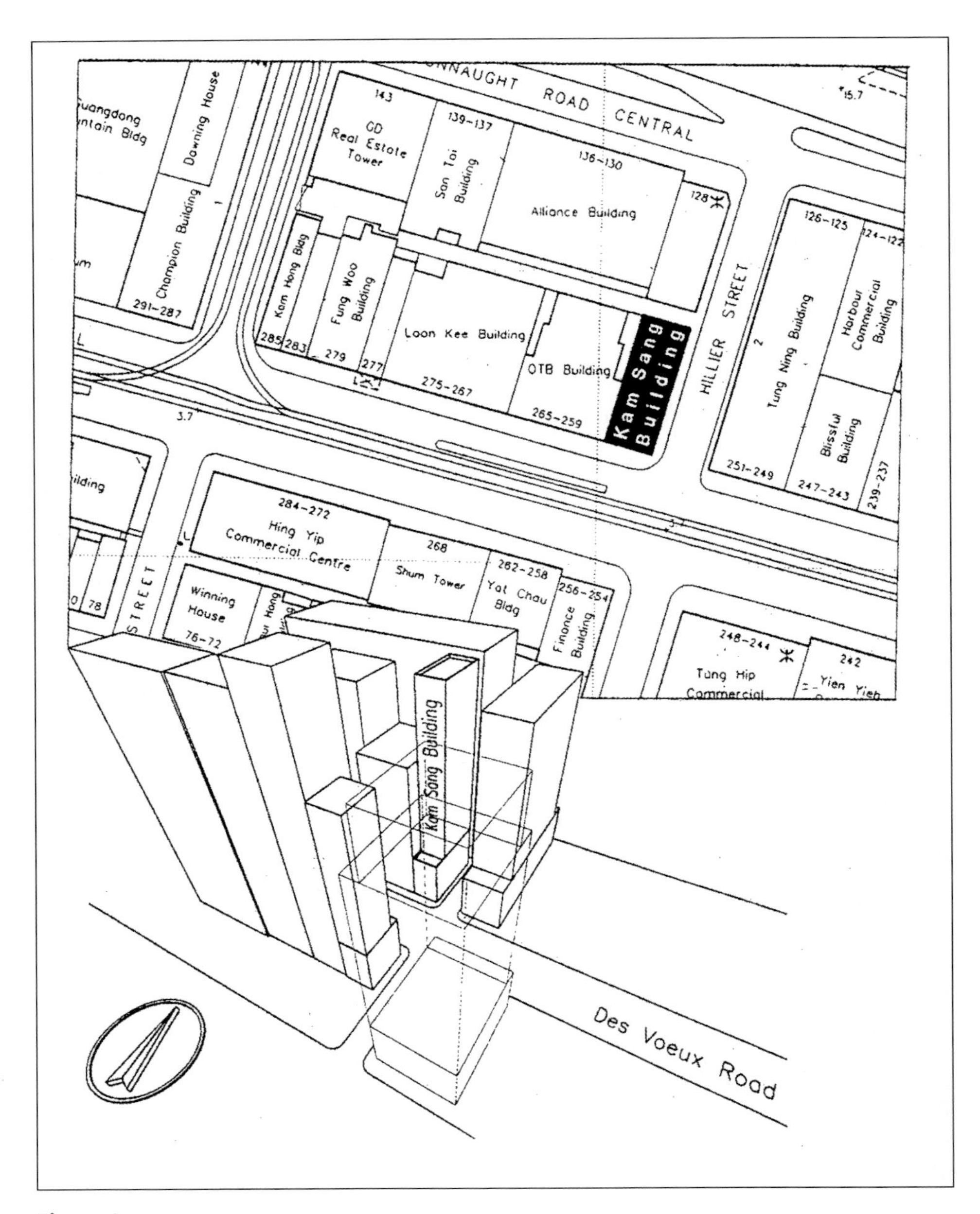

Figure 6

separated from Kam Sang Building by a 3m-wide service lane. Therefore, the OTTV contributed by the rear façade (north elevation) is negligible. In the following sections, the shadow cast on the east, south and west façades of Kam Sang Building at 8:00a.m., 10:00a.m., 12:00p.m., 2:00p.m. and 4:00p.m. were recorded to see the extent of the effect caused by adjacent buildings.

Observation

From Figure 7, it can be seen that the building under investigation was shaded by the surrounding buildings during most of the day. The portion of façade area being shaded ranged from 23.54% to 100%. This area is dependent on the width of the surrounding streets and the height of the surrounding buildings. It is clear that the narrower the streets, the larger the area will be that is shaded by the surrounding buildings.

Although Kam Sang Building is bound by Des Voeux Road, of which at 23m wide is probably wider than most streets in Hong Kong on average, the area of the façade being shaded is significant.

The above investigation shows that the assumption made in the method of OTTV calculation, where the building under question is self-standing regardless of its surroundings, would probably provide an inaccurate answer.

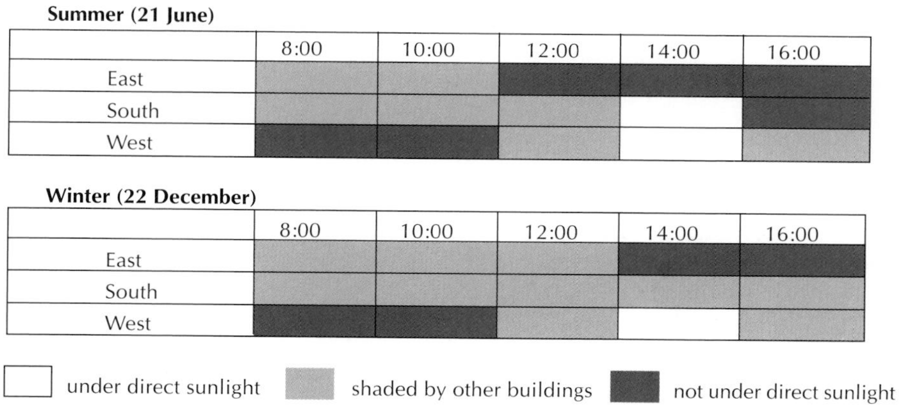

Summer (21 June)

	8:00	10:00	12:00	14:00	16:00
East					
South					
West					

Winter (22 December)

	8:00	10:00	12:00	14:00	16:00
East					
South					
West					

under direct sunlight   shaded by other buildings   not under direct sunlight

Figure 7

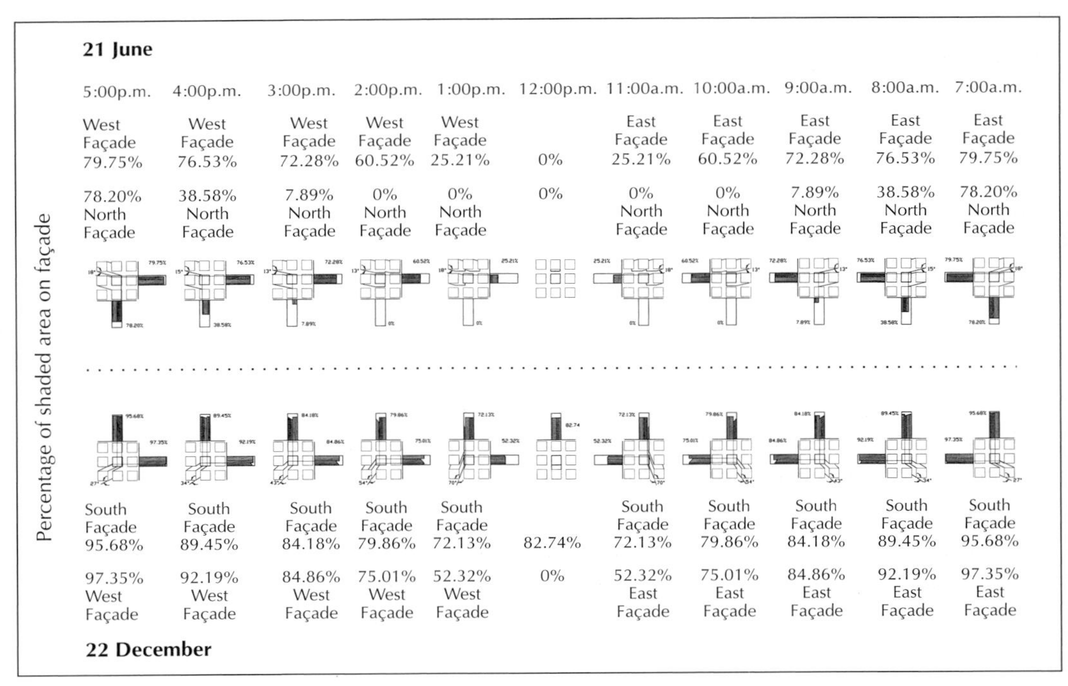

Figure 8

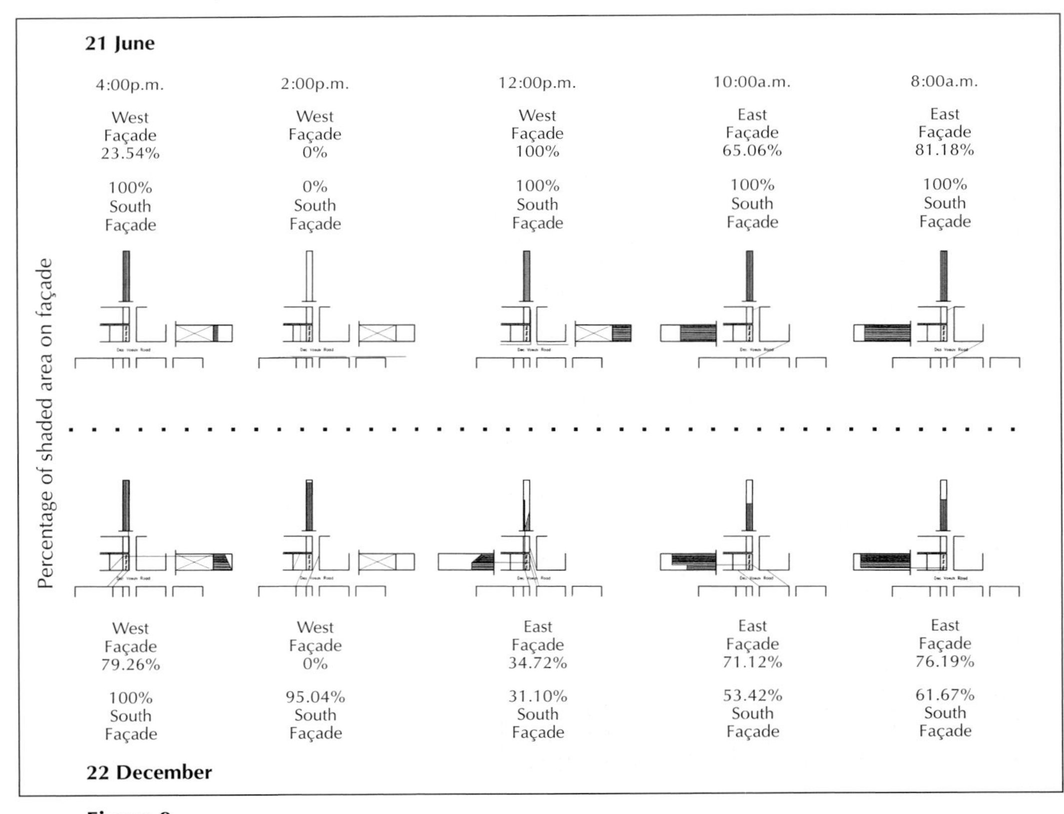

Figure 9

CASE STUDIES

Titus Square

Address	26 Nathan Road and 4 Middle Road, Tsim Sha Tsui
Building Type	Office
No. of Storeys	28

Titus Square is an office tower located in Tsim Sha Tsui. Similar to many office buildings in Hong Kong, the lower floors (G/F to 5/F, in this case) are designed for retail use and the upper floors for office use. Here, 6/F, 7/F and 14/F are for mechanical services and are not taken into account for OTTV calculation.

A curtain wall of 8mm coated glass and light grey tiles are selected as the basic external finishes for the fenestration and opaque wall respectively. The roof is reinforced concrete slab with insulation board and concrete tile.

Located between Nathan Road and Middle Road, the building is shaded from direct sunlight by Far East Mansion, Sheraton Hotel and Peninsula Hotel. This poses a difficulty in applying the standard method of OTTV calculation to accurately reflect the real situation.

Plate 2

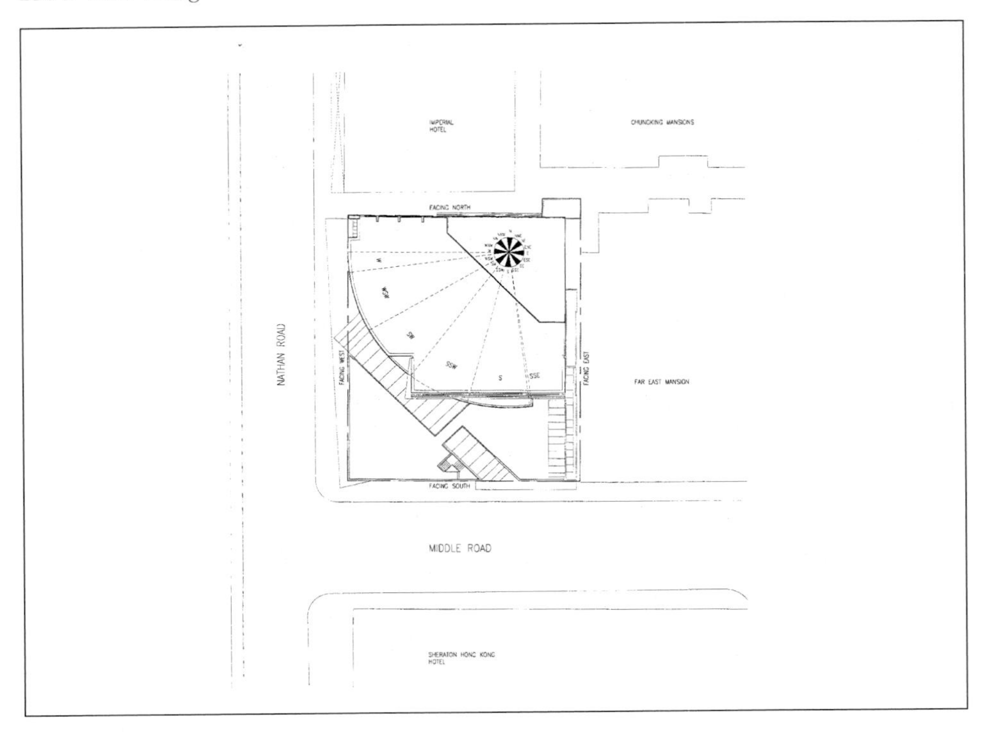

Figure 10 Orientation Key Plan

Summary of OTTV of building envelope

Façade		
Orientation	**Gross area (m²)**	**Gross heat gain**
E	2029.32	22733.91
S	1564.63	60337.69
SSE	7.55	155.84
SSW	190.87	3956.62
W	882.31	29260.59
SW	466.5	15151.39
WSW	497.59	18128.62
N	2290.19	10234.89
Subtotal	7928.96	159959.55
Roof		
	719.68	14887.05
Subtotal	719.68	14887.05

Tower Walls OTTV 20.17W/m^2
Tower Roofs OTTV 20.69W/m^2

Tower OTTV **20.22W/m^2**

In the following sections, the south façade of the building is studied. The following is a summary of wall construction.

Opaque Wall

W1 Curtain wall glass on 230mm reinforced concrete beam
8mm silver coated glass
75mm air gap
75mm mineral wool insulation
230mm reinforced concrete beam
10mm plaster with white paint

W2 Curtain wall spandrel glass panel only
8mm silver coated glass
75mm air gap
75mm mineral wool
10mm plaster with white paint

W5 125mm reinforced wall
5mm light grey tile on external surface
10mm cement sand render
450mm reinforced concrete wall
Plain concrete internal surface

W7 125mm vitreous enamelled panel
5mm enamelled tile on external surface
125mm reinforced concrete wall
10mm plaster with white paint

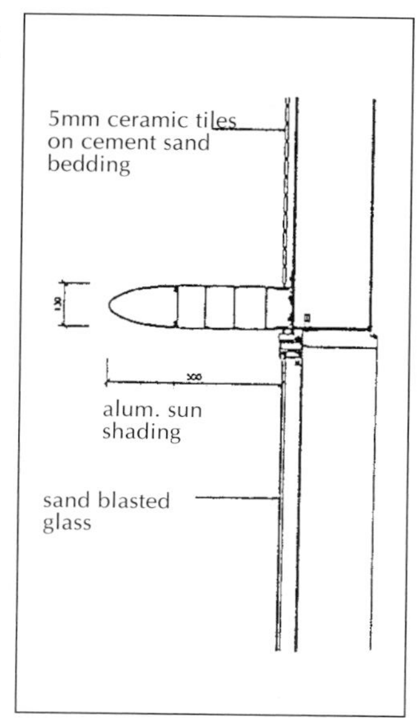

Figure 11 Typical Section

Glass

F2 Curtain wall glass
8mm thick
Silver coated
SC 0.35

F4 Sandblasted clear glass
12mm thick
SC 0.5
No external shading device

F5 Clear glass
12mm thick
SC 0.5
No external device

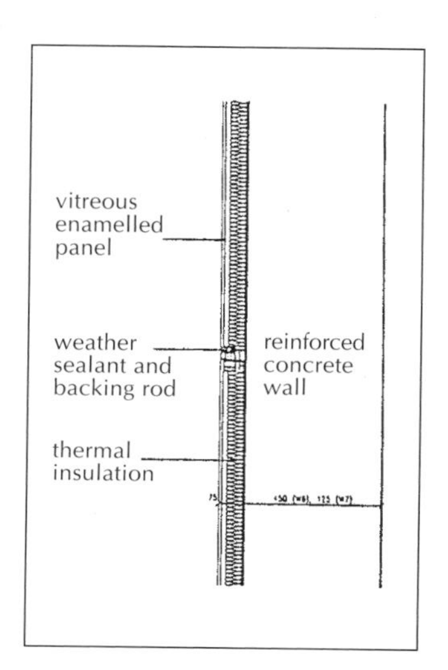

Figure 12 Typical Section

Building (Energy Efficiency) Regulation
Form OTTV 1
Calculation Of U Value Of Composite Wall/Roof And Details Of Other Values

Façade Orientation Facing: South

Wall code no.		W1	W2	W5	W7
Location of wall		Curtain wall on RC beam 4/F-27/F	Curtain wall on glass panel 4/F-27/F	125 RC wall 4/F-27/F	125 vitreous enamelled panel, 4/F-27/F
External finish material		Silver coated glass	Silver coated glass	Light grey tile	Enamelled panel
Conductivity	W/mk	1.05	1.05	1.5	1.5
Density	kg/m³	2500	2500	2500	2500
Thickness	m	0.008	0.008	0.005	0.005
Absorptivity	(a)	0.25	0.25	0.58	0.58
Surface film resistance	m² k/W	0.044	0.044	0.044	0.044
Resistance	m² k/W	0.008	0.008	0.003	0.003
Intermediate component - A		Air gap	Air gap	Cement render	RC wall
Conductivity	W/mk			0.72	2.16
Density	kg/m³			1860	2400
Thickness	m	0.075	0.075	0.01	0.125
Resistance	m² k/W	0.153	0.153	0.014	0.058
Intermediate Component - B		Mineral wool	Mineral wool	RC wall	
Conductivity	W/mk	0.039	0.039	2.16	
Density	kg/m³	50	50	2400	
Thickness	m	0.075	0.075	0.125	
Resistance	m² k/W	1.92	1.92	0.058	
Intermediate component - C		RC beam			
Conductivity	W/mk	2.16			
Density	kg/m³	2400			
Thickness	m	0.23			
Resistance	m² k/W	0.106			
Internal finish material		White paint on plaster	White paint on plaster	Plain concrete	White paint on plaster
Conductivity	W/mk	0.38	0.38		0.38
Density	kg/m³	1120	1120	1120	1120
Thickness	m	0.01	0.01	0.01	0.01
Absorptivity	(a)	0.3	0.3	0.3	0.3
Surface film resistance	m² k/W	0.299	0.299	0.299	0.299
Resistance	m² k/W	0.026	0.026	0.026	0.026
U value of composite wall	W/mk	0.391	0.408	2.392	2.323
Area of wall	m²	160.93	332.42	16.37	48.8
Density of composite wall	kg/m²	587	35	331	324
Class of composite wall		5	2	3	3
Equivalent temperature difference		1.4	5.01	3.6	3.6

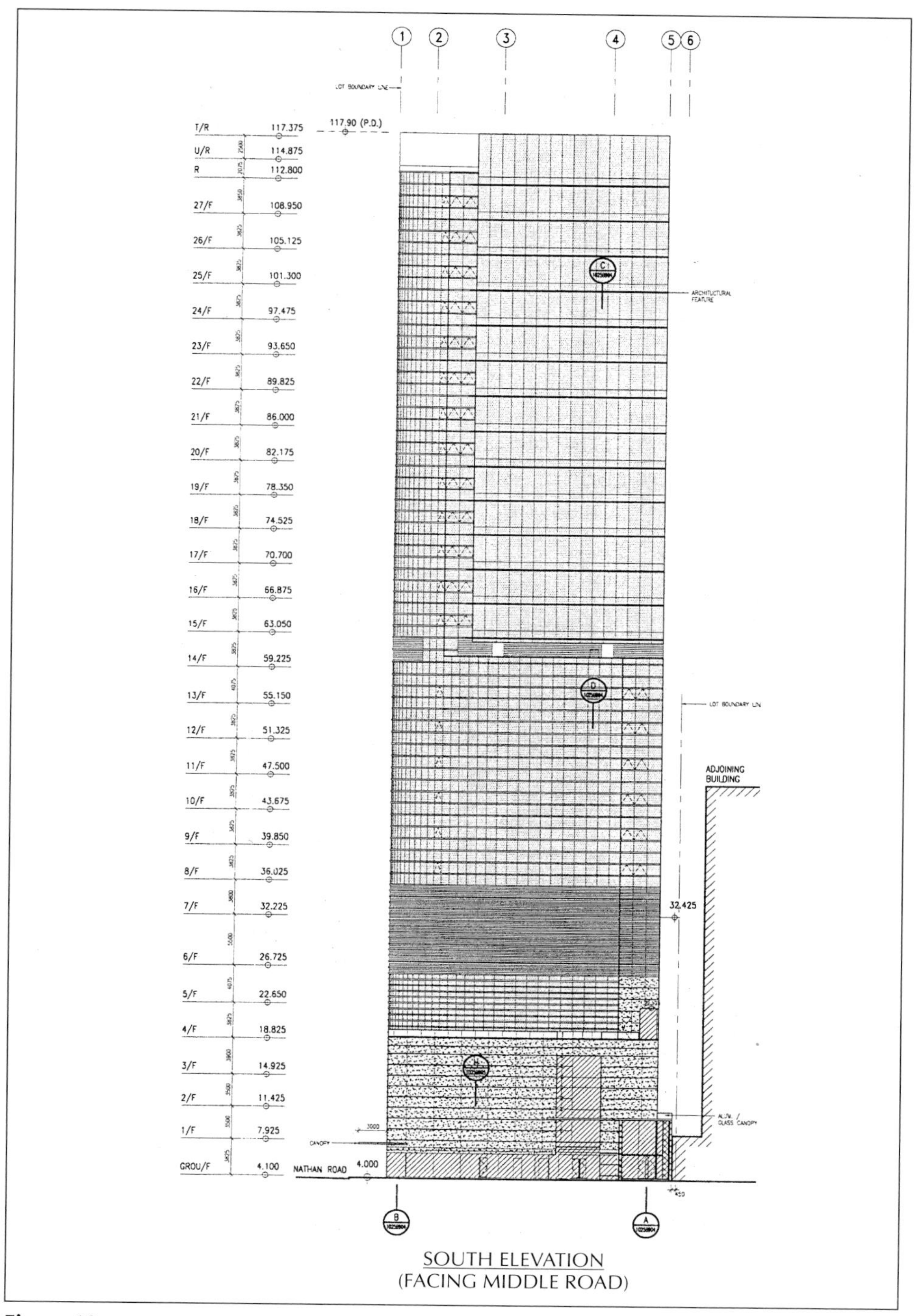
T/R 117.375
U/R 114.875
R 112.800
27/F 108.950
26/F 105.125
25/F 101.300
24/F 97.475
23/F 93.650
22/F 89.825
21/F 86.000
20/F 82.175
19/F 78.350
18/F 74.525
17/F 70.700
16/F 66.875
15/F 63.050
14/F 59.225
13/F 55.150
12/F 51.325
11/F 47.500
10/F 43.675
9/F 39.850
8/F 36.025
7/F 32.225
6/F 26.725
5/F 22.650
4/F 18.825
3/F 14.925
2/F 11.425
1/F 7.925
GROU/F 4.100
117.90 (P.D.)
NATHAN ROAD
4.000
32.425
ADJOINING BUILDING
SOUTH ELEVATION
(FACING MIDDLE ROAD)

Figure 13

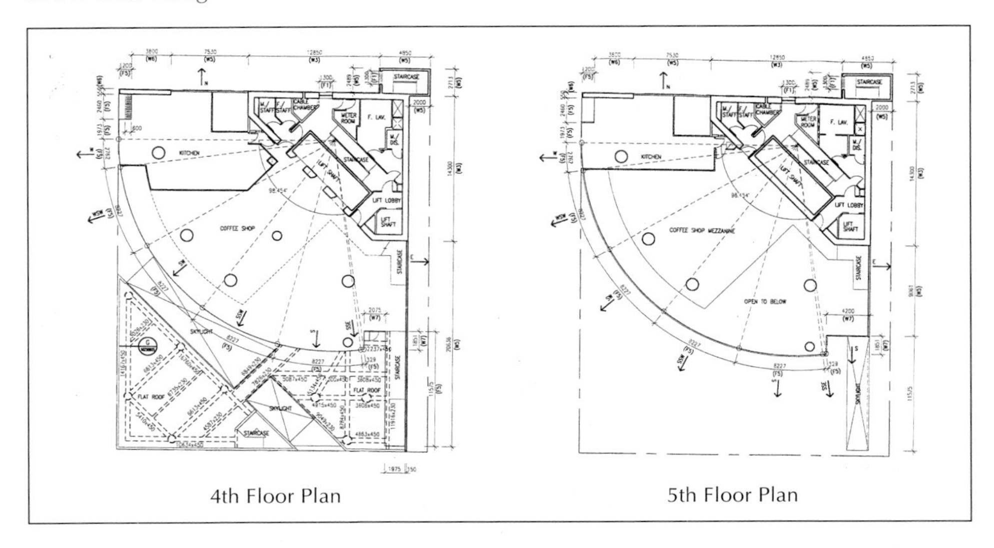

4th Floor Plan

5th Floor Plan

Figure 14

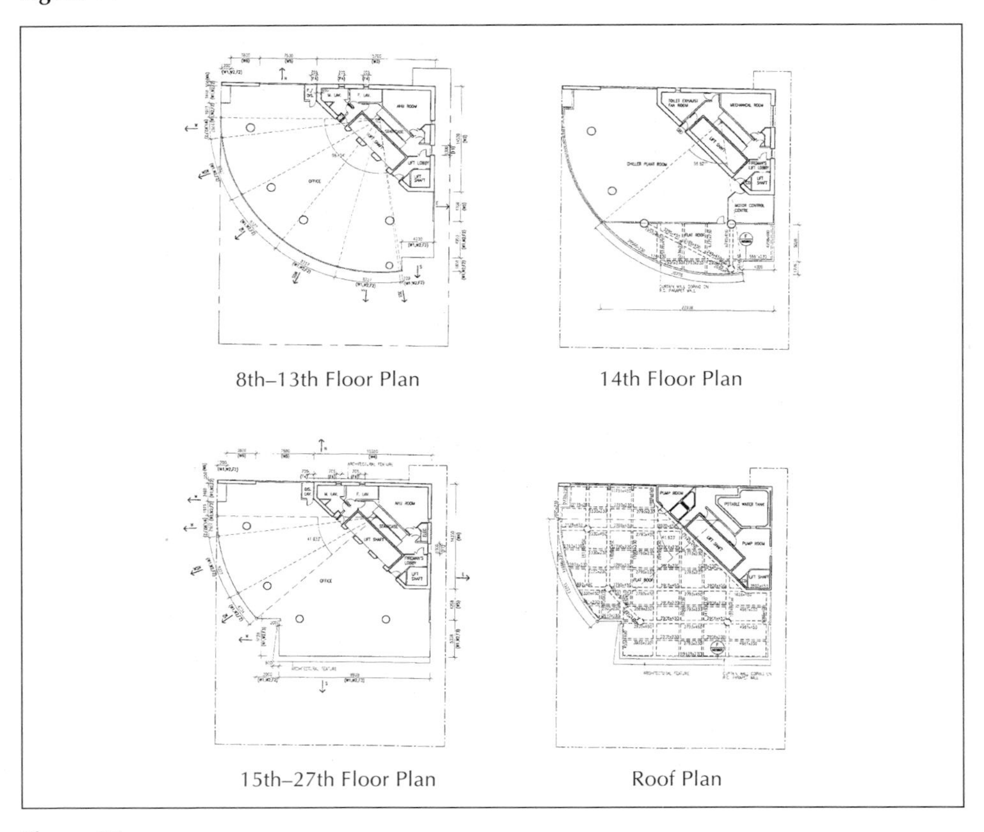

8th–13th Floor Plan

14th Floor Plan

15th–27th Floor Plan

Roof Plan

Figure 15

Building (Energy Efficiency) Regulation
Form OTTV 2
Window/Rooflight Schedule

Façade Orientation Facing: South

Window code no.	F1	F2	F3	F4	F5
Location of window	4/F - 27/F Unshaded	8/F - 13/F 15/F - 27/F Unshaded	15/F - 27/F Shaded	8/F - 13/F 15/F - 27/F Unshaded	G/F - 5/F Unshaded
Glazing type	Wired	Coated	Coated	Sandblasted	Clear
Thickness	0.012	0.0008	0.008	0.012	0.012
SC	0.5	0.35	0.35	0.5	0.5
Type of shading device			Aluminium		
Length of overhang			0.765		
Height of window			3.12		
Overhang projection factor			0.25		
ESM for overhang	1	1	0.823	1	1
ESM for sidefin	1	1	1	1	1
Area of glazing	0	130.03	803.5	0	72.55

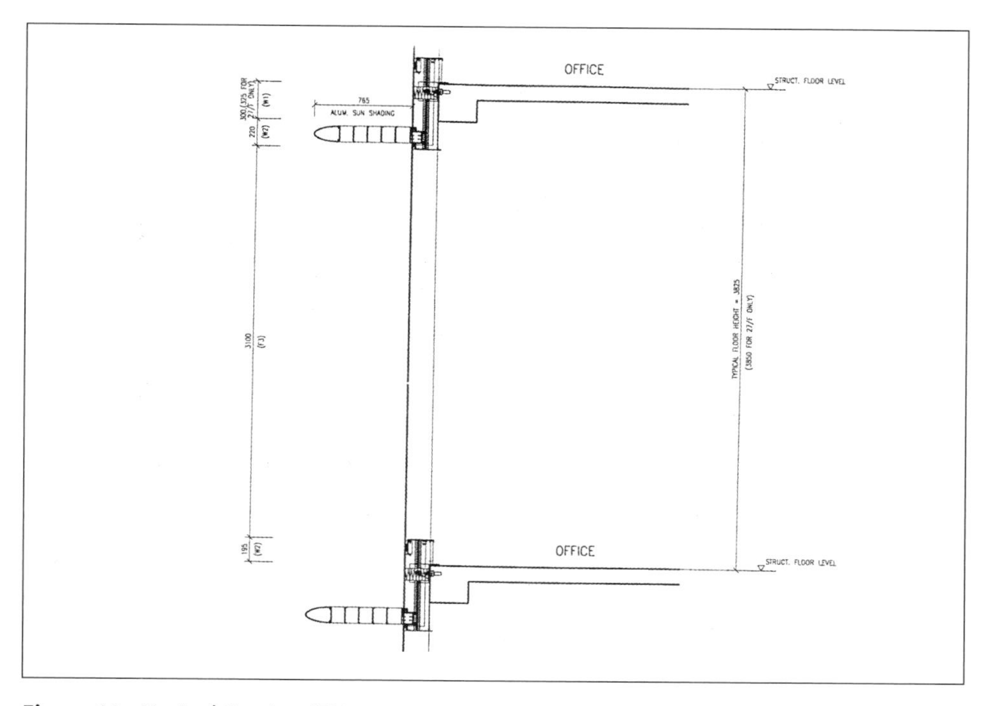

Figure 16 Typical Section (F3)

Building (Energy Efficiency) Regulation
Form OTTV 3
Calculation Of OTTV Of Individual Façades In Building Envelope

Façade Orientation Facing: South

Opaque wall

Code	Description	Aw	U	a	Tdeq	Sum
W1	Curtain wall on RC beam 4/F - 27/F	160.933	0.391	0.25	1.40	22.02
W2	Curtain wall on glass panel 4/F - 27/F	332.419	0.408	0.25	5.01	169.87
W3	600 RC wall 4/F - 27/F	0	1.309	0.58	1.4	0
W4	450 RC wall 4/F - 27/F	0	1.759	0.58	1.4	0
W5	125 RC wall 4/F - 27/F	16.374	2.392	0.58	3.6	81.78
W6	450 vitreous wall 4/F - 27/F	0	1.721	0.58	1.4	0
W7	125 vitreous wall 4/F - 27/F	48.828	2.323	0.58	3.6	236.84

Subtotal 558.554m^2 Heat gain 510.51W

Fenestration

Code	Description	Afw	SC	ESM	SF	Sum
F1	4/F - 27/F unshaded	0	0.5	1	0	0
F2	8/F - 13/F and 15/F - 27/F unshaded	130	0.35	1	191	8690.5
F3	15/F - 27/F shaded	803.5	0.35	0.823	191	44206.6
F4	8/F - 13/F and 15/F - 27/F unshaded	0	0.5	1	0	0
F5	G/F - 5/F unshaded	72.5	0.5	1	191	6923.8

Subtotal 1006.0m^2 Heat gain 59820.9W

Gross area 1564.554m^2 Gross heat gain 60331.41W

OTTV (South Façade) 38.56W/m^2

Observation

Though glass was extensively used for the construction of the elevations, the OTTV for the southern façade is just 38.56W/m^2 and for the whole tower 20. 22W/m^2. The reasons for this can be summed up as follows:

- *Overhang* — In the typical wall section of the tower (accounting for 202. 628m^2 out of 302.477m^2 of the eastern façade glazing area, and 803.50m^2 out of 1006.077m^2 of the southern façade glazing area), a 765mm-long aluminium overhang is incorporated. It reduces the heat gain by a factor of 0.823 (ESM = 0.823).
- *SC* — Coated glass with a 0.35 SC was employed in the typical wall section. Compared with tinted glass with a SC of 0.7, the use of coated glass halves the heat gain.

Plate 3

St. Thomas The Apostle Church

Address	Tsing Luk Street, Tsing Yi
Building Type	Church / Kindergarten
No. of Storeys	7

St. Thomas The Apostle Church is a church complex incorporating a kindergarten, church halls, an administrative office, function rooms and priest quarters. Reinforced concrete with brick tiles and tinted glass are selected as the basic external finishes for the opaque wall and fenestration respectively. The roof is constructed with reinforced concrete slab and concrete tiles.

Because the Building (Energy Efficiency) Regulations apply to only commercial buildings and hotels, some compartments of the church complex are exempt from the regulations. Therefore, only calculations for the church hall and chapel, assembly hall and function rooms are required by the Buildings Department. For areas such as the kindergarten and priest quarters, calculation of OTTV is not necessary.

The assembly hall is studied in the following sections.

Plate 4

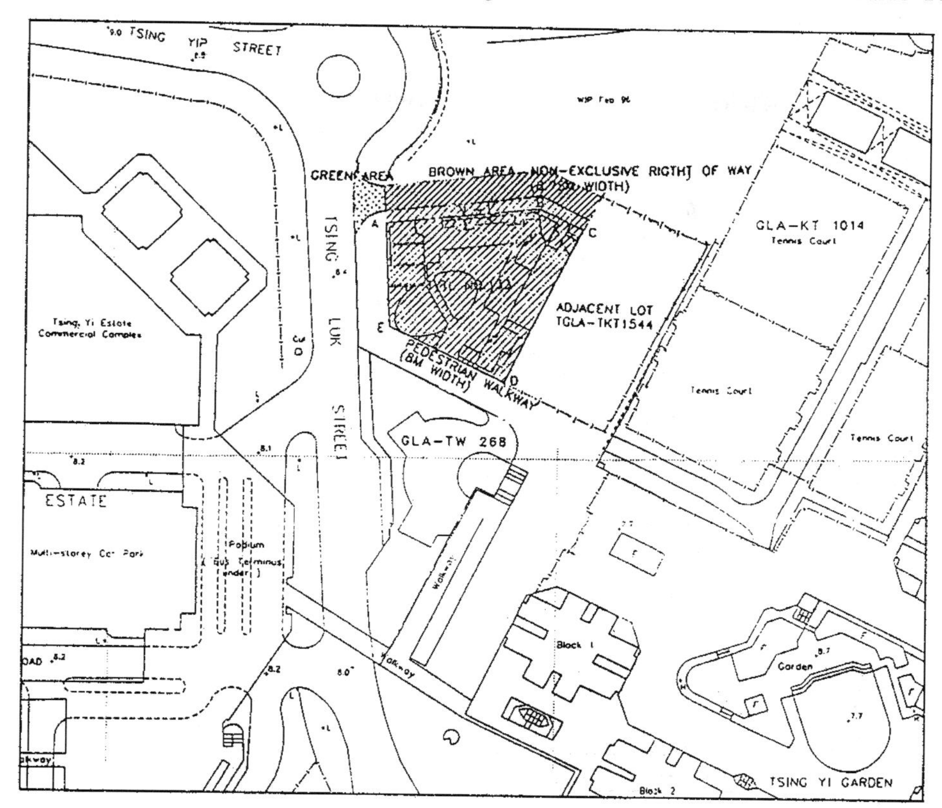

Figure 17 Orientation Key Plan

Summary of OTTV of building envelope (assembly hall)

Façade		
Orientation	**Gross area (m²)**	**Gross heat gain**
N	125.66	5663.68
SE	106.76	4039.31
WSW	66.5	320.81
Subtotal	298.86	10023.8
Roof		
	186.82	499.91
Subtotal	186.82	499.91

Tower Walls OTTV 33.54W/m²
Tower Roofs OTTV 2.68W/m²

Tower OTTV 21.67W/m²

In the following sections, the north façade of the assembly hall on the 3/F is studied.

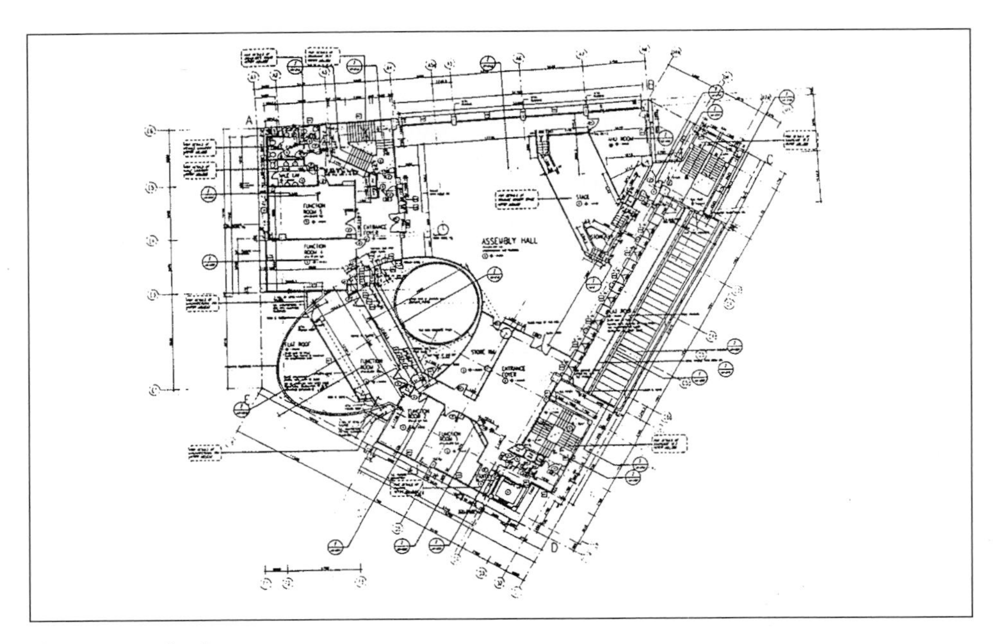

Figure 18 3/F Plan

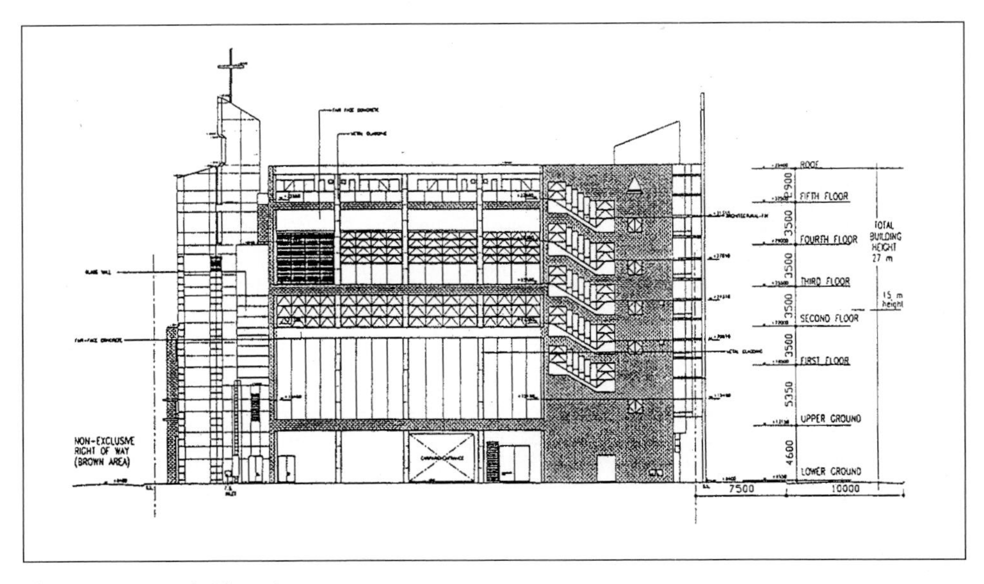

Figure 19 North Elevation

Building (Energy Efficiency) Regulation
Form OTTV 1
Calculation Of U Value Of Composite Wall/Roof And Details Of Other Values

Façade Orientation Facing: North

Wall code no.		**W1**	**W2**	**W5**	**W7**
Location of wall		800mm column	1050mm column	400mm beam	150mm RC wall
External finish material		Aluminium cladding	Brick tile	Uncoloured concrete	Uncoloured concrete
Conductivity	W/mk	160	0.95		
Density	kg/m^3	2800	1900		
Thickness	m	0.002	0.0127		
Absorptivity	(a)	0.1	0.88		
Intermediate component - A		Air gap	C/S plaster	RC	RC
Conductivity	W/mk		0.72	2.16	2.16
Density	kg/m^3		1860	24000	2400
Thickness	m	0.05	0.01	0.4	0.15
Intermediate Component - B		RC	RC		
Conductivity	W/mk	2.16	21.6		
Density	kg/m^3	2400	2400		
Thickness	m	0.8	1.05		
Intermediate component - C		C/S plaster		Air gap	
Conductivity	W/mk	0.72			
Density	kg/m^3	1860			
Thickness	m	0.01		0.05	
Internal finish material		Gypsum plaster	Gypsum plaster	Gypsum plaster	Gypsum plaster
Conductivity	W/mk	0.38	0.38	0.38	0.38
Density	kg/m^3	1120	1120	1120	1120
Thickness	m	0.01	0.01	0.01	0.01
Absorptivity	(a)	0.3	0.3	0.3	0.3
U value of composite wall	W/mk	0.75	1.13	1.81	2.28
Area of wall	m^2	10.08	4.56	9.51	31.7
Density of composite wall	kg/m^2	1936.8	2573.93	971.2	371.2
Equivalent temperature difference		1.70	1.70	1.70	2.05

Building (Energy Efficiency) Regulation
Form OTTV 2
Window/Rooflight Schedule

Façade Orientation Facing: North

Window code no.	**F1**
Location of window	Tower unshaded
Glazing type	Tinted
Thickness	0.006
SC	0.7
ESM for overhang	–
ESM for sidefiin	–
Area of glazing	69.75

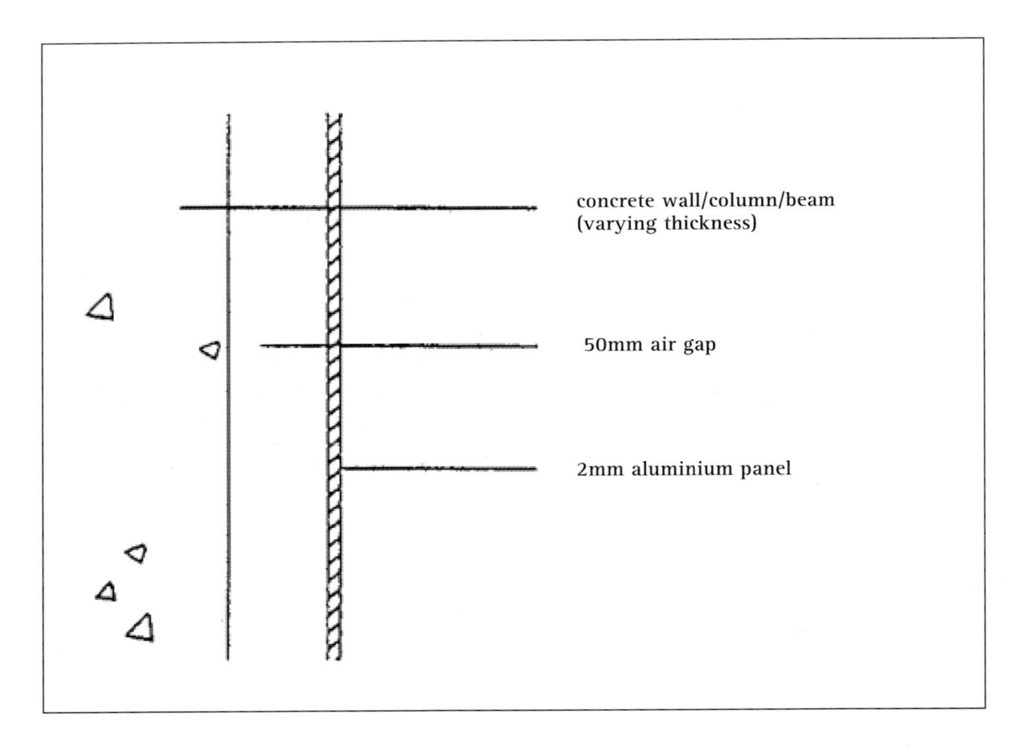

Figure 20 W1

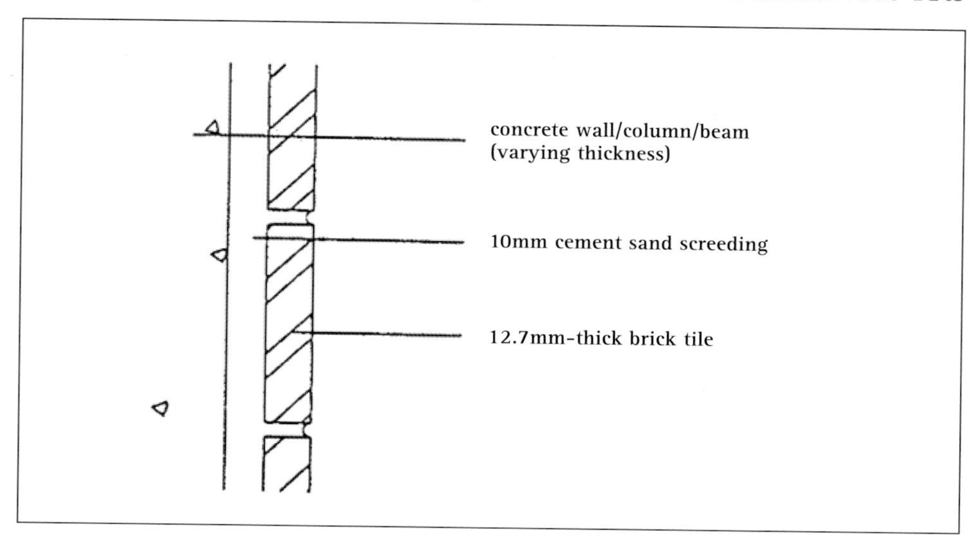

Figure 21 W2

Building (Energy Efficiency) Regulation
Form OTTV 3
Calculation Of OTTV Of Individual Façades In Building Envelope

Façade Orientation Facing: North

Opaque wall

Code	Description	Aw	U	a	Tdeq	Sum
W1	1050mm column	10.08	0.75	0.1	1.7	1.29
W2	400mm beam	4.56	1.13	0.88	1.7	7.71
W3	800mm column	9.51	1.81	0.65	1.7	19.02
W4	400mm beam	31.7	2.28	0.65	2.05	96.31

Subtotal 55.85m² Heat gain 124.33W

Fenestration

Code	Description	Afw	SC	ESM	SF	Sum
F1	Unshaded	69.75	0.7	–	104	5077.80

Subtotal 69.75m² Heat gain 5077.80W

Gross area 125.60m² Gross heat gain 5202.13W

OTTV (North Façade) 41.42W/m²

Observation

This situation is different to most other buildings in which the whole building is considered as a single envelope. For this project, under the Building (Energy Efficiency) Regulations, only the church hall and chapel, assembly hall and function rooms are included in the OTTV calculation. This gives rise to some occasions where the rooms that require calculation are surrounded by compartments of which calculation is not required. This means that some external walls or roofs of the rooms under calculation are not exposed to external air.

In the actual calculation process, the heat gain of these walls is assumed to be zero. This probably causes significant error in the results obtained since the OTTV is the total heat gain divided by the total wall and roof area. Should the wall area on one side of the room be assumed to be zero, the OTTV can increase significantly provided that the heat gain of this wall is not high. Therefore, the whole building should be included in the OTTV calculation when there is at least one compartment that is eligible.

Plate 5

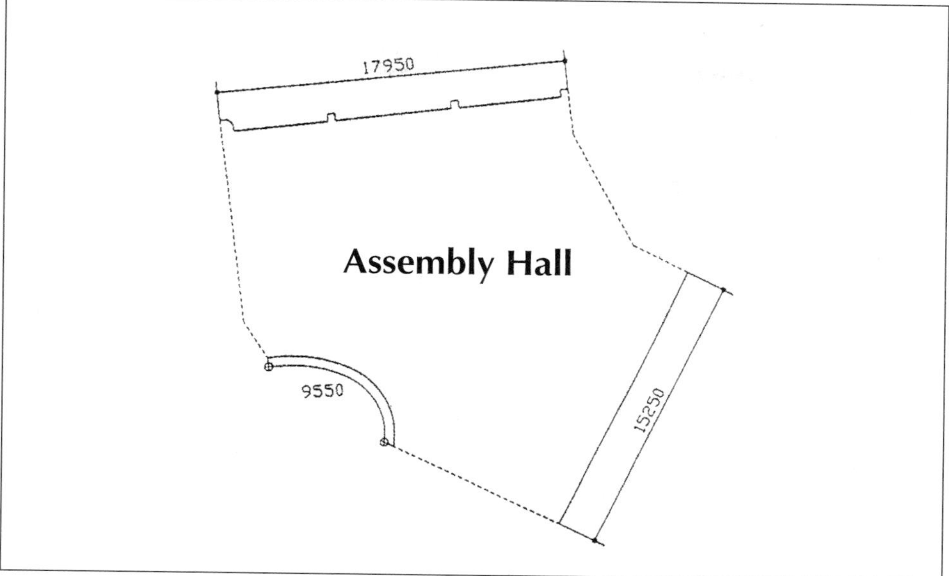

Figure 22 Diagram of Assembly Hall

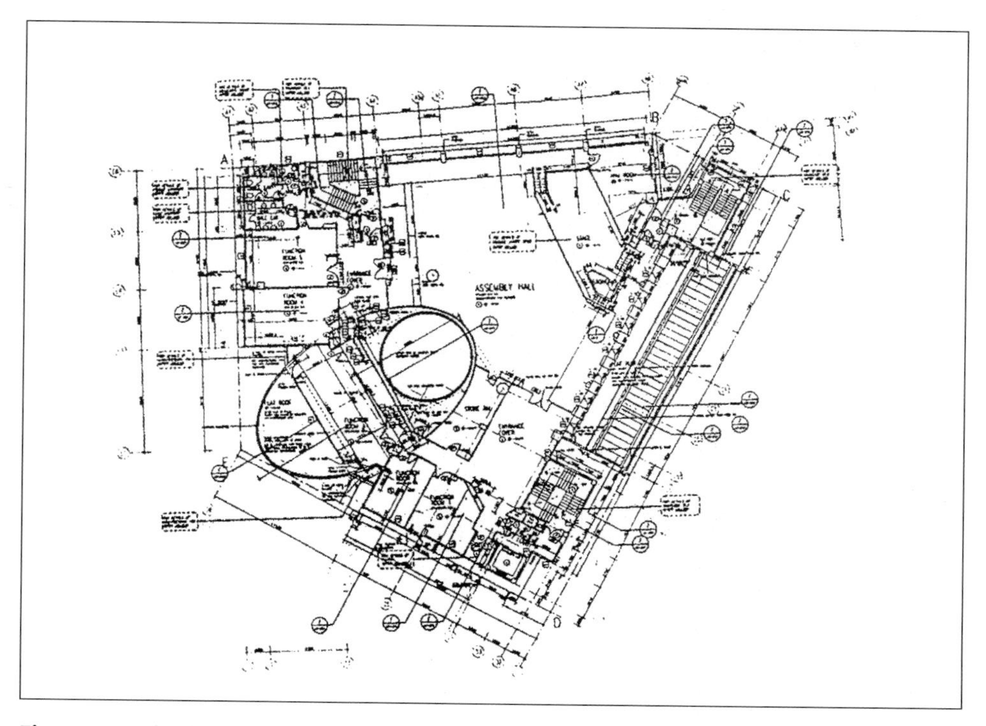

Figure 23 Third Floor Plan

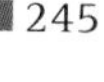

KCRC Kowloon Station Extension

Address	Hung Hom, Kowloon
Building Type	Transport Terminal
No. of Storeys	1

This project is the extension of the existing KCRC Station, with the existing buildings lying to the west of the extension as shown on the orientation key plan (Figure 24). The building is conceived as a large shell housing all other accommodations inside. For the exterior wall, tinted glass and GRC panels are used as the major finishes. Panels with galvanized steel external finish and intermediate mineral wool are extensively used on the roof.

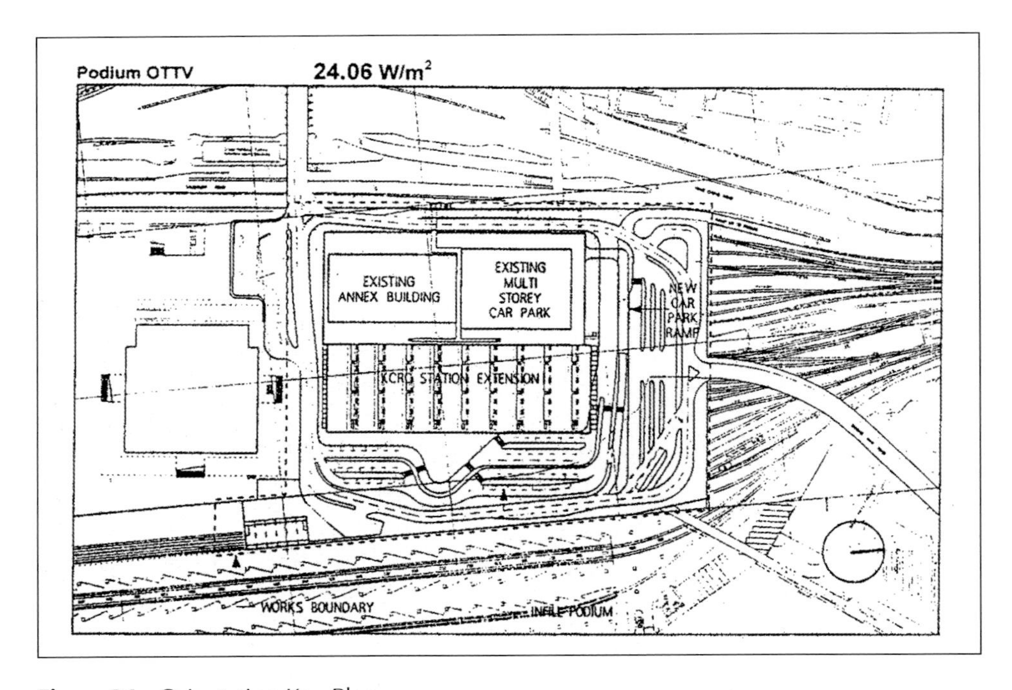

Figure 24 Orientation Key Plan

Summary of OTTV of building envelope

Façade		
Orientation	**Gross area (m²)**	**Gross heat gain**
East	1724.17	87987.52
North	945.75	13691.34
West	1724.16	65988.57
South	949.39	33434.75
Subtotal	5347.47	201978.09
Roof		
Roof	11185.55	236347.7
Roof facing north	1329.9	20774.49
Roof facing south	1329.9	2672.1
Subtotal	13845.35	259794.29

Podium Walls OTTV 33.77W/m²
Podium Roofs OTTV 18.76W/m²

Podium OTTV 24.06W/m²

Plate 6

Plate 7

In the following section, the south façade of the KCRC Kowloon Station Renovation and Extension is studied.

Building (Energy Efficiency) Regulation
Form OTTV 1
Calculation Of U Value Of Composite Wall/Roof And Details Of Other Values

Façade Orientation Facing: South

Wall code no.		W1	W2	W3	W4
Location of wall		Existing building	Existing building	Louvre wall	Parapet
External finish material		GRC panel	GRC panel	GRC panel	GRC panel
Conductivity	W/mk	0.44	0.44	0.44	0.44
Density	kg/m^3	1300	1300	1300	1300
Thickness	m	0.025	0.0015	0.025	0.025
Absorptivity	(a)	0.65	0.65	0.65	0.65
Intermediate component - A		Reinforced concrete	Air gap	Reinforced concrete	Reinforced concrete
Conductivity	W/mk	2.16		2.16	2.16
Density	kg/m^3	2400		2400	2400
Thickness	m	0.175	0.035	0.175	0.175
Intermediate component - B			Vapour barrier		
Conductivity	W/mk		-		
Density	kg/m^3		-		
Thickness	m		-		
Intermediate component - C			Concrete block		
Conductivity	W/mk		0.44		
Density	kg/m^3		1300		
Thickness	m		0.14		
Intermediate component - D			Air gap		
Conductivity	W/mk		-		
Density	kg/m^3		-		
Thickness	m		0.025		
Internal finish material		GRC panel	Plaster	GRC panel	GRC panel
Conductivity	W/mk	0.44	0.38	0.44	0.44
Density	kg/m^3	1300	1120	1300	1300
Thickness	m	0.025	0.025	0.025	0.025
Absorptivity	(a)	0.65	0.65	0.65	0.65
U value of composite wall	W/mk	2.79	1.14	2.79	2.79
Area of wall	m^2	172.68	109.8	133.86	115.19
Density of composite wall	kg/m^2	485	229.5	485	485
Equivalent temperature difference		3.28	2.72	3.28	3.28

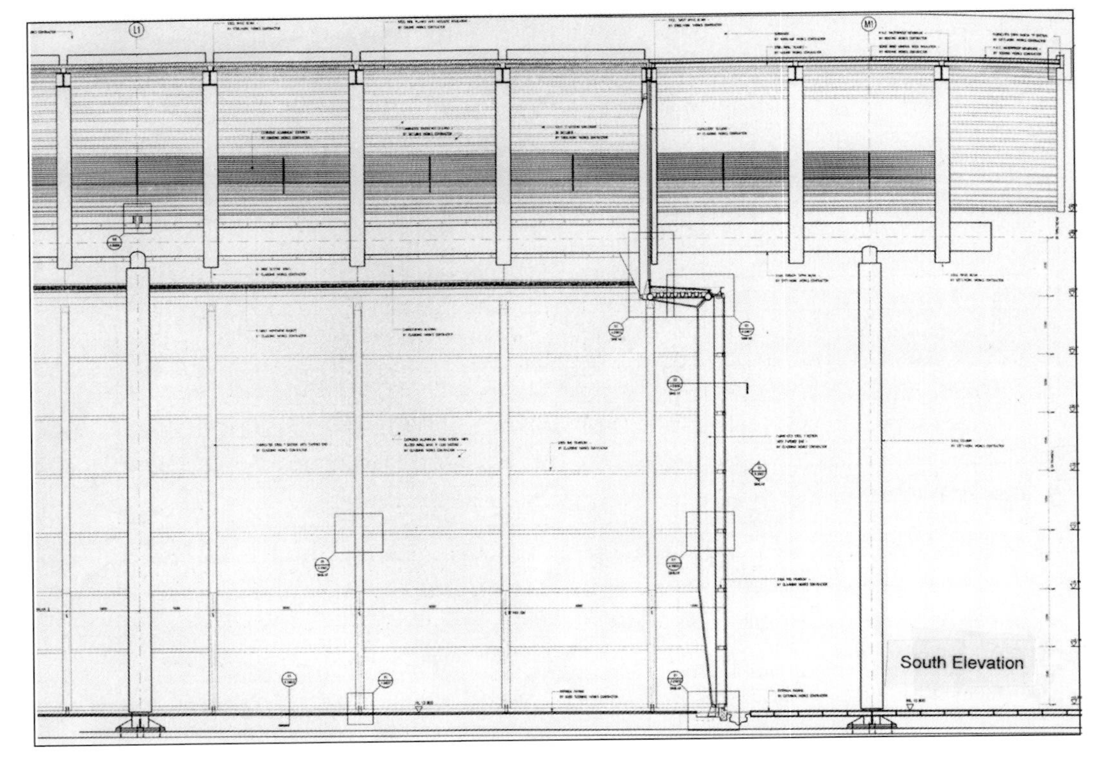

Figure 25

Building (Energy Efficiency) Regulation
Form OTTV 2
Window/Rooflight Schedule

Façade Orientation Facing: South

Window code no.	**F1**
Location of window	Curtain wall
Glazing type	Laminated
Thickness	0.013
SC	0.45
Type of shading device	Solid overhang
Length of overhang	
Height of window	
Overhang projection factor	
ESM for overhang	0.877
ESM for sidefin	1
Area of glazing	417.86

Building (Energy Efficiency) Regulation
Form OTTV 3
Calculation Of OTTV Of Individual Façades In Building Envelope

Façade Orientation Facing: South

Opaque wall

Code	Description	Aw	U	a	Tdeq	Sum
W1	Existing building	172.68	2.79	0.65	2.15	673.28
W2	Existing building	109.80	1.14	0.65	3.60	292.90
W3	Lourve wall	133.86	2.79	0.65	2.15	521.92
W4	Parapet	115.19	2.79	0.65	2.15	449.13

Subtotal 531.53m² Heat gain 1937.23W

Fenestration

Code	Descritpion	Afw	SC	ESM	SF	Sum
F1	Curtain wall, laminated	417.86	0.45	0.88	191.00	31605.26

Subtotal 417.86m² Heat gain 31605.26W

Gross area 949.39m² Gross heat gain 33542.49W

OTTV (South Façade) 35.33W/m²

▮ CONCLUSION

Following the Code of Practice for OTTV in Buildings, various factors like the dimensions of windows, materials and detailing have been studied and compared. Obviously, OTTV figures decrease as the dimensions of windows increase. This is, however, in contrast to using windows for views and aesthetics, where both functions have environmental values not directly measurable by OTTV standards.

Certain types of glass like reflective glass or tinted glass give better OTTV performance. Their use is more related to construction costs, internal view and appearance of the building. An increased capital outlay has to be made for better OTTV performance.

When the overhanging sunshading device is enlarged, the OTTV is reduced. However, the overhang should be proportional to the size of the window opening to achieve the best effect. Again, these overhanging devices have construction cost implications and aesthetic significance.

The effect of adjacent buildings is not included in OTTV calculations.

Although this is a variable factor that is uncontrollable, this factor can be predicted to a certain extent. For example, new buildings erected in groups will provide shading for a number of years. Another example is that of for a large site with more than one building but under one management, the effect of which will be more controllable. This investigation shows that the existence of adjacent buildings does have a practical effect on OTTV and should not be ignored. Non-functioning devices to reduce OTTV can increase initial building costs, and are also a factor in indirect environmental considerations.

The case studies demonstrate how Hong Kong buildings have been designed in response to OTTV requirements under the present building code. Sunshades, overhangs, tinted and reflective glass are common techniques used.

▮ ACKNOWLEDGEMENTS

The preparation of this chapter was based on research funded by the University of Hong Kong's Committee on Research and Conference Grants (A/C code: 337/058/0046). The co-researcher was Stephen Siu-yu Lau.

Many thanks are due to All Arts Ltd., Foster Asia (Hong Kong) Ltd. and Rocco Design Ltd. for their assistance in the case studies.

Part D: Environmental Legislation

The call for environmental protection is appealing to all. The problem lies in the implementation. In addition to business incentives and educational promotion, legal provisions are required to regulate and enforce environmental protection. This part considers the law related to environmental planning and pollution control. The chapters in this part provide an overview of the issues from a governmental perspective, and a general coverage on the major pieces of environmental legislation. As the Environmental Impact Assessment (EIA) Ordinance is the central piece of new environmental law that affects major developments, it deserves detailed analysis from both the management and implementation points of view.

In 'Environmental Planning and Impact Assessment of Major Development Projects in Hong Kong', Elvis Wai-kwong Au introduces the ways that the

Hong Kong government has been exploring to meet the rising community expectations for a better environment. The chapter presents the development of an environmental planning system by the government to deal with the environmental implications of major property developments at the earliest possible stage. Au goes through the historical development of different phases of the EIA process leading to the EIA Ordinance in Hong Kong in 1998. Since 1990 and before the EIA Ordinance, major development projects such as the Airport Core Programme were subject to the scrutiny of the environmental monitoring and audit system. The chapter explains the role of the new EIA Ordinance and the statutory framework for regulating and enforcing the EIA process under the Ordinance. Au also points out that the EIA Ordinance can be an aid to environmentally sound management that facilitates business decision-making by private developers and the government.

In 'Impact of Major Environmental Legislation on Property Development in Hong Kong', Edwin Hon-wan Chan provides an overall view of the major pieces of environmental law that affect property development in terms of development rights, hurdles for the development process and restriction on site construction works. He begins by introducing the two main classes of environmental policy instruments that promote and enforce environmental protection: those that rely on 'market-based' and 'command and control' environmental management philosophies. The relevant pieces of legislation are discussed under the two major areas of land use planning and pollution control. The discussion is backed up by two legal case studies that illustrate the lengthy and uncertain process in securing land use rights and development approvals when environmental issues are involved. The case studies and specific reference to provisions in pollution control ordinances also help to explain the advantages and limitations of the two environmental management philosophies. The chapter concludes with an emphasis on the potential of voluntary compliance and incentive schemes in promoting environmental protection.

In 'Environmental Impact Assessment Ordinance – An Introduction', Chi-sun Poon and Xiang-dong Li focus their discussion on one new major piece of environmental legislation in Hong Kong: the EIA Ordinance. The chapter explains the types of projects affected and the Environmental Permit required under the Ordinance. It provides an outline of the requirements for preparing an EIA report. With a case study, the authors contribute their practical experiences by explaining the procedures involved in an EIA study and by illustrating the impact of EIA at the construction and operational stages of a property development project.

In 'Building Control to Enhance the Environment', Wong Wah Sang discusses the importance of the Building Code in moulding the physical environment, and the possible improvement of the Building Code to enhance environmental awareness. Several questionnarire surveys were conducted through professional institutes as well as users to search for aspects of improvement, which are grouped under Gross Floor Area, Saleable Floor Area, environmental issues, planning and site matters, and internal layouts.

11

Environmental Planning And Impact Assessment Of Major Development Projects In Hong Kong

Elvis Wai-kwong Au

INTRODUCTION

In the last few decades, the economic growth in Hong Kong has been phenomenal, turning the once relatively small fishing village into a modern metropolis, ranked as the tenth largest economy in the world. It has been facing the classic conflicts between development and the environment. The 1980s saw the emergence of new environmental laws, policies and programmes aimed at arresting the deterioration of the environment and preventing new problems from occurring. Part of the environmental management framework is a system and a set of procedures for environmental planning and impact assessment of new development activities.

Developments in Hong Kong are characterized by their high-rise, high-density patterns. Many areas in Hong Kong are hilly and the coastline is rugged and indented, thereby creating a number of airsheds and water basins of limited dispersive capacity. The scarcity of land and the associated high concentration of activities give rise to different types of land use conflicts and a unique set of environmental problems. With rising community expectations for a better environment, the government is looking for ways to harmonize the conflicts between development and the environment. An important part of such an effort is the emphasis the government places on better planning of new projects and a proper application of the environmental impact assessment (EIA) process.

▮ ENVIRONMENTAL PLANNING SYSTEM IN HONG KONG

In response to these challenges, an environmental planning system has been developed over the years to deal with the environmental implications of major developments at the earliest possible stage.

In the 1985 Polmet Conference, Ashcroft argued that the general approach to environmental planning in Hong Kong was based on pragmatism and flexibility and the system was largely developed as an extension of the government's consultative system. Since then, the environmental planning system had become more comprehensive, proactive, sophisticated and diversified. Over the years, an environmental planning and EIA system has been established with the following key components:

- the Environment Chapter of Hong Kong Planning Standards and Guidelines (HKPSG): this chapter was comprehensively revised in 1990 to provide guidance for planners, architects and engineers in planning and designing major developments in Hong Kong;
- an integration of environmental planning into forward planning through environmental studies in territorial development strategy, subregional planning and local plan-making process; and
- an EIA process that is embodied in the government's public works approval process, and which is applied to major private sector projects through the government's approval mechanism.

The key environmental inputs to different stages of forward planning are given in Table 1. The development of the environmental planning and EIA system is presented in Table 2.

Table 1 Environmental input to different stages of forward planning

Level of planning	Objectives of environmental input	Means of input
Strategic planning through Territorial Development Strategy	address key strategic environmental issues, environmental carrying capacities, environmentally suitable areas for developments	the Environment Chapter of HKPSG, strategic environmental assessment
Subregional planning and feasibility studies of new towns	address the environmental acceptability of major development strategies and plans	the Environment Chapter of HKPSG, environmental planning studies at the regional level; environmental assessment of plans, site search, and EIA studies
Outline Development Plans, Outline Zoning Plans, and Layout Plans	address environmental compatibility of land uses, environmental acceptability of plans, and environmental facility.requirements	the Environment Chapter of HKPSG, environmental assessment of plans, environmental advice at planning committees
Project planning and implementation	address the environmental impacts of project design and implementation, and monitor the actual environmental impacts	project EIA, environmental monitoring and audit, the Environment Chapter of HKPSG

HISTORICAL DEVELOPMENT OF EIA IN HONG KONG

The first generation of the application of the EIA process formally began with the promulgation of the National Environmental Protection Act of 1969 in the United States. Under this law, proponents of development projects that involved US federal land, federal tax dollars or federal jurisdictions were required to file an environmental impact statement. Since then, various forms of statutory or discretionary EIA procedures have been established in many countries.

Against this backdrop of a world trend, since the late 1970s and early 1980s, the EIA process in Hong Kong has been applied through administrative rather than statutory means. It is important to note that all developments in Hong Kong are controlled one way or the other by lease conditions imposed in crown leases. Because of this, the Hong Kong government was able, where

Table 2 Development of envrionmental planning and EIA system in Hong Kong

1985	An Environment Chapter was added to the HKPSG incorporating environmental consideration into planning.
1986	A Government Circular on the Environmental Review of Public Sector Projects was promulgated.
1988	A revised Government Circular on the Environmental Review of Major Development Projects was issued, covering new towns and new projects.
1990	• The Environmental Chapter in the HKPSG was comprehensively revised and updated. • An administrative arrangement was established with the Hong Kong Industrial Estate Corporation to review applications for tenancy within industrial estates.
1992	• A Government Circular on the Public Access to EIA reports was issued. • A revised Government Circular on the EIA of Major Public Sector Development Projects was issued, strengthening the administrative procedures for EIA and incorporating provisions for environmental monitoring and auditing. • The Policy Address stated that major papers submitted to the Executive Council must include paragraphs on 'environmental impact assessment'. • The Public Works Sub-Committee under the Finance Committee of the Legislative Council required that all papers seeking funds approval from this subcommittee must include paragraphs on 'environmental implications'.
1994	An EIA subcommittee under the Advisory Council on the Environment was formed to advise on EIA reports.
October 1994	The Policy Address included a commitment to introduce legislation on EIA.
29 January 1996	The EIA Bill was gazetted.
May 1996	A generic environmental monitoring and audit manual was distributed.
July 1996	The strategic environmental assessment of the Territorial Development Strategy was presented to the Advisory Council on the Environment.
4 February 1997	The EIA Ordinance was enacted.
February 1997	A policy on off-site ecological compensation under the EIA process was promulgated.
June 1997	The Technical Memorandum on EIA was approved by the Legislative Council.
November 1997	The regulations on fees and appeal board were approved by the Provisional Legislative Council.
First half of 1998	The EIA Ordinance came into operation.

necessary, to impose EIA requirements through its rights as ground landlord of almost all land in Hong Kong. The development of EIA can be broadly classified into the following phases.

Phase 1: Trial-and-test Period (1979–85)

Effort was made to formulate a set of administrative EIA procedures for public and private sector projects, with regard to development pressures and social characteristics at that time. The application of the EIA process by the pioneers in the then Environmental Protection Agency resulted in integration of environmental considerations into the site selection, design, construction and operation of major development projects such as a new coal-fired 1 700MW power station (Reed and Woolley, 1982), and other development projects such as the Tin Shui Wai development and the Chek Lap Kok airport. In this period, the number of EIAs conducted was less than that for later periods. From 1979 to 1985, a total of 23 EIAs were completed. In 1985, the emphasis on the integration of environmental factors into the land use planning process led to the formulation of a set of environmental standards and guidelines incorporated into the HKPSG.

Phase 2: Systematic Application Of An Administrative System (1986–91)

Over the last decade, EIA in Hong Kong has developed from an ad hoc requirement imposed on a small number of government and private infrastructure projects, to a set of systematic administrative procedures to be followed by proponents of all major or environmentally significant development projects (private and public) in Hong Kong. The Hong Kong government issued in 1986 an internal directive entitled 'Environmental Review of Major Development Projects', setting out the screening process and EIA requirements for public works. The administrative EIA procedures were subsequently laid down in two documents: the joint Planning, Environment and Lands Branch Technical Circular No. 2/92 and Works Branch Technical Circular No. 14/92, entitled 'Environmental Impact Assessment of Major Development Projects', for government projects; and Advice Note 2/92, entitled 'Application of the Environmental Impact Assessment Process to Major Private Sector Projects', issued by the Environmental Protection Department (EPD). During this period, a total of about 80 EIAs were completed, covering a wide range of projects such as roads, sewerage and sewage treatment works, major residential developments and waste disposal facilities.

Phase 3: Focused Approach Towards Project Implementation (1991–93)

The number of EIAs increased significantly during this period. From 1992 to

1994, there were about 239 EIAs completed or ongoing, compared to about 80 EIAs in the six-year period between 1986 and 1991. There had been a strong demand from the public, district boards, the Legislative Council and other governmental advisory bodies for a more thorough consideration, before the commencement of construction, of environmental impacts of development projects. The EIA process underwent three major developments in this period. First, EIA was formally stated as a planning tool for decision-makers. The procedures also included a conflict resolution mechanism to resolve disagreements. Second, there was also a new requirement to make EIA reports available to the public for inspection. Third, a proper system of environmental monitoring and auditing was introduced to track the actual performance of projects.

Phase 4: Wider Application Of The EIA Process To Policies And Works Programmes (1993–94)

Since the 1990s, the EPD had contributed to a number of strategic planning studies, including the Territorial Development Strategy Review. In his policy address in October 1992, the then Governor of Hong Kong introduced a new requirement for EIAs to be included in papers submitted to the Executive Council. Because of this initiative, policy submissions to the Executive Council have to include an environmental implication section to assist in better decision-making. Likewise, environmental implication sections have become a prerequisite for all requests by public sector projects for funding approvals considered by the Public Works Sub-Committee of the Legislative Council's Finance Committee. 'Environmental Implication' statements have since been included in papers seeking funding and policy approval.

Phase 5: Towards A Formal And Effective Statutory EIA System (1994 Onwards)

The EIA Ordinance was enacted on 4 February 1997 and came into operation on 1 April 1998.

ENVIRONMENTAL MONITORING AND AUDIT SYSTEM FOR MAJOR DEVELOPMENT PROJECTS

The 1990s saw the Hong Kong government embark on a mammoth infrastructure programme, the Airport Core Programme (ACP), worth HK$158 billion (or US$20.3 billion), to meet Hong Kong's transport needs well into the next century. The ACP projects comprised ten interlinked infrastructure projects to be built within a period of five years, with completion dates focused on mid-1997. To ensure that environmental benefits were maximized while avoiding any adverse impacts or at least limiting them to within acceptable standards, the government initiated a series of the most comprehensive EIA studies ever undertaken in Hong Kong. The most important outcome of these EIA studies was a comprehensive, project-specific and site-specific package of environmental mitigation and enhancement measures incorporated at the planning stage. However, though some adverse construction impacts were unavoidable, there was a strong need to monitor the implementation of these projects and to follow through the measures recommended in the EIAs in order to reduce the gap between promise and performance.

During the implementation of major development projects, environmental monitoring and audit programmes must be established and reviewed to track and reduce adverse impacts and to ensure mitigation requirements, identified by the EIA study, are implemented in a timely fashion. The programmes for these projects are established through various approaches including contractor self-audit, environmental teams advising the contract management, and environmental project offices (ENPOs) auditing a number of projects in a study area. The EPD has established two independent ENPOs, each comprising a dedicated consultancy team to provide expert advice on cumulative construction impacts arising from multi-billion-dollar civil engineering projects.

In 1996 and 1997, more than 60 environmental monitoring and audit programmes were implemented to ensure satisfactory environmental performance of major projects. Such programmes are now standard requirements for major developments in Hong Kong.

▮ THE NEED FOR AN EIA ORDINANCE

Despite the major contributions and proven achievements of the EIA system, a statutory framework was required for the prediction and assessment of potentially adverse environmental impacts from development projects undertaken by the public and private sectors, and to make implementation of preventive and mitigatory measures enforceable.

Under the existing mechanisms, EIA plays an important role in preventing adverse environmental consequences that may arise from major development projects. On many occasions, through implementation of preventive measures such as improving location and design, adverse environmental impacts can be minimized, thus saving subsequent remedial work which can be more costly and disruptive. More than 300 EIAs have been completed since 1983 and about 100 EIAs are currently being undertaken. Of these, about 60% are public sector projects while the remaining 40% are private sector or public corporation projects.

The current EIA procedures are, however, administrative in nature. There has been difficulty in attaining full implementation of the mitigation measures recommended in EIA studies. Under the current administrative arrangements, there is no satisfactory and effective mechanism to enforce EIA recommendations.

In view of these drawbacks in the administrative arrangements, a policy commitment was included in the Governor's Policy Address in 1994 to introduce legislation on EIA. The EIA Bill was gazetted and introduced to the Legislative Council in January 1996. After one year's public discussion, the EIA Ordinance was enacted on 4 February 1997 and came into operation on 1 April 1998.

FRAMEWORK OF THE EIA ORDINANCE AND TECHNICAL MEMORANDUM ON EIA

Application And Coverage

The EIA Ordinance applies to both public and private sectors. The Director of Environmental Protection (DEP) is responsible for regulating and enforcing the EIA process set out in the Ordinance. Its main provisions are highlighted in the following paragraphs.

The Ordinance provides for the designation of development projects that require an Environmental Permit; such projects are listed as Designated Projects in the Schedules attached to the Ordinance. The projects in Schedule 2 are individual projects, whereas those in Schedule 3 are complex developments involving a number of constituent projects. The Secretary for Planning, Environment and Lands is empowered to amend the list of Designated Projects by order published in the Gazette.

For residential developments, the Ordinance covers those that are to be built within Buffer Zone 1 or 2 of Deep Bay, or in other ecologically sensitive areas such as sites of special scientific interest, or in country parks, as well as those residential developments proposing more than 2 000 flats that are located in unsewered areas.

The EIA Process And Public Involvement

A person who is planning a Designated Project is required to apply to the DEP for an EIA study brief with a description of the project profile, and then prepare an EIA report in accordance with the brief. The DEP will coordinate a review of the findings of the EIA report and if it meets the requirements set out in the study brief, it will be displayed for public inspection for one month. The Advisory Council on the Environment will be notified and may select the project for detailed consideration. Having taken into account relevant environmental concerns and the comments received, the DEP will decide whether the EIA report should be approved, whether an Environmental Permit should be issued, and if so, whether and what conditions should be imposed. The EIA procedures are shown in the flow chart in Figures 1 and 2. Streamlined arrangements have been provided to allow a project proponent to apply directly for an Environmental Permit if the impact of the project has been sufficiently addressed in a previous EIA or, on review of the project profile, the DEP is satisfied that the project is unlikely to cause a significant environmental impact.

Statutory Time Limits

Statutory time limits have been set for each step of the EIA process. There are time limits for the DEP to respond to applications and reports, so too for responses by the public and the Advisory Council to project profiles and EIA reports.

Offences

The Ordinance renders it an offence to carry out a Designated Project without an Environmental Permit or not in accordance with Permit conditions. Projects that had been approved under other legislation or had commenced construction or operation before the EIA Ordinance came into effect are exempted. The maximum penalty stipulated in the Ordinance is a HK$5 million fine plus two years' imprisonment. There are due diligence defences for non-compliance with Permit conditions. If the offences result in environmental damage, the DEP, with the consent of the Secretary for Planning, Environment and Lands, may issue a cessation order and take action to remedy the damage and recover the costs.

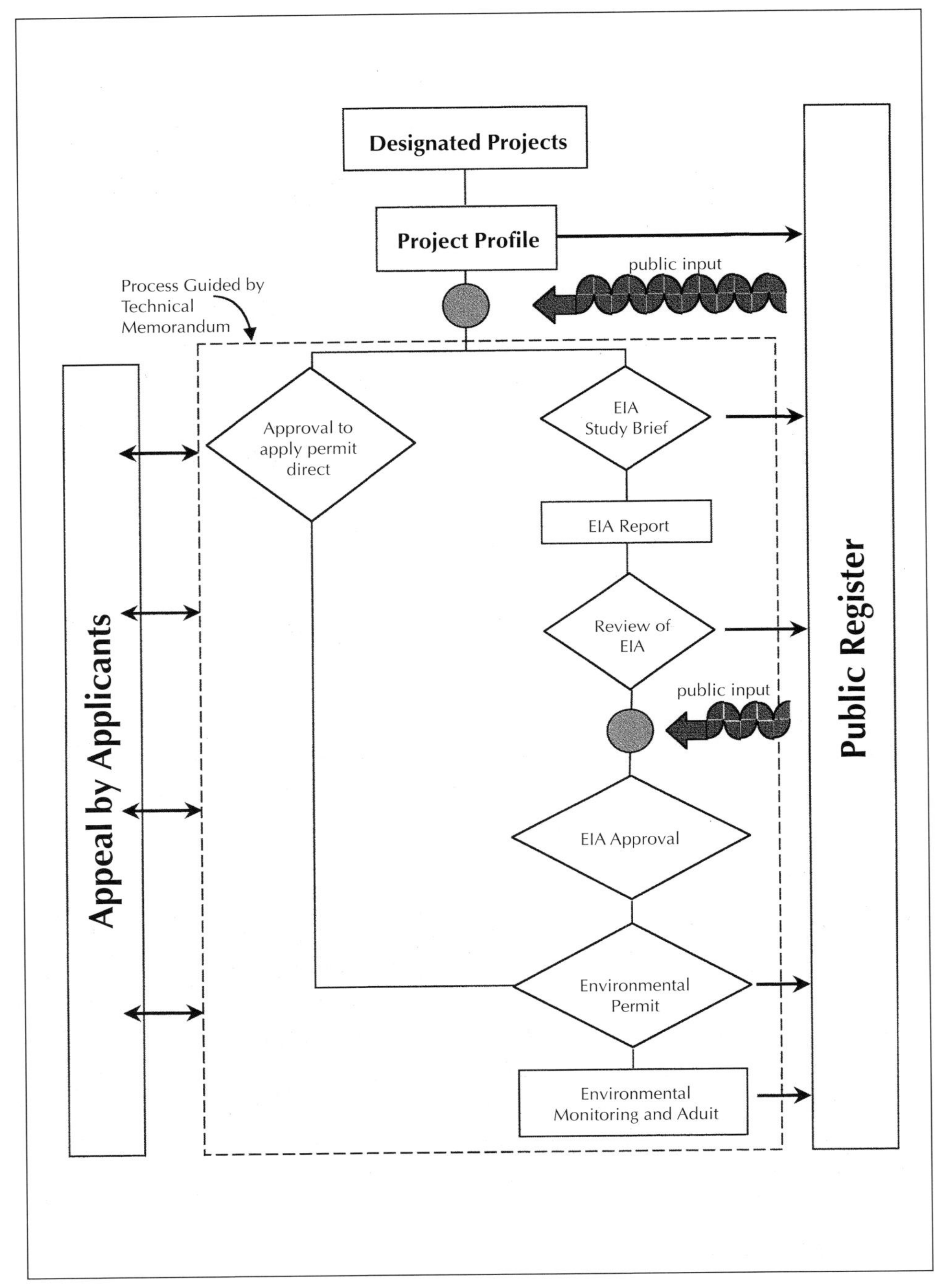

Figure 1 EIA process Under the EIA Ordinance

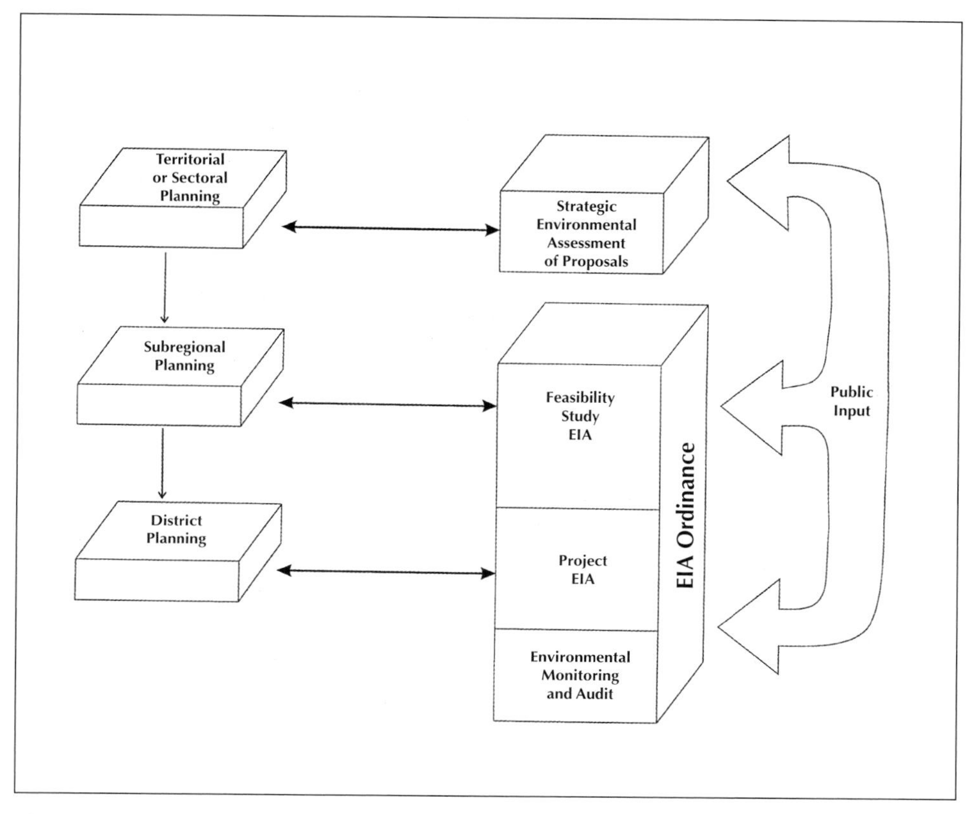

Figure 2 EIA Ordinance and strategic evnironmental assessment in Hong Kong.

Technical Memorandum On EIA

The Ordinance provides for the issue of Technical Memorandums, subject to negative approval by the Legislative Council, to set out the detailed operational requirements on the structure and content of EIA studies, to guide the DEP in the issue of an Environmental Permit, and to provide mechanisms for taking advice from other government authorities and for resolving conflicts. A Technical Memorandum was drawn up within the government and in consultation with interested professional organizations, and was approved by the Legislative Council in June 1997.

Appeal Mechanism

To ensure suitable checks and balances, the Ordinance has provisions for appeal by applicants against key decisions made by the Director or the Secretary. The appeals would be dealt with by an independent Appeal Board set up under the Ordinance.

▌ AN AID TO ENVIRONMENTALLY SOUND MANAGEMENT OF DEVELOPMENT PROJECTS

In its report 'Environmental Assessment – A Business Perspective', the World Business Council for Sustainable Development stated that:

> Environmental assessment of business decisions leads to better decisions for business.
>
> The business itself is usually the most important 'user' of its own environmental assessments.
>
> At the heart of sound environmental management is the assessment of effects, real or potential, on the environment as a consequence of business activities, and the planning and implementation of measures to avoid or mitigate damage.

Such business decisions include those made by the government or the private sector with regard to major development projects. The EIA Ordinance and its associated Technical Memorandum on the EIA process would be an aid to environmentally sound management of major development projects in four ways: transparent requirements and guidelines; clearly defined timeline for responses by stakeholders; designing environmental safeguards into development projects; and proactive tools to reduce or manage environmental risks.

For a proponent or an implementation agent to comply with the Ordinance, the following key steps and requirements would need to be observed:

- Check if a project is a Designated Project. At an early planning stage, steps should be taken to check if a proposed project, albeit in a very preliminary form, falls under Schedule 2 or 3 of the EIA Ordinance. If it does, then the provisions and requirements in the Ordinance should be considered and followed.
- Check the register to see if there is already an approved EIA report that adequately addresses the impact of the Designated Project at issue and if the report findings are still relevant. If there is, then there are provisions for seeking permission to apply directly for Environmental Permits without duplicating the effort in undertaking EIA studies.
- Apply for an EIA study brief at the planning stage of the project. During the process of drawing up the study brief, there is a provision for the public to comment on the project profile, which would be advertised in

newspapers. The aim is to identify public concerns as early as possible and to address them during the EIA study to help avoid late focus on major environmental issues.

- Undertake the EIA study at the planning stage in accordance with the study brief and the Technical Memorandum, and submit the EIA report for approval. If the report is found to be technically adequate, the report will be exhibited for the public to comment on, and consultation with the Advisory Council on the Environment may also be made in parallel. These steps are to ensure that major public concerns are identified and resolved well before the project commences.
- Once the EIA report is approved, apply for an Environmental Permit. The Environmental Permit conditions would include those measures and conditions that have already been agreed on during the EIA process, but would not duplicate conditions that are to be imposed under other pollution control ordinances.
- Design, construct and operate the project in accordance with findings and recommendations of the EIA study as well as the Environmental Permit conditions. To help ensure compliance with the agreed measures, there would also be requirements for monitoring and auditing the environmental performance of the project.
- If there are variations to the project that result in a need for variation of Permit conditions, apply for a variation of Environmental Permit conditions. Simple variations of Permit conditions could be dealt with within 30 days of the submission of application.

▮ STRATEGIC ENVIRONMENTAL ASSESSMENT IN HONG KONG

Since 1988, major development plans have been required to be subjected to EIA studies. More than a dozen major environmental assessments of plans or new town developments have been carried out. The provision for applying strategic environmental assessment to policies and strategies was established in October 1992's Policy Address. Papers on major policies submitted to the Executive Council have to contain an environmental implication section to set out clearly the likely environmental costs or benefits that would arise from implementing the proposal. Environmental implication sections are required for the following types of proposals: proposals for new policies or strategies; amendments to existing ones; specific matters that involve environmental issues; proposals or projects for which suitable EIAs have already been carried out; and environmental strategies, policies and proposals.

From 1994 to 1996, a total of about 200 Executive Council papers included environmental implication sections. Of these, roughly 50% involved environmental issues and required detailed environmental input. This provision allows environmental concerns and issues to be addressed at an early stage when the opportunity to influence directions and options is greatest.

Among all these cases, the most comprehensive application of strategic environmental assessment was for the Territorial Development Strategy Review. The strategic environmental assessment of this Strategy commenced in 1992 and was conducted in stages. The study was completed in December 1995, and was presented to the Advisory Council on the Environment in July 1996. The findings were included in the public consultation digest of the Territorial Development Strategy Review. The assessment analysed the environmental implications of more than 20 different alternative development options for different rates and extents of economic and regional development. Throughout the process, the findings of the strategic environmental assessment influenced the strategy formulation, with a number of environmentally damaging options discarded or significantly modified. The results led to a more comprehensive study on sustainable development for Hong Kong.

▮ CONCLUSION

A comprehensive environmental planning and EIA system, as can be seen in Figure 3, has been developed in Hong Kong over the years, and would continuously be improved for use by those involved in the development process and applied for a better environmental future for the present and future generations.

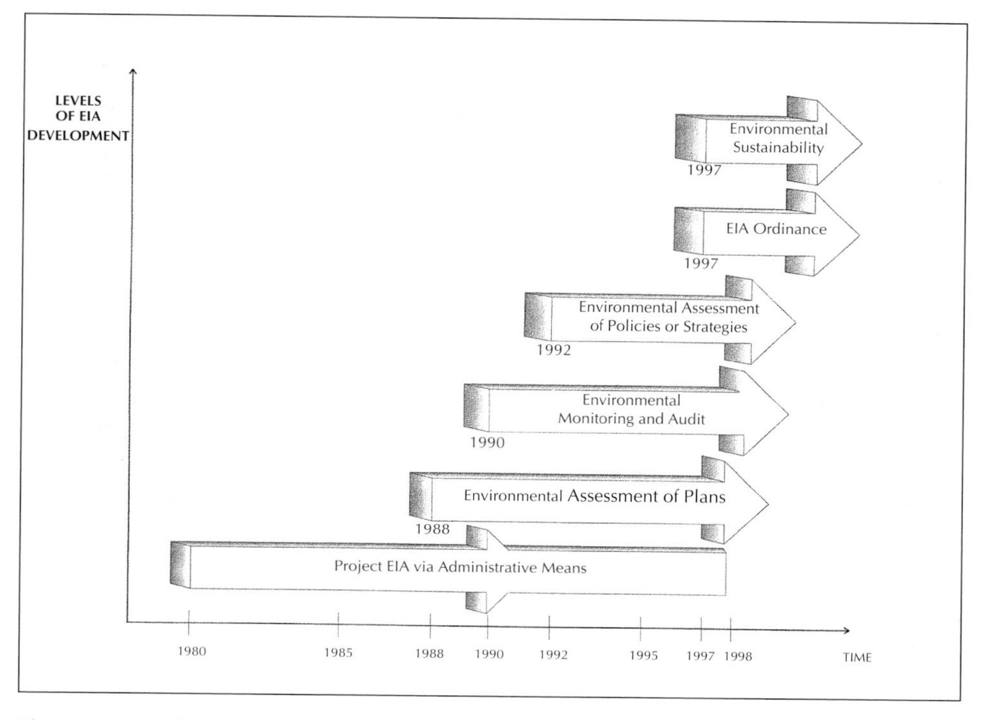

Figure 3 Development of Hong Kong's EIA System

▮ REFERENCES

Ashcroft, B.C. 1987. Environmental planning in Hong Kong. In *Polmet '85: Pollution in the urban environment*, eds. Chan, M.W.H. et al. London: Elsevier Applied Science Publishers.

Au, W.K. 1991. Planning against pollution – an art of the possible. In *Polmet '91: Pollution in metropolitan and urban environment*, ed. Boxall, J., Hong Kong Institute of Engineers.

—— and P.J. Baldwin. 1994. Application of the EIA process in Hong Kong – towards a more effective and formal system. *IAIA '94*, Canada (June 1994).

Hong Kong government. *Environmental impact assessment ordinance* and *Technical memorandum on environmental impact assessment process*.

Reed, S.B. and B. Woolley. 1982. Assessing the environmental impact of a new power station for Hong Kong. *Journal of the Institute of Energy* (June 1982).

Disclaimer: The views expressed in this chapter are those of the author, and do not necessarily represent those of his employer or his organization.

12

Impact Of Major Environmental Legislation On Property Development In Hong Kong

Edwin Hon-wan Chan

INTRODUCTION

In the last decade, Hong Kong has, in economic terms, grown to a status equivalent to developed countries. People's expectations for the quality of life have become higher. To support the growth, massive infrastructure construction and property development have been carried out or planned. Property development has become one of the major economic activities in Hong Kong. Trailing behind property development for a better-built habitat are the issues of environmental protection and preservation of nature.

The need to be environmentally conscious in development activities may be coming from a self-awareness by citizens in Hong Kong. It is also a response to the international concept of sustainable development and ISO 14000 requirements. The latter is one of a new wave of quality assurance systems following the ISO 9000 that has swept across all parts of the world. Many development projects in Hong Kong involve foreign investment and technological cooperation. Apart from funding and know-how, the foreign parties also inject international concepts and demands in environmental protection.

Environmental issues are a global concern. The UN and some economic blocs such as the European Community and ASEAN have introduced environmental protection model laws or directives to member countries. Those countries unwilling to comply are also pressured to conform by economic and trade sanctions. The same pressure is also put on Asian countries and metropolitan cities, including Hong Kong. In response, the Hong Kong government published a White Paper, 'Pollution in Hong Kong – a time to act', aiming to clean up and protect our environment in Hong Kong before it was too late. Several reviews had been conducted, with progress reports published by the government. Following the White Paper and subsequent reviews, a lot of internationally familiar environmental laws such as the Air Pollution and the Environmental Impact Assessment laws have been promulgated in Hong Kong. This chapter intends to outline the major pieces of environmental legislation affecting property development, and examines their impact in Hong Kong.

POLICY INSTRUMENT AND CONTROL FRAMEWORK

Policies to enhance the quality of the environment will no doubt receive support from all fronts, both globally and locally. For many proposed environmental policies, the challenging problems concern practical issues, especially on how to achieve effective integration with other economic and land development activities. Different types of environmental policy instruments will have diverse and significant impacts on a society in terms of compliance cost, restriction to trade and economic activities, and also the use of limited natural resources. At a workshop held in Hong Kong in 1996 to discuss environmental law in Asia, one of the recommendations was that '[s]election of environmental policy should be based on an assessment of each instrument's (i) effectiveness in limiting environmental damage, (ii) cost-effectiveness, (iii) administrative ease, (iv) fairness, and (v) political acceptability' (Barrow and Cottrell, 1996). Hence, a policy instrument will have to take into account issues of politics, economics and enforcement.

The best approach for promoting environmental policies is through education to induce self-awareness and voluntary participation by citizens in improving the environment. One effective way is to have transparency in the policy-making and implementation so that laws are made with the consent and cooperation of those being governed. Public consultations provide channels for public participation in law-making. Environmental laws enacted with the support of the public would encourage more citizen-initiated legal actions that would help to increase effective enforcement. Environmental education of this nature will have to start with the general masses, however, this will take generations to bear fruit.

Apart from employing education to facilitate an environmental policy, there are two main classes of environmental policy instruments (Barrow and Cottrell, 1996). First, the 'market-based' environmental management philosophy advocates designing and setting up specific measures that encourage positive changes in environmentally damaging activities. It is a more performance-based requirement, as an affected party has more flexibility in deciding ways to modify his/her behaviour in order to comply with the management measures. Because the affected person has a choice, it is perceived as a voluntary approach. Economists believe that the compliance cost of this approach is more cost-effective because the affected party has the initiative and flexibility to seek innovative and cost-saving solutions. Second, the 'command and control' philosophy advocates designing and setting up specific measures to exercise direct control over activities. The idea is to have tailor-made control measures to suit the situation of each affected party. However, very often, a set of requirements would apply uniformly to all those subject to the same type of control. A particular cost-effective solution for one party may not necessarily be the best option for another party. From an economic point of view, this approach does not necessarily allow an environmental policy to be achieved at the lowest possible cost.

Regulators prefer the latter approach because it provides certainty by rules and achieves fairness for administration purposes. For the same reasons, the approach sacrifices flexibility and is often perceived as enforcement by direct control. Although education and promoting voluntary action could cut down the reliance on rigid environmental laws, they may not produce the desired results, especially not in a relatively short period. To avoid ambiguity and the possibility of favouritism in the administration of environmental control, the 'command and control' approach requires the support of detailed rules and regulations for direct control. Once a certain set of environmental policy instruments are adopted, the forms of control measures will be reflected by an appropriate set of laws and regulations.

Environmental law is very diverse. It covers many fields and affects all walks of life. In the following, we will discuss the major pieces of environmental law and regulations that have a significant impact on property development in Hong Kong.

▮ ENVIRONMENTAL IMPACT ASSESSMENT ORDINANCE

The new Environmental Impact Assessment (EIA) Ordinance cap 499 has been in effect since April 1998. Before the enactment of the EIA Ordinance, the Environment Chapter of the Hong Kong Planning Standards and Guidelines, which was comprehensively revised in 1990, provided useful guidance for planners, architects and engineers in planning and designing major property developments in Hong Kong. It is still the most important guide for practitioners in carrying out design and construction works to comply with environmental requirements. Principles and practice of the EIA Ordinance are discussed in detail in other chapters of this book and are not repeated here. In essence, the Ordinance provides the framework for assessment of the environmental impact of projects. Only certain projects are subject to the control of the EIA Ordinance.

Impact Of EIA Ordinance On Property Development

The EIA system in Hong Kong will have a major impact on the economic and social aspects of our society, and will certainly affect property development. Developments within the definition of 'designated projects' are required to have EIA reports submitted to the Director of Environmental Protection for approval. The reports may encounter several rejections, amendments and resubmissions before the final approval is obtained. Mitigation measures will also affect the design and construction stages of a development project. The EIA process will incur additional costs and delay to a development project (see Case Study 1 below). If the system is supported with a good information system exploring the latest information technology and using data of existing EIA reports, a better decision on design can be made, and the process to comply with environmental requirements can be more predictable. The cost of mitigation measures may be reduced because alternative design or construction methods could be explored. The extra expenditure by developers may be offset by the goodwill gained in promoting environmental friendliness, which could be a valuable asset for the corporate image of an organization.

Continuing this line of thinking, there are other economic benefits of environmental sensitivity for developers involved in international projects or local projects funded by international organizations. These international organizations such as the World Bank have indicated that one of the important elements in considering financial assistance or underwriting infrastructure projects is the quality of EIA reports (Smith and van der Wansem, 1995). The EIA Ordinance, supported with a clear Technical Memorandum and guidelines, will surely help international projects in Hong Kong seek financial support from international banks and organizations. As one of the attributes of ecological balance, the EIA system, though requiring extra expenditure, will no doubt bring a more sustainable environment and better health to Hong Kong citizens in the long run. According to research carried out by Gilpin (1995), the life span of citizens is longer in places where EIA is practised. As the EIA Ordinance has been implemented for only a short while, more time is needed to confirm its benefits and drawbacks. Monitoring and reviewing processes from government authorities and pressure groups will serve a meaningful purpose.

Case Study 1: Noise Mitigation Measures For Roadworks

Noise mitigation measures were implemented in relation to a roadworks project in Tsuen Wan to prevent some 850 premises from being exposed to noise levels exceeding the legally permissible limit. The HK$274.1 million project included the construction of a flyover across Castle Peak Road to link up Sha Tsui Road with Tsuen King Circuit, in order to ease traffic congestion there and to provide an additional vehicular access point to the area.

An EIA study for the project, completed in 1994, concluded that some 850 premises would be exposed to noise levels exceeding the noise limit laid down in the Hong Kong Planning Standards and Guidelines unless mitigation measures were provided. A government spokesperson said that to mitigate the noise impact, direct technical remedies including low-noise road surface, road covers and roadside noise barriers would be provided. In addition, indirect technical remedies in the form of window insulation and air-conditioners for some 480 affected premises, including a kindergarten/church, would also be provided. The Traffic and Transport Committee of the Tsuen Wan District Board accepted in principle the proposed noise mitigation measures. The Advisory Council on the Environment endorsed the EIA report, and the Executive Council approved the indirect technical remedies.

(Reference: Extract of a press release obtained from the Web-site *http://www.pelb.gov.hk/index.htm* of the Secretary of Planning, Environment and Lands Bureau, Hong Kong Special Administrative Region, 24 March 1998.)

▮ LAND USE PLANNING

A lot of environmental damage is caused by, and justified in the name of, development for the necessity of industrial growth, providing infrastructure and housing for society. This is even more so for metropolitan cities like Hong Kong with scarce land resources. Environmental protection is applicable not just to the countryside and other sensitive receivers. It is precisely because of the conflicting demands and problems experienced in a metropolitan city that we should explore integrated environment management policies. This was an obvious conclusion from the 1996 workshop held in Hong Kong on environmental law in Asia: 'Land use planning—and especially industrial and infrastructure siting—should be a starting point for environmental management' (Barrow and Cottrell, 1996). The report of the workshop considered the situation in Hong Kong: 'Land use planning is a vastly under-utilised tool in the region for lessening the environmental consequences of development and this has undermined success in reducing environmental damages from existing source' (Barrow and Cottrell, 1996).

The environmental control framework in Hong Kong is moving towards a more integrated approach to ensure that environmental factors are taken into account at all stages of planning and project development. For territorial and subregional planning, the Environmental Protection Department (EPD) is involved to formulate environmental input to the Territorial Development Strategy carried out by the Planning Department. The strategy will identify future strategic growth areas to meet long-term development needs. It also involves comprehensive assessment of the environmental impact and development potential of these areas. At the local planning level, environmental factors recommended in the Environmental Chapter of the Hong Kong Planning Standards and Guidelines have to be considered when planning applications are submitted by planners, architects and engineers.

With the EIA Ordinance coming into effect, an Environmental Permit becomes a prerequisite for gaining planning approval for 'designated projects'. A city cannot sustain continuous development on greenfield land or redevelopment sites. To improve the quality of the environment in old urban areas and to encourage rehabilitation of buildings, the government is carrying out integrated planning and development studies for strategic growth areas by selecting priority projects and exploring means to enhance the financial viability of these projects. It aims to set up the Urban Renewal Authority (URA), which will have the statutory power to speed up the pace of urban renewal and implement more comprehensive development schemes.

Conservation and creation of a quality environment go hand-in-hand through the stages of land use planning. When a desired land use pattern is established, planning controls are put into effect through the gazetting of statutory plans including Outline Zoning (OZ) Plans or Development Permission Area (DPA) Plans to be complied with by all.

Town Planning Ordinance

The main piece of legislation governing land use planning and supporting environmental management is the Town Planning Ordinance (TPO) cap 131. The objective of the Ordinance is 'to promote the health, safety, convenience and general welfare of the community by making provision for the systematic preparation and approval of plans.' Draft OZ or DPA plans prepared under the Ordinance may show or make provisions for 'country parks, coastal protection areas, sites of special scientific interest, green belts or other specified uses that promote conservation or protection of the environment.' This is to ensure that wildlife habitats are preserved, and an environment of high ecological value is protected.

Green Belt Zoning

The green belt zones in Hong Kong cover a large percentage of rural land to preserve natural vegetation. These zones provide passive recreation outlets for the public and act as a buffer and reserve zone between urban and rural land uses, and between development and designated conservation areas. To preserve existing natural landscapes, the Planning Authority prohibits any extensive clearance of existing natural vegetation causing adverse visual impact on the surrounding environment. Only in exceptional circumstances will a new development proposal in a green belt zone be approved. The proposal must be justified on very strong planning grounds, supported by an appropriate environmental mitigation proposal.

Conservation Area Zoning

Conservation areas consist of ecological landscapes and topographical features in rural areas. These zones also protect water-collecting areas against pollution and erosion. They act as buffers to protect country parks or sites of special scientific interest. The planning intention for this type of zoning is to retain the existing natural characteristics of the area. Any redevelopment of an existing house is restricted to the original plot ratio, site coverage and building height. New developments in these areas are not allowed unless they are required to support the conservation of the areas' natural features and scenic qualities.

Site Of Special Scientific Interest (SSSI) Zoning

These sites are to preserve existing, sensitive wildlife habitats from human disturbance. They provide breeding grounds for a diverse community of wildlife. These areas are important ecological sites and natural environment that should be preserved. The planning intention is to prevent any detrimental activities within the SSSI. The URA would approve only developments serving the special need within the areas, such as for educational purposes.

Planning Disputes

Planning permission from the Town Planning Board (TPB) under section 16 of the TPO is required for developments within the aforementioned zoned areas. Within these special areas, only agricultural use and uses related to the purpose of conservation are permitted. Each application will be assessed on its individual merits based on guidelines issued for each of the zones. Applicants may be granted approval, with or without conditions, or disapproval. An applicant could ask for a review of his/her application if it is disapproved by the TPB. If it is rejected again in the review process, an appeal may be made to the Town Planning Appeal Board (TPAB) under section 17B of the TPO for the case to be heard by an independent body. An EIA with evaluation of the ecological/ environmental impacts of the development proposal is always recommended for an application to the TPB or appeal to the TPAB. If the TPAB's decision is not acceptable, legal battles could follow in the form of an application for judicial review by the High Court and consequent appeal to the law courts of a higher level. The process could be very lengthy and the outcome uncertain after incurring heavy costs (see Case Study 2).

Case Study 2: Nam Sang Wai

Nam Sang Wai (NSW) is next to the Mai Po Nature Reserve (MPNR), which is surrounded by two buffer zones (Buffer Zones 1 and 2). The explanatory statement of the 1991 DPA Plan covering NSW read: 'There are areas where private initiatives may wish to provide comprehensive low-rise, low density residential developments . . . due regards should be given to minimising the environmental drainage and traffic impact of these developments on the surrounding areas.'

In August 1992, a developer proposed a comprehensive development comprising a golf-course and 2 550 residential units in buildings of two to eight storeys at NSW within Buffer Zone 2 in Deep Bay. As an attraction and environmental mitigation measure for the proposed development plan, the developer offered to develop the nearby Lut Chau site within Buffer Zone 1 into a nature reserve and provide a wildlife habitat in NSW. The proposed development was rejected by the TPB for the following reasons:

- *the proposed development was not in line with the planning intention for the area, which was primarily to protect and conserve the landscape and ecological value of the area and its scenic quality necessary to sustain MPNR;*
- *the intensity of the proposed development was excessive for low-density residential development in the rural area; and*
- *the drainage impact assessment of the proposed development, in particular the impact of the loss of fishponds and its impact on wildlife, was inadequate to demonstrate that the development would not have an adverse impact on the area.*

In August 1994, the developer appealed to the TPAB. The TPAB allowed the appeal and approved the plan by a majority for the following reasons:

- *the proposal fully complied with the planning intention of the DPA plan;*
- *the intensity of the proposed development was less than that already permitted;*
- *in intrinsic landscape terms, the proposal presented a substantial improvement and that in intrinsic ecological terms, the proposal did not present a threat to the MPNR and in fact enhanced the habitats in the Deep Bay area.*

Adding to the legal complication was the fact that during the planning application and appeal stages, two policy planning guidelines were issued, in September 1992 and November 1993, for the area. The 1992 guidelines introduced for the first time the concept of buffer zones. The purpose was to preserve the MPNR. Buffer Zones 1 and 2 have a different intention. They serve to protect the MPNR with the 'gradation concept' of zoning.

The TPB sought a judicial review but this was dismissed by the High Court in April 1995. The TPB appealed to the Court of Appeal, which upheld the appeal and quashed the decision of the TPAB. The Court of Appeal accepted the TPB's argument that the TPAB had misunderstood the explanatory statement of the DPA plan and the subsequent guidelines. It ruled that 'the planning intention to be derived from those documents was that the existing landscape, including the fishponds, was to be preserved more or less intact, and that there was to be no large-scale residential development.'

The developer appealed to the Privy Council in London, which upheld by a majority in favour of the developer in December 1996. The Privy Council ruled that:

- *the Appeal Board [TPAB] was not bound to follow the explanatory statement or the guidelines although they were material considerations to be taken into account and could not be disregarded;*
- *there was no misunderstanding on the part of the Appeal Board [TPAB]. If it had been the intention to preserve the whole of buffer zone 2 in its existing condition, it would have been easy enough to say so by designating the whole area a conservation zone. Not only was that not done, part of the area had already been designated residential.*
- *even if there were some misunderstanding on the part of the Appeal Board [TPAB], it could not avail the TPB . . . . The Appeal Board had clearly stated that they regarded the proposed development as an improvement in landscape terms. They dealt fully with every aspect of the TPB's appeal. The TPB chose to fight the battle of the fishponds. They lost on planning grounds.*

(Reference: Henderson Real Estate Agency Ltd. *v* Lo Chai Wan, Privy Council Appeal No. 54 of 1996. Italics are extracted from *Hong Kong Law Report and Digest* [1997], pp. 258-70.)

In the NSW case, although the developer eventually obtained approval for the proposed development, the whole procedure took about five years. During that period, not only had the property market changed, but the process resulted in also additional legal costs and professional design fees being incurred. With changing sentiment in society for greater environmental protection, EIA requirements and mitigation measures during the development's implementation stage could become more stringent. The dispute in the above case was mainly on the interpretation of planning intention, which was hidden in the explanatory statements of the statutory plans and subsequent planning guidelines. It begs the question of why the planning intention could not be spelt out more clearly in the first place. The planning requirement was performance-based, with general terms defining planning intention and inviting private initiatives to propose comprehensive development. It resembles the 'market-based' environmental management policy that allows flexibility, but it could also be fertile ground for planning disputes that may defeat the economic benefit attached to the policy.

When ambiguity arises, it becomes questionable whether the planning professionals and executives in government departments can help to clarify the original planning intention.

Interpretation of planning guidelines through legal process may not necessarily be the best option. In the NSW case, after a lengthy legal battle all the way up to the final law court, it did not seem to have set a landmark precedent with major legal principles established to avoid future planning disputes. For such situations, a prescribed planning intention with clearer development control measures enforceable by the regulators would appear more useful for all parties concerned.

Enforcement Against Unauthorized Developments

The planning intention to protect our environment is reflected in the land use zoning plan. To achieve the intended purpose, any unauthorized developments must be stopped. The Buildings Ordinance cap 123 provides a very effective control against unauthorized developments involving construction works because building plans have to be submitted to the Buildings Department for approval. The Buildings Department will enforce against any unauthorized developments by forcing the removal of unauthorized building works under powers contained in the Buildings Ordinance. Although the Planning Department can enforce planning control under section 23 of the TPO by issuing an Enforcement Notice, Stop Notice or Reinstatement Notice against unauthorized developments not involving construction works, the enforcement provisions are less effective. Unauthorized developments, such as change of agricultural land into vehicle repair yards, and container storage and port backup uses in the rural parts of the New Territories, are causing serious problems to the environment, traffic, safety, flooding and health. Because the land use planning and environmental policy for the rural areas have not been able to follow the rapid pace of change in the economy and the needs of the rural community in the last decade, the government has adopted a sensible and sensitive approach in enforcement and prosecution. Rezoning proposals made to the TPB to regularize unauthorized developments may be considered as an integral part of the effort to clean up environmental black spots (SPEL, 6 December 1996).

Country And Marine Parks

The special land use zones under section 4(1)(g) of the TPO include SSSI, country parks, marine parks and marine reserves, conservation areas, and green belts. Developments in country park and marine park zones are subject to the control of the Country Parks Ordinance (CPO) and Marine Parks Ordinance (MPO). The CPO provides for the designation, control and management of country parks so as to conserve their nature for recreational and educational purposes. Country parks and special areas are designated on the advice of the Country and Marine Parks Board created under the CPO. In 1994, there were 21 country parks comprising in total 40 833 hectares, and 14 special areas with a total of 1 639 hectares (Hong Kong government, 1994). They are administered by the Agriculture and Fisheries Department. The provisions in the MPO cap 476 concerning property development are very much the same as those of the CPO. Indeed, the same Country and Marine Parks Board advises the Lands Department on development control in marine parks.

Many country and marine parks overlap with the land use zoning plans made under the TPO. With the advice from the Country and Marine Parks Board, the Lands Department coordinates development controls in the country parks and special areas to avoid duplication. The objectives are to ensure minimum disturbance to the country park zones and a right balance between public interest and that of private landowners. Development in country parks without prior approval of the Lands Department is prohibited. Under section 16 of the CPO, if a proposed use of any leased land within a country park would substantially reduce the enjoyment and amenities of the country park, the leasee may be requested to discontinue or modify the use or be prohibited from proceeding with the proposed use. If the requirements of a notice are not complied with, the leased land may be resumed in accordance with the Lands Resumption Ordinance for public interests. The Country and Marine Parks Board will approve only small developments required to serve the recreational and educational functions of a country park. For substantial new development, even just next to a country park, the developer is required to go through a lengthy procedure and undertake an uphill struggle in order to get approval (see Case Study 3).

Case Study 3: Shalotung, Pat Sin Leng

Since 1978, the Pat Sin Leng Country Park has been designated a country park under the CPO. It is subject to protection against any disturbing effects from new development. The country park zoning did not include the area of a village of over 300 years old at the Shalotung Valley, which was outside the then land zoning system and was used as a rural settlement. Since 1979, a developer had been buying the private land in the village from the original villagers. From then on, the developer had been proposing development plans including a golf-course and low-density residential development for the area. The development proposal involved 89 hectares of land, of which about 31 hectares fell within the Pat Sin Leng Country Park that needed to be leased from the government. Most of the proposed structures would be constructed on the private land in the village area, and the 31 hectares within the Country Park would be part of a proposed golf-course. Apart from the land grant required from the District Lands Office, the development also required approval from the Country Parks Board of the Agriculture and Fisheries Department.

In 1987, the government agreed in principle to allow the proposed development. From then on, the proposal attracted public attention and objection from environmentalists. Most of the villagers were happy to sell their land and support the development. The government also had limited power to refuse residential development on the private land of the village that could have environmental impact on the Country Park. A whole package of the proposed development, including the 31 hectares of Country Park for the golf-course plus guarantees from the developer for public access through the golf-course, was considered by various government panels/boards/committees. With reference to the prescribed objectives of the CPO, the government undertook a balancing exercise to consider potential harm against the advantages of the proposed development, and concluded the proposed whole package of development to be more favourable in the public interest.

In 1992, the government's approval was recommended to the Executive Council for a final decision. The main concern of environmental groups was that, if approved, the case would set a precedent for the government to allow development in all country parks. Such scepticism was reinforced by the fact that in 1992, a development proposal was submitted by another developer for a similar development at the other end of the Pat Sin Leng Country Park. The environmental groups filed an application for judicial review on the Country Park Board's decision for the Shalotung case. In April 1992, the High Court allowed the environmentalists' application to overturn the Country Park Board's decision, on a legal technicality.

In September 1992, the developer submitted a revised development scheme so that none of the proposed development fell within the Country Park. However, environmental groups still objected to the revised scheme because rich animal and plant life in the Shalotung Valley would be under threat. In 1995, villagers in Shalotung Valley were trying to bulldoze six hectares of their private land, claiming it as a return to farming. This was also objected to by the environmental groups in order to preserve the ecological value of the Valley. In December

1996, the Advisory Council on Environment issued a joint statement saying: 'The risks to habitat, however small from the proponents' perspective, are not acceptable' (*South China Morning Post*, 17 December 1996). It prevented the villagers from destroying the ecologically sensitive valley. For nearly 20 years, the proposed development in Shalotung had been the centre of bitter dispute among landowners, developers and environmentalists.

(References: 20 December 1987, 13 January 1992, 7 March 1992, 24 September 1992, 24 January 1993, 13 June 1995, and 17 December 1996, *South China Morning Post,* Hong Kong.)

In the Shalotung case, it is interesting to note that development on the private land of the village was allowed under the land use zoning plans. Because the area was bordering a country park, stringent environmental requirements had prohibited the proposed development. If environmental requirements were important factors, the village should have been zoned as part of the Country Park or under other type of environmental sensitive zone. To deter development by imposing environmental issues during the development proposal stage appears to be an afterthought to make up for the deficiency in environmental planning at the outset. Another important issue in the case was to achieve a delicate balance between the rights of the private landowners, including the original villagers in Shalotung, and the public interest represented by the active environmentalists. The public expectation to enjoy the adjoining Country Park had overridden the rights of property owners to deal freely with their land. The case shows that environmental issues are not confined by physical ownership boundaries, and can be made the justification to interfere with the rights of one's neighbours.

▌ POLLUTION CONTROL ORDINANCES

Noise Pollution

The Noise Control Ordinance cap 400 and its subsidiary regulations are the main legal provisions for noise control. The major impacts of noise pollution control law on property development are encountered not only during the EIA stage but also at site layout and building design. Guidance on detailed requirements and standards is set out in the Environmental Chapter of the Hong Kong Planning Standards and Guidelines. It may affect the orientation and siting of building blocks and the façade/window treatment of the buildings. Sometimes, purpose-built noise barriers may be required under the EIA report.

The Noise Control Ordinance has more direct influence on property development during the construction stage. Because machinery and equipment are used in the construction processes, construction sites have always been a significant source of noise pollution. Regulations for specific controls such as the Noise Control (Air Compressors) and Noise Control (Hand Held Percussive Breakers) have been promulgated. These regulations require machinery to comply with the statutory noise standards and be issued with a noise emission label. It is necessary for those who may cause any noise in construction to obtain a Construction Noise Permit before the construction work starts. Construction work is grouped into either general construction work or percussive piling work. Both are regulated by a construction noise permit system. Restrictions on the operating hours of the piling works and general construction works require sophisticated site management from contractors in scheduling work to avoid noisy operations during sensitive hours, and the contract period should allow for such restrictions. Sometimes, a Construction Noise Permit could be critical to the completion of a construction project, and applicants could appeal to the Appeal Board if the Board unreasonably refuses to issue a permit.

Water Pollution

The main legislation controlling water pollution in Hong Kong is the Water Pollution Control Ordinance (WPCO) and its many subsidiary regulations. The Ordinance enables the government to declare water control zones covering the whole of Hong Kong and to carry out the necessary work to prevent pollution on a cost-recovery basis. Within the water control zones, a licence is required to discharge effluent. The criteria for a licence and its limits are set out in the Technical Memorandum of Effluent Standards. The 1993 amendment of the WPCO allows for control of the connection of waste water from private premises to the public sewerage, and also requires that private communal sewage treatment plants are properly operated and maintained. The WPCO has a major impact on the development planning for sizeable developments in places where connections to public sewerage are not feasible. Detailed requirements of design and construction of drainage/sewage are in the Buildings (Standards of Sanitary Fitments, Plumbing, Drainage Works and Latrines) Regulations made under the Buildings Ordinance.

Air Pollution

The Air Pollution Control Ordinance (APCO) cap 311 is the main piece of legislation to control pollution of the atmosphere so as to achieve the air-quality objectives of Hong Kong environmental policy. Similar to noise control, the major impacts of the APCO on property development are experienced during the EIA stage and during site layout design. The assessment report may restrict the use of a site, require the relocation of a building or change of building height. A proposed development involving fuel burning may be restricted because of difficulties in obtaining licences. Under section 10 of the Ordinance, the Air Pollution Control Authority (the Authority) may issue an air pollution abatement notice to the owner of a premise or to the person carrying out the activities at a premise, requiring him/her to cease or reduce the emission of air pollutants from the process. The construction process is not within the definition of Specified Process in Schedule 1 of the Ordinance, which requires a licence in order to carry out the process. However, some construction operations produce dust, smoke and other airborne toxic waste that are major sources of air pollution. If the effect is to cause safety problems and to disturb the normal activities of the public, an abatement notice may require the construction operation to cease or be modified. The Ordinance affects property development mainly at the construction stage because the construction process resembles factory manufacturing, involving the use of equipment and plants. For these reasons, there are several regulations made under the Ordinance to specifically control construction activities.

The purpose of the APC (Dust and Grit Emission) Regulation is to control dust emissions from works that have dust-emission potential. Because of the specific nature of the construction process, the general requirements of the APC (Dust and Grit Emission) Regulation and the abatement notice of the Ordinance do not serve the purpose of pollution control on construction works. The APC (Construction Dust) Regulation was introduced in June 1997, aiming at controlling the major source of air pollution from construction sites by requiring contractors to implement specified dust control measures. They include the installation of dust control systems, enclosing dusty materials and stock piles or spraying them with water or dust suppression chemicals, treating unpaved surfaces, and implementing good on-site housekeeping measures. Contractors have to notify the Authority before they can undertake certain specified construction work. The Regulation sets out control standards to be met in carrying out the work, and penalties for failure to give proper notice or meet the standards.

Asbestos Control

Control of asbestos in buildings relies on a combination of several pieces of legislation. They include the APC (Asbestos) (Administration) Regulations, Waste Disposal (Chemical Waste) (General) Regulations, Factories and Industrial Undertakings (Asbestos) Special Regulations, and the Factories and Industrial Undertakings (Notification of Occupational Disease) Regulations. The enforcing authorities for the APCO are the EPD and the Labour Department. The Labour Department is mainly concerned with the health and safety of workers. The Factories and Industrial Undertakings Regulations lay down various control measures and requirements for protective equipment, storage and distribution of asbestos materials and conditions for working with asbestos. The APC (Asbestos) (Administration) Regulations provide for the registration of asbestos consultants, supervisors, laboratories and contractors.

Since May 1993, the EPD has been operating a 'cradle-to-grave' licensing system to control the manufacture/import of asbestos, and keeps track of its sale and use, removal and disposal. Asbestos is known to be hazardous and is classified as chemical waste under the Waste Disposal Ordinance. The material was seldom used in construction in the last decade, and the control of asbestos is mainly concerned with the removal of asbestos. Under the APCO, the owner of a premise that contains suspected asbestos materials shall engage a registered asbestos consultant to carry out an investigation on the material and submit an asbestos investigation report. If the investigation confirms the existence of asbestos-containing materials, the owner should submit to the Authority an asbestos abatement plan before he/she carries out any intended abatement work. Only registered asbestos contractors are allowed to carry out asbestos removal works, and a licence is required for the disposal of asbestos in a landfill site.

Waste Disposal

The Waste Disposal Ordinance cap 354 and its subsidiary regulations provide a comprehensive framework for the management of waste from the point of creation to the final disposal of the waste. The Ordinance is enacted to protect the health and welfare of the public from any adverse effects associated with the storage, collection, treatment and disposal of all wastes (Hong Kong government, 1989). Construction waste takes up a large proportion of the overall solid waste in Hong Kong. According to the EPD Final Report 1991, the main sources of solid waste in construction works are building demolition, site clearance, excavation, building renovation and roadwork. Most of the construction wastes are inert granular that may become pollutants on land and in drainage/sewerage systems. Waste is classified into two categories: municipal waste and other specialized waste. Construction waste belongs to municipal waste. Asbestos waste is classified as 'chemical waste' under the Waste Disposal (Chemical Waste) (General) Regulations and is subject to more stringent controls under those Regulations. Depending on how the construction wastes are managed, it can be a valuable commodity for reclamation purposes, while it could also be an expensive nuisance in taking up valuable landfill space.

Waste management involves too many variables. The benefits are tangible and intangible, and have to be considered and valued within an overall view by the government in representing the public interest. The overall strategy for construction and demolition waste is similar to the municipal waste strategy – avoid, minimize, recycle and dispose. Construction waste is required to be properly collected and treated either by the relevant collection authorities (e.g. the EPD, Regional Council or Urban Council) for a fee or by licensed persons in the private sector. The relevant authority controls the use of dumping/ landfill sites. A charge is imposed for waste disposal at landfills, and any illegal dumping will be penalized. As with other pollution control ordinances in Hong Kong, waste disposal control is based on the 'polluter pays' principle and is backed up by a haul of sanctions and penalties for non-compliance.

Building Control

With advice from the EPD, the Buildings Department helps to enforce legislative control on environmental matters under the Buildings Ordinance cap 123. The Buildings Department may disapprove submitted building plans if they contravene any enactment, which includes all environmental control ordinances. The Building (Refuse Storage Chambers and Chutes) Regulations and the Building (Standards of Sanitary Fitments, Plumbing, Drainage Works and Latrines) Regulations made under the Buildings Ordinance allow the Buildings Department to require storage and handling facilities for solid waste and foul drainage, and waste treatment facilities in a building. These Regulations provide detailed requirements for the design of refuse chutes and private drainage works within buildings.

Before the enactment of the APC (Construction Dust) Regulations, the Building (Demolition Works) Regulations made under the Buildings Ordinance provided for *inter alia*, the prevention of dust nuisance and the removal of construction waste caused in the process of demolition. Contractors were and are still required to erect dust screens to protect the works and notify the EPD of the demolition works for monitoring through the Buildings Department. Under the (Construction Dust) Regulations, it is a mandatory requirement of a contractor to notify the EPD prior to commencement of works, failing which the contractor will be liable to a penalty.

The Building (Energy Efficiency) Regulations were first promulgated in 1995. They are aimed at regulating commercial and hotel buildings, so that they are designed to achieve energy efficiency. The criteria are measured against the overall thermal transfer value. The external walls and roofs are designed to achieve at least the minimum standard required to avoid waste of energy. The Regulations have been criticized in that the requirements are too rigid in terms of choice of materials, colour scheme and stifle innovative options in building design. The Regulations should have allowed more freedom to building designers to achieve energy efficiency through an overall design involving the combination of building shading, orientation, building services systems and property management. The Hong Kong Building Environmental Assessment Method (HK-BEAM) discussed at the end of this chapter could provide an insight into this approach.

ADMINISTRATIVE CONTROL

Most of the environmental laws on pollution control in Hong Kong are based on the 'command and control' philosophy that requires direct supervision and enforcement by regulators. The EPD is responsible for the implementation of most of the measures in the aforementioned pollution control legislation. It aims to provide a 'one-stop shop' service to integrate pollution control activities. It is also responsible for processing construction site permits and licences, and carrying out multidisciplinary inspection visits to premises. The EPD has recently set up Local Control Offices (LCO) to support its control activities. The LCOs help to strengthen communication with local communities and enhance the EPD's position for enforcement work at a district level.

As part of the EPD structure, the Professional Persons Environmental Consultative Committee (ProPECC) was set up in October 1991 and is chaired by the Director of Environmental Protection. The Committee consists of representatives from relevant professional institutes and business organizations. It provides a forum for the exchange of views on all environmental matters related to the duties and practices of members of the institutes/organizations. One of the most important contributions from the Committee is the publication of ProPECC Practice Notes, which have been widely distributed to professionals and organizations in relevant fields. The Practice Notes provide guidance on technical standards required and administrative procedures for practice.

▮ VOLUNTARY COMPLIANCE

As explained at the beginning of this chapter, the most effective and fundamental cure to prevent environmental damage is to attend to the root of the problem through educating our current and future generations to protect our environment and to support sustainable development. All environmental laws require the full support and understanding of the public to achieve the targets set out in the legislation. If such support could be developed to enable voluntary involvement, it would be a very positive sign of hope for our environment. Professional institutions and business organizations involved with property development are beginning to take responsibility towards our environment. Modelled on the UK Building Research Establishment's Environmental Assessment Method, the HK-BEAM was set up in the early 1990s and is operated by the Centre of Environmental Technology Ltd., the executive arm of the Private Sector Committee on the Environment. It is developed and funded by the Real Estate Developers Association with the technical assistance of the Department of Building Services Engineering of The Hong Kong Polytechnic University, the Welsh School of Architecture and the ECD Energy and Environment Ltd. The scheme is a voluntary initiative by property owners to assess the overall environmental performance of new and existing commercial buildings. It sets the criteria for good environmental performance in buildings. The aims of the HK-BEAM are shown in the Appendix.

The ISO quality management system has already gained recognition worldwide. The ISO 14000 series is a voluntary environmental management system. The purpose is to create global management standards for controlling and preventing pollution to our environment in the respect of air, water and land. The ISO 14000 system would be the best environmental management system to promote voluntary compliance. Certification under the ISO 14000 system could be established as an incentive scheme. Potential polluters holding the ISO certificate could apply for partial exemption from compliance with statutory environmental requirements. After benchmarking exercises to evaluate the equivalent credit weight of the ISO 14000 certificate, it could also be utilized by potential polluters as trade-offs to pay less under the 'polluters pay' principle.

CONCLUSION

In general, most of the pollution control measures affect construction activities in one way or another. Among environmental legislation, EIA and land use planning laws have the most significant impact on the planning and design of large-scale property development. On the other hand, most of the pollution control measures affect property development mainly during the construction stage. The requirements under the pollution control ordinances can be met with extra time and costs for the project. It often implies a longer time required for approvals and issuing of licences, and this becomes critical to the development programme. To lessen the impact of the pollution control legislation on property development, specific regulations and clear practice guidelines reinforced with an efficient administration system will help the construction industry to plan ahead and avoid time being wasted in speculating on government requirements. As illustrated by the case studies, although the 'market-based' environmental management philosophy provides incentives and flexibility for compliance, it is not always the best option for all situations. Certainty that comes with statutory provisions can provide strong support for regulators to enforce pollution control, and in some cases, certainty could avoid unnecessary planning disputes.

For a long-term solution to protect our environment, environmental policy should explore education to promote self-awareness of environmental protection. Independent assessment and certification by voluntary organizations will provide recognition of the commitment of the property owner/developer for environmental friendliness and quality. It sets an excellent example for the 'market-based' environmental management philosophy to comply with environmental policy. Being a voluntary initiative, it often carries a mission with objectives that are above the standards set by legislation. Being a self-regulated and self-imposed task, it is hoped that the incentive to comply and to achieve the goal will be higher. It is a very positive model to generate a sense of responsibility towards our environment beyond what the legislation imposes upon us. All legislation will be in vain if all those being governed have no heart for it.

ACKNOWLEDGEMENTS

Part of the research work in this chapter was supported by a research grant from the Hong Kong Polytechnic University.

▮ REFERENCES

Aglionby, A. 1997. Environmental risk avoidance: officer's and director's liabilities. The HKELA/Pearson Professional Conference, Hong Kong.

Bachner, B. 1996. Economics and the regulatory environment 1996. *The Hong Kong Environmental Law Association Newsletter,* Vol. 3.3.

Barrow, B. and J. Cottrell. 1996. Making environmental law in Asia more effective. A published participants' report at a regional workshop, 4–8 March 1996, Hong Kong. Hong Kong: Centre of Urban Planning and Environmental Management.

Centre of Environmental Technology Ltd. 1996. *HK-BEAM (new offices), an environmental assessment for new air-conditioned office premises, version 1/96, and 1996.*

——. 1996. *HK-BEAM (existing offices), an environmental assessment for existing air-conditioned office premises, version 2/96, and 1996.*

Chan, E.H., M.W. Chan and B. Chung. 1996. Pilot study report: global comparison of building control systems. Research monograph presented at the Fourth World Congress of Building Officials, November 1996, Hong Kong, p. 13.

COM. 1992. Towards sustainability – an EC programme of policy and action in relation to the environment and sustainable development. *COM (92) Final,* 27 March 1992.

Finch, E. 1992. Environmental assessment of construction projects. *Construction Management and Economics* 10, no. 1 (January 1992).

Gilpin, A. 1995. *Environmental impact assessment: cutting edge for the twenty-first century.* Cambridge: Cambridge University Press.

Hong Kong government. 1989. *White Paper: Pollution in Hong Kong – a time to act.* Hong Kong: Government Printer.

——. 1994. Chapter 10: Conservation. In *Hong Kong planning standards and guidelines* (1994 edition), Hong Kong.

Secretary of Planning, Environment and Lands Bureau (SPEL). 1996. Web-site *(http://www.pelb.gov.hk/index.htm)* of the Planning, Environment and Lands Bureau of the Hong Kong SAR government. Searched on 24 March 1998.

Smith, D.B. and Mieke van der Wansem. 1995. *Strengthening EIA capacity in Asia: environmental impact assessment in the Philippines, Indonesia, and Sri Lanka.* Washington, D.C.: World Resources Institute.

Uff, J., H. Garthwait and J. Barber. 1994. *Construction, law and the environment.* London: Willy Chancery.

Wong, F. 1997. Implementing CLP's environmental management system. The HKELA/Pearson Professional Conference, Hong Kong.

Yip, S. and J. Knight. 1996. The EIA bill and nature conservation: policy, criteria and stewardship. *The Hong Kong Environmental Law Association Newsletter*, Vol. 3.1.

APPENDIX

The following are the aims of the HK-BEAM, as stated in the publication *HK-BEAM versions 1/96 and 2/96:*

- *To reduce the long-term impact that buildings have on the environment.*
- *To raise awareness of the large contribution that building makes to global warming, acid rain and depletion of the ozone layer, as well as local environmental issues.*
- *To promote and encourage energy-efficient buildings, building services systems and equipment.*
- *To reduce the unsustainable uses of increasingly scarce resources such as water, timber, and other natural materials.*
- *To improve the quality of the indoor environment and hence health and well-being of the occupants.*
- *To provide recognition for buildings where the environmental impact has been reduced.*
- *To set targets and standards that are independently assessed and so help to minimise false claims and distortions.*
- *To enable developers, operators and users to respond to a demand for buildings that have less impact on the environment, and to help stimulate such a market.*

13

Environmental Impact Assessment Ordinance – An Introduction

Chi-sun Poon and Xiang-dong Li

INTRODUCTION

The Environmental Impact Assessment (EIA) Ordinance was enacted in early 1997 to provide for assessing the impact on the environment of certain project proposals, for the protection of the environment. The Technical Memorandum on Environmental Impact Process, which sets out the principles, procedures, guidelines, requirements and criteria of the assessment process, was also published by the government to complement the implementation of the Ordinance. Some of the materials in this chapter are adapted from the Ordinance and the Technical Memorandum.

DESIGNATED PROJECTS

There are two types of projects specified in the Ordinance as designated projects.

Designated Projects Requiring An Environmental Permit

The Ordinance specifies in its Schedule 2 that for certain projects to be designated projects, an Environmental Permit from the Environmental Protection Department (EPD) must be obtained before their commencement. Normally, an EIA study is required to be carried out before the project proponent can apply to the EPD for an Environmental Permit. However, the Ordinance also provides flexibility to the project proponent to apply to the EPD for permission to apply directly for an Environmental Permit if sufficient justifications can be provided. The types of designated projects include:

Roads, railways and depots
Airports and port facilities
Reclamation, hydraulics and marine facilities, dredging and dumping
Energy supply
Water extraction and water supply
Sewage collection, treatment disposal and reuse
Waste storage, transfer and disposal facilities
Utility pipelines, transmission pipelines and substations
Waterways and drainage works
Mineral extraction
Industrial activities
Storage, transfer and trans-shipment of fuels
Agriculture and fisheries activities
Community facilities
Tourist and recreational developments
Major residential and other developments
Major decommissioning projects

(A more detailed listing can be obtained from the government's Web site, www.inf.gov.hk/epd/index.htm, or from the Schedule of the EIA Ordinance.)

Major Designated Projects Requiring An EIA Report

Additionally, the Ordinance stipulates in Schedule 3 that before the commencement of the following major designated projects, thorough EIAs must be conducted.

The projects include:

- Engineering feasibility study of an urban development project with a study area covering more than 20 hectares or involving a total population of more than 100 000.

- Engineering feasibility study of a redevelopment project with a study area involving an existing or new population of more than 100 000 people.

PREPARATION OF PROJECT PROFILE

The Ordinance requires the proponent of a designated project to prepare a description of the project (project profile) in accordance with the Technical Memorandum. The purpose of the project profile is to enable the EPD to determine (1) the scope of the environmental issue associated with the designated project that should be addressed in the EIA study, if such a study is required by the EPD, or (2) whether the applicant can proceed directly to apply for an Environmental Permit. The project proponent may then apply to the EPD for an EIA study brief, or for permission to apply directly for an Environmental Permit.

Contents of a typical project profile

Basic information
Outline of planning and implementation programme
Possible impact on the environment
Major elements of the surrounding environment
- Outline existing and planned sensitive receivers and sensitive parts of the natural environment that might be affected by the proposed project
- Outline major elements of the surrounding environment and existing and/or relevant past land use(s) on site that might affect the area in which the project is proposed to be located

Environmental protection measures to be incorporated in the design, and any further environmental implications
- Describe measures to minimize environmental impacts
- Comment on the possible severity, distribution and duration of environmental effects
- Comment on any further implications

EIA STUDY BRIEF

Upon receipt of the application by the proponent, if the EPD decides that an EIA study is required, the EPD will issue to the applicant an EIA study brief. The study brief would normally set out the following:

- the scope of environmental issues to be addressed;
- the requirements that the EIA study will need to fulfil;
- the necessary procedural and reporting requirements;
- the methodologies or approaches that the EIA study needs to follow;
- other matters that the EIA study must take into account.

▮ PERMISSION TO APPLY DIRECTLY FOR AN ENVIRONMENTAL PERMIT

If the project proponent applies for permission to apply directly for an Environmental Permit, the EPD, after considering the project's details and other environmental issues, may approve or reject such an application. Normally, the EPD will approve an application if it is satisfied that the environmental impacts of the proposed project have been adequately assessed by a previous EIA study, or the environmental impact of the project is unlikely to be adverse and the mitigation measures described in the project profile are adequate.

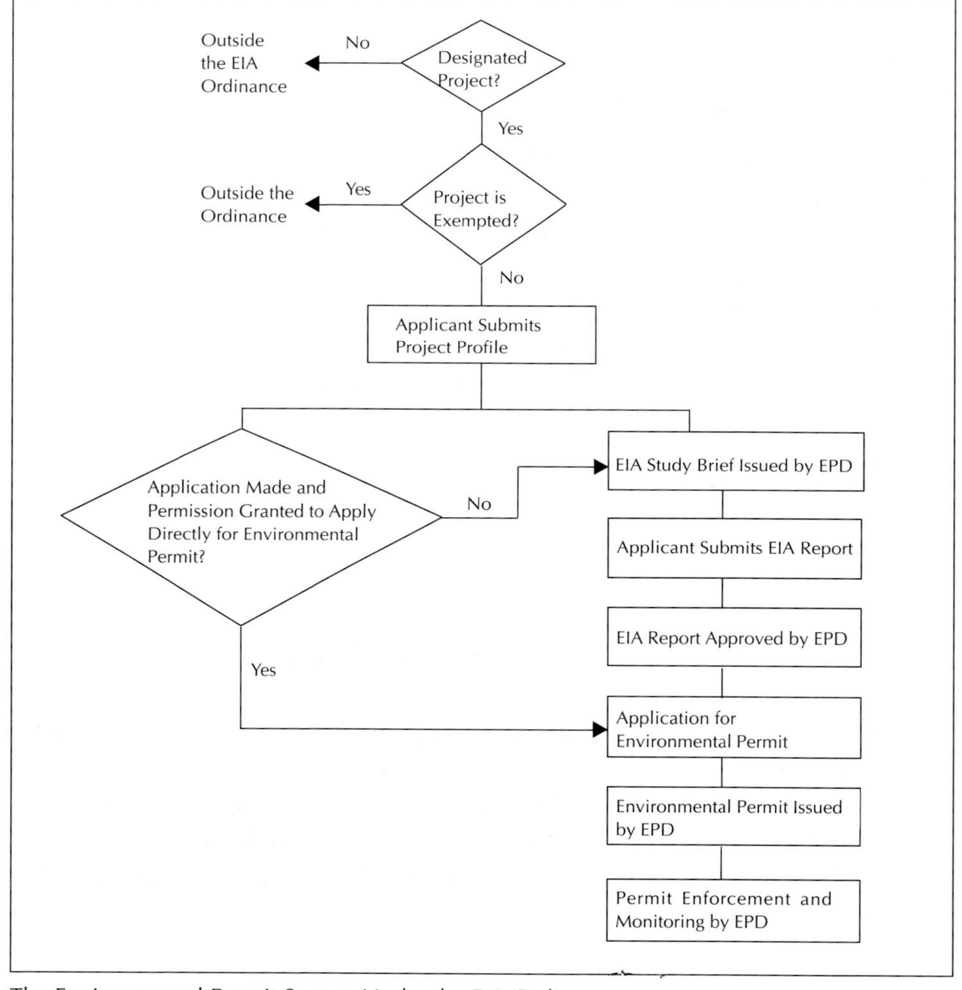

The Environmental Permit System Under the EIA Ordinance

PREPARATION OF AN EIA REPORT

If an EIA study is required, the EIA study would need to be carried out, usually by professional environmental consultants on behalf of the project proponent, in accordance with the EIA study brief and the Technical Memorandum. The EPD requires the project proponent to submit an EIA report that should provide a detailed assessment of the likely environmental impact and environmental benefits of the project for the EPD's approval before the proponent can make an application for an Environmental Permit. The approved EIA report will be placed on a register and made accessible to the public. The contents of a typical EIA report are shown as follows.

Contents of a typical EIA report

Executive summary in English and Chinese
Introduction
Description of the project
Environmental legislation, policies, plans, standards and criteria
Description of the environment
Description of assessment methodologies
Identification of environmental impacts
Prediction and evaluation of environmental impacts
Mitigation of adverse environmental impacts
Definition and evaluation of residual environmental impacts
Environmental monitoring and audit
Conclusions and recommendations
Schedule of recommended mitigation measures
Appendix — Responses to comments received

APPLICATION FOR AN ENVIRONMENTAL PERMIT

A proponent who wishes to carry out a designated project listed in Schedule 2 of the Ordinance must obtain an Environmental Permit from the EPD before the project can commence. The EPD may approve or reject an application for an Environmental Permit. When the EPD approves an application, it may issue the Environmental Permit with conditions outlined in Schedule 4 of the Ordinance.

Matters that may be specified in an Environmental Permit

1. The design, alignment, plan, layout or visual appearance of a designated project.
2. The physical scale, scope or extent of a designated project.
3. The methods for carrying out a designated project, or the timing, phasing or order of the stages of a designated project.
4. The amount, rate or quality of a discharge, emission or deposit of pollutants or wastes from a designated project, including the colour or temperature of, or amount or concentration of, a contaminant, impurity or other substance in the discharge, emission or deposit.
5. Limits on the strength, severity or level of the environmental impact of a designated project.
6. The mitigation of the environmental impact of a designated project, and the timing, phasing or order of mitigation measures, including:
   a. pollution control or environmental protection equipment, processes, systems, practices or technologies;
   b. equipment, processes, systems, practices or technologies for the prevention, reduction, reuse, recovery and recycling of wastes or waste water.
   c. equipment, processes, systems, practices or technologies for the management of wastes, including the storage, treatment or disposal of wastes;
   d. acoustic barriers and enclosures, noise insulation equipment, or equipment, processes, systems, practices or technologies for the avoidance, prevention, reduction, minimization or control of noise;
   e. equipment, processes, systems, practices or technologies for the avoidance, prevention, reduction, minimization or control of air pollution; or
   f. processes, systems, practices, procedures or technologies for the conservation, preservation or protection of flora, fauna, ecological habitats, sites of special scientific interest or of cultural heritage importance, or resources.
7. Pollution control, environmental protection or other mitigation measures to be carried out outside the site of a designated project.
8. Compensation or restoration measures for the conservation, preservation or protection of flora, fauna, ecological habitats or other ecological resources, including replanting, relocation, re-establishment or rehabilitation measures, to mitigate an adverse environmental impact of a designated project.
9. Landscaping measures to mitigate the environmental impact of a designated project.
10. Programmes or exercises for monitoring the environmental impact of a designated project or the effectiveness of measures to mitigate its environment impact, whether such impact may occur within or outside its physical boundary or site, and the review and audit of data and information derived from such programmes or exercises, including specification of:
    a. the parameters or impact to be monitored;
    b. the frequency of monitoring, or the procedures, practices, methods or equipment to be used for monitoring, including the maintenance and calibration of such equipment, quality assurance and laboratory accreditation procedures;
    c. the standards or criteria to be used for evaluating and auditing monitoring data;

(to be continued)

d. plans and procedures for action in response to the results of such monitoring programmes or exercises, including action to intensify or increase monitoring, inspect or investigate revealed or indicated problems, or take remedial measures to address such problems;
e. the nature, format or frequency of the reporting of the results and findings of monitoring or action plans and procedures.

11. Equipment, methods, processes, systems, procedures or practices for the construction, operation, use, implementation or decommissioning of a designated project.
12. Requirements for the training, qualifications or expertise of personnel involved in the carrying out of a designated project.
13. The preparation of management plans, procedures manuals or other materials and documents for guiding and regulating the carrying out of a designated project.
14. Environmental studies, investigation or information to be obtained and submitted during the carrying out of a designated project.
15. The release to the public of reports on monitoring or auditing work or other reports or information in relation to the assessment or carrying out of a designated project.
16. The requirements for carrying out environmental monitoring by accredited laboratories, or environmental audit by qualified personnel.
17. The requirements of the implementation and completion of mitigation measures to be checked and certified by qualified personnel, and for the submission of certified reports on the status of the implementation of mitigation measures.

The Ordinance also sets out the procedures for the surrender of the Permit, the issue of further Environmental Permits upon change of responsibility, and application for a variation of or cancellation of an Environmental Permit.

The Environmental Permit may impose requirements for the monitoring and/or auditing of the environmental impacts of the project for verification of the prediction previously made in the EIA study or the effectiveness of the measures proposed to mitigate its environmental impacts.

APPEAL

The Ordinance provides for the proponent rights to appeal to an independent Appeal Board if he/she is aggrieved by the decision of the EPD with regard to:

- the requirements of the EIA study brief;
- the decision not to issue an Environmental Permit;
- the decision not to permit a direct application for an Environmental Permit;
- the decision not to approve an EIA report.

The Ordinance also lays down the procedures on how the appeal is conducted.

PARTICIPATION OF THE PUBLIC IN THE EIA PROCESS

Under the EIA Ordinance, the public has ample opportunities to be involved in the EIA process. First, the proponent of a designated project is required to advertise the availability of the project profile and the EIA report in an English and a Chinese newspaper. The public can comment on the project profile and the EIA report via the EPD. Second, the EPD is required to keep a register containing project profiles, EIA study briefs and EIA reports of the designated projects, and the register is open for inspection by the public.

BUILDING WORKS AFFECTED BY THE EIA ORDINANCE

Generally, the major focus of the EIA Ordinance is on civil engineering projects, specific industrial projects and other projects that may give rise to significant environmental impacts. Building works that are associated with common residential, industrial or commercial activities that do not pose much environmental concern are excluded, except for:

- a residential or recreational development, other than New Territories exempted houses, within Deep Bay Buffer Zone 1 or 2;
- a large residential development of not less than 2 000 flats and not served by the public sewerage network;
- building work in a country park or a special conservation area.

CASE STUDY: SHEUNG SHUI SLAUGHTER HOUSE

This case study illustrates the scope and procedures for carrying out an EIA.

The Proposed Development

The proposed slaughterhouse lies between Road 2801 (Po Wan Road) and the Kowloon-Canton Railway in Sheung Shui. The site area is approximately 14 hectares and the total plant room working floor area will be 38 000 m^2, excluding the car park, unloading area and waste water pre-treatment plant. Detailed drainage design will be required during the construction phase as the entire site is situated on low-lying land with the potential of flooding in the rainy season. The proposed plant is designed to accommodate approximately 15 000 pigs and 2 400 heads of cattle. The slaughtering capacity per day will be an average of 5 000 pigs and 400 heads of cattle.

The project area will comprise four quite separate sections by the functions within. They are as follows:

1. *Lairage* (three storeys)
   An agricultural-style environment where pigs and cattle will be kept prior to slaughter. The area allowed for this is approximately 25 000 m^2.
2. *Slaughter Hall and By-products Plant* (two storeys)
   This is considered to be the business end of the development. Here the pigs and cattle will be slaughtered, dressed and by-products separated.
3. *Administration and Amenities Block* (three storeys)
   This will be situated at the 'clean' end of the proposed development, comprising facilities such as offices, shower facilities, changing rooms, mess rooms, kitchen, and laundry.
4. *Kowloon-Canton Railway Goods Yard*
   This will be situated to the west of the slaughterhouse and will contain facilities such as unloading platforms, holding pens, lorry waiting areas, and mess rooms for handling staff.

EIA Study

Background

The Architectural Services Department (ArchSD) commissioned an environmental consultancy company (Mott MacDonald Hong Kong Ltd.) to carry out an EIA of the proposed Sheung Shui Slaughter House, the ancillary by-products plant, and on-site and off-site activities associated with the construction and operation of the proposed slaughterhouse.

Purpose Of The EIA

The purpose of the EIA was to assess the likely environmental impacts of the proposed development of the Sheung Shui Slaughter House, and to determine suitable measures to be incorporated into the design, construction and operation of the scheme to mitigate any adverse impacts to a level that is considered to be acceptable and in compliance with statutory controls.

The EIA study examined all environmental aspects of construction and operation in order to identify the nature, extent and location of environmental impacts as far as possible in view of the available information. All issues stated in the study brief from the EPD were addressed. The quality of the environment is described in terms of the following:

- noise impact;
- air quality/odours impact;
- water quality impact;
- visual impact and land use.

Objectives Of The EIA

The objectives of the EIA are as follows:

1. to describe the proposed development and the requirements for carrying out the development;
2. to identify and describe the elements of the community and environment that are likely to be affected by the development;
3. to identify, assess and evaluate the net environmental impact (i.e. after practicable mitigation) and cumulative effects expected to arise during the operation and construction phases of the development in relation to neighbouring land uses and water bodies;
4. to propose infrastructure provision or mitigation measures in order to minimize pollution, environmental disturbance and nuisance during the construction and operation of the development;
5. to identify, assess and specify methods, measures and standards to be included in the detailed design and in the contractual agreement between the government and the proponent, which are necessary to mitigate these impacts and reduce them to acceptable levels;
6. to design and specify the environmental monitoring and audit requirements necessary to ensure the effectiveness of the environmental protection measures adopted;
7. to carry out any necessary environmental survey and baseline monitoring work to achieve the objectives; and
8. to quantify, where necessary, by use of models or other predictive methods, the environmental impacts arising from the construction and operation of the development.

The environmental assessment was conducted in two stages. The Initial Environmental Assessment Report involved preliminary evaluation of the environmental impacts arising from the construction and operation of the proposed plant. Those impacts requiring more detailed evaluation were identified, and these were fully evaluated in the detailed EIA study.

Major Conclusions And Recommendations

Visual Impact

The site for the slaughterhouse has already been allocated and there is no need for other land users to be relocated. This development will have a significant visual impact on the local environment. However, this must be seen in the context of the adjacent land use of a sewage treatment plant. The EIA study recommended screen planting to mitigate the visual impact, and this should be in place as early as possible as this will also provide an opportunity to mitigate construction-stage impacts. Planting around the border of the site could start well in advance of other construction works.

Noise

There should be no need for construction at night, and the daytime construction noise should be within the standards recommended by the EPD. Although the noise impact during construction will not be significant, it is recommended that good construction management practice should be adopted.

During the operational phase, the fabric of the buildings will contain the process noise occurring inside the main structure. However, problem areas are likely to be the meat despatch area, the pig waiting lairage, and the railway goods yard.

Noise prediction modelling results have indicated that acceptable noise levels can be achieved after the installation and operation of the recommended mitigation measures, such as design of building face direction and installation of acoustic panels at various noise sources. Additionally, operational procedures can also be adopted that would also help to reduce noise levels.

The standard for road traffic noise level is not likely to be exceeded, and no mitigation of this noise is required. However, rail noise has been identified as a major contributor to the overall noise level predicted, and it is recommended that noise barriers be constructed at the yard.

Air Pollution

The EIA study has concluded that air pollution during the construction phase will not be a major issue.

Odours from plant operations have, however, been the key issue of this study. A comprehensive study of odour emissions has been carried out, including an odour panel to identify levels of emissions, and computer modelling. The study has concluded that the plant must be enclosed, with the provision of ventilation systems and odour control units to meet the EPD standards.

Conventional odour control units will be suitable for most of the plant operations to achieve an efficiency of more than 90%. However, the rendering plant will need a much higher level of odour removal efficiency (about 99.5%), and incineration of process exhaust gases may be required to achieve this. It should be noted that the assessment has used a very high value for odour generated by the rendering plant for a 'worst scenario' analysis. This assumption may not be realistic in practice. It is recommended that further work be carried out during the development of the detail design to obtain further data of odour emissions from plants to decide the design of odour control units. It will be more cost-effective to design a plant with a lower level of odour than to install expensive odour removal equipment.

There will also be a boiler at the plant, but the emissions from its operation will be well within the required standards. Vehicle exhaust emissions also will not exceed the required standards.

Water Pollution

The EIA study has recommended control measures to avoid unacceptable or undesirable water pollution impact during the construction stage.

It has been identified that all liquid waste can be treated to achieve the discharge contents prescribed in the Technical Memorandum by the EPD. Alternative treatment and process designs with a well-balanced effluence were identified. The discharge will be further treated at the Shek Wu Hui Sewage Treatment Works.

Land Use Implications

The impact assessment of the air and water quality has identified that there are land use/allocation implications because several options, in terms of odour control mitigation measures, result in different levels of odours around the site.

For water-quality impacts, the size of the waste water pre-treatment system will be larger than the allocated area of 250m^2, but the plant can be contained in the revised area of 0.4 hectare, which the government has indicated is available for the proposed plant.

Overall Conclusion

The general conclusion of the EIA study was that the construction and operation of the slaughterhouse at the proposed Sheung Shui site can meet the environmental protection statutory requirements, provided that the considerations and mitigation measures are taken in the design, construction and operation stages. It is particularly important to adopt design/build and operate/transfer contracts for this development project.

■ CONCLUSION

EIA is a structured and systematic approach to the assessment and control of possible environmental impacts arising from proposed development projects. The enactment of the EIA Ordinance in 1997 marked a new era in EIA study in Hong Kong. Professionals who are involved in the planning, development, design and construction of engineering and building projects in Hong Kong must be familiar with the requirements of the new system. The case study of this chapter illustrates the scope and procedures for carrying out an EIA. The impacts at both the construction and operational phases of the proposed project have to be considered and evaluated. An EIA study would normally follow the study brief issued by the EPD, and covers various environmental aspects such as noise, air, water, waste, visual and land use issues. An EIA study does not only assess the possible environmental impacts of the project, but must also evaluate and recommend possible mitigation measures to address such impacts.

■ REFERENCES

Architectural Services Department, Hong Kong government. 1995. *Sheung Shui Slaughter House environmental impact assessment (final report).* Mott MacDonald Hong Kong Ltd. in association with EBC Hassell.

Environmental Impact Ordinance (Chapter 499) of Laws of Hong Kong.

Environmental Protection Department, Hong Kong government. 1997. *Technical memorandum on environmental impact assessment process.*

14

Building Control To Enhance The Environment

Wong Wah Sang

INTRODUCTION

Le Corbusier (1927) advocated regulating lines in architecture to serve 'to make very beautiful things'. An orderly control is important, but now building regulations in Hong Kong as the prime legislative control have no such equal intent. Greenstreet (1996) investigated building codes in the United States and found that statutory control added heavy responsibilities on to designers and tended to discourage innovative designs for buildings. Though all societies exert control to attain standards for the common good, excessive sophistication often becomes heavy legislative obstacles to achieving innovative designs.

As stated by Alan Downward (1992), building control was developed from concerns ranging from fear of fire to the health of occupants. In the UK, concern was from poor sanitation and drainage to provision of air spaces around buildings. The development has been still under British standards for health, safety and quality. In Hong Kong, the urban context is characteristic of a high-density fabric where people are always competing for space to build different physical forms to accommodate various activities in order to sustain the city.

Anson Chan (1999), the Chief Secretary for Administration, stated that the 'challenge is to provide safe and well-managed accommodation for our community in a very crowded environment'. Architects and related professionals as well as property developers generally follow the government's rules for design and development of the built environment. Occasionally, some modifications are allowed, provided that they do not contravene the general provisions contained in the Building Code. However, as society progresses, the underlying principles behind the Code have to be adjusted to aim at a sustainable environment. In view of the above, it is the objective of this chapter to explore and discuss deficiencies in the present system of building control in order to make positive suggestions for improvement or alterations to cope with the advancement of the modern metropolis of Hong Kong.

BACKGROUND OF RESEARCH

Buildings in Hong Kong usually conform to the provisions of the Buildings Ordinance and Regulations, reflecting a physical interpretation of the Code on the external façade. The Code in Hong Kong concentrates on density, safety, health and environment as well as other allied legislation. As stated by Gordon Siu (1999), the Secretary for Planning, Environment and Lands, 'one of the Government's principal policy objectives is to ensure that buildings in Hong Kong are functionally designed, are environmentally friendly and are safe'. Each of these aspects is addressed by various Building Regulations or Codes of Practice to affect both the internal environment and external expression of buildings. Due to the high property prices in Hong Kong, these regulations are often exploited directly or indirectly to achieve maximum development potential. In such design process, other architectural issues, such as environmental concern, cultural expression, site context, and other creative and innovative ideas, may become factors of secondary importance.

In line with this, Lung (1998) commented that Hong Kong had been developing with 'borrowed time' and a 'borrowed place'. People worked mainly out of greed and self-profit. However, new aspects of development might be opened up after the return of sovereignty of Hong Kong to China in 1997. Not mentioned by Lung was the possibility of improving building control to cater for cultural vibrancy as well as environmental consciousness.

Hall (1996) stated that people were concerned about the quality of their immediate environment. The government has the authority to decide on building control. However, there should be a balance between private and communal interests depending on circumstances. An equilibrium state with a balance between control and freedom is important, and this will depend on local criteria. In Hong Kong, the high-density urban fabric produces a unique set of criteria to be tested by survey, from different perceptions in the urban environment.

Hall also looked at the specification of density as a means to control the physical environment of an area. This can be expressed as number of buildings, rooms or persons per unit. This is a measure that is easily calculated and communicated. However, in Hong Kong, the built environment is quantified by 'gross floor area' (GFA), which forms the basis for architects to achieve maximum development potential in each project. The definition of GFA is stated by the Building Authority (BA). However, there is still some grey area, depending on whether functions stated for certain areas are to be included or not in GFA calculation. The BA and the Lands Office may also take different views in some aspects of this GFA calculation. So it appears that a redefinition or modification of the present GFA calculation is necessary to cope with modern needs for innovative design. The unified views of different government authorities are also necessary to give consistency to a development.

With expectations for a sustainable world in the twenty-first century, these values of environmental concern, cultural expression, local site character and other creative and innovative ideas have to be brought back to our built environment. Improving the Code or adding special clauses to the Lease Conditions or Town Plan are all possibilities to create better buildings.

▮ SURVEY ON BUILDING REGULATIONS AFFECTING DESIGN OF BUILDINGS

As architects are the professionals who interpret and execute architectural design work, a survey on building regulations affecting the design of buildings was carried out in the second quarter of 1999 among all members of the Hong Kong Institute of Architects (HKIA). Opinions were collected as a database for possible future improvement of building regulations. Sixty-one questionnaires were returned. Most respondents were employed in the private sector and had more than five years' practical experience.

The following are the findings of the survey.

Calculation Of GFA (See Figure 1)

Most respondents agreed to have the following items discounted in the calculation of plot ratio under the Buildings Ordinance:

1. External wall surfacing and finishes
   - External wall finishes such as tiles, plaster
   - Cladding and the space between the external wall and the cladding
2. Environmental protection devices
   - Sunshading device
   - Noise shield
   - Heat insulating material for the external wall
3. Internal parts relating to building services
   - Lift shaft
   - Services areas e.g. pipe ducts (electrical cable ducts, water pipe ducts, A/C pipe ducts, etc.)
4. Amenity facilities
   - Window flower box that projects more than 500mm
   - Observation deck and related independent transportation route for the public
   - Podium garden
   - Incorporated Owners' Committee meeting room

Most respondents disagreed to have the following items discounted in the calculation of plot ratio under the Buildings Ordinance:

1. Lift lobby
2. Staircase
3. Exposed structure such as columns and beams on the external side of the building
4. True bay window with same floor level as the rest of the room
5. Balcony

Environmental Issues

1. Under Building (Planning) Regulation (1998) 33, the area of the enclosed verandah, etc., is included in the calculation of the lighting and ventilation of rooms. Most respondents agreed that this was appropriate.
2. Practice Note for Authorized Persons and Registered Structural Engineers (PNAP) 172 (5/95) imposes energy-efficiency requirements through the Building (Energy Efficiency) Regulation (1995). It requires commercial or hotel buildings to be designed to have suitable overall thermal transfer

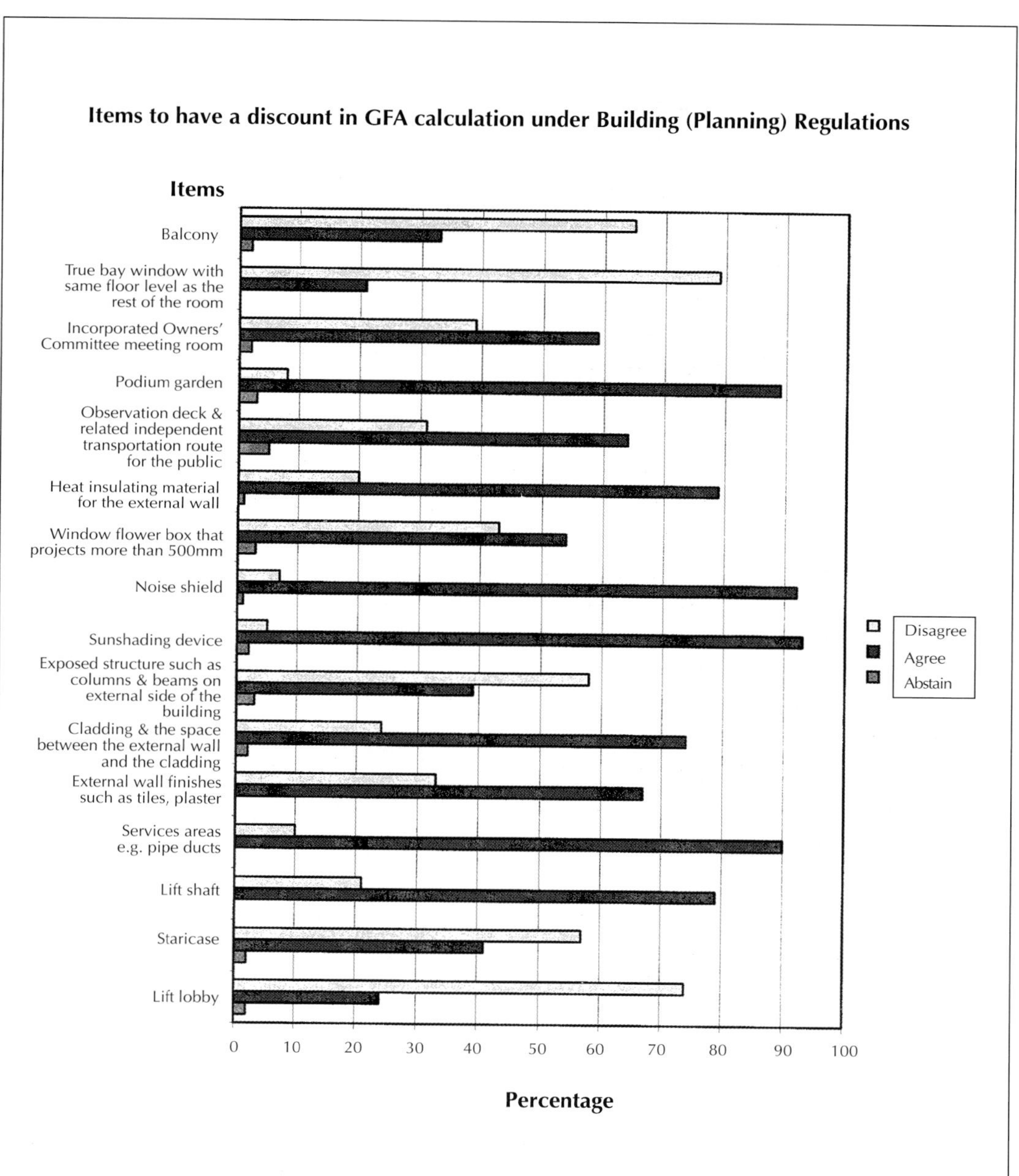

Figure 1 Survey on Building Regulations Affecting Design of Buildings

values (OTTV). Most respondents agreed that this should be applied to residential building as well.

3. Most respondents disagreed that the lighting and ventilation regulations should consider the orientation and surrounding land use.
4. Most respondents disagreed that the prescribed-window regulations should consider the orientation and surrounding land use.
5. A prescribed window facing a street of 4.5m or more in width is deemed to satisfy the lighting/ventilation requirement. However, a prescribed window facing an adjacent building is subject to the criteria of the lighting angle/rectangular horizontal plane, which very often requires the building to be much further apart than 4.5m. Most respondents thought that this was unfair.

Planning And Functions

1. Under Buildings Ordinance (1998) 16(1)(g), the BA may refuse to give approval for any plans of building works whereby the carrying out of the building works shown thereon would result in the building differing in height, design, type or intended use from the one previously existing on the same site or those in the immediate neighbourhood. Most respondents agreed that this should be modified.
2. Most respondents agreed that it would be fair for a small class A or B site to become a large class B or C site when combining with an adjacent site.
3. Most respondents agreed that more definitions should be included in warehouse regulations (e.g. the definition of megamarkets, shopping areas and showrooms).
4. Most respondents agreed that facilities for the disabled should be provided in the common areas of residential buildings.

Street-related Matters

1. Under Building (Planning) Regulation 5(2), the BA may require the provision of an access lane or access road within the site of any new building. Most respondents disagreed that more conditions should be added to this regulation.
2. Under Building (Planning) Regulation 26, where the width of an existing street in front of any new building is less than 4.5m, no part of such building shall be nearer to the centre line of the street than 2.25m. Most respondents were satisfied with the existing regulation.

3. Most respondents agreed that an open space next to a site should be considered as a street when considering the classification of the site.
4. Most respondents agreed that Building (Planning) Regulation 28 regarding the service lane was outdated.

Internal Space Arrangement

1. Under Building (Planning) Regulation 24, every room used or intended to be used for the purpose of an office or for habitation in any building shall have a height of not less than 2.5m measured from floor to ceiling: provided that there shall be not less than 2.3m measured from the floor to the underside of any beam. Most respondents disagreed that this should be extended to other types of building.
2. Under Building (Planning) Regulation 39(3)(d), the main staircase of every building which exceeds one storey in height shall not have more than 16 steps in any flight without the introduction of a landing. Most respondents were satisfied with the existing regulation.
3. Most respondents thought that 3m was the most comfortable clear ceiling height, including finishes (e.g. false ceiling), for a residential building.

Kitchen

1. Most respondents disagreed that the definition of kitchen, kitchen fixtures, fitment and equipment requirements and its interior fitout standards needed to be revised.
2. Most respondents agreed that open kitchens should be allowed in domestic buildings.

▮ SURVEY ON USERS' OPINION ON THEIR HOMES

'Building an ideal home' was one of the main subjects in the 1999 Policy Address by Mr Tung Chee-hwa. A pleasant and safe living environment is now one of the targets of Tung's policy, but what are the users' views on their homes?

To establish a database from users' point of view about residential buildings, 100 questionnaires were collected through random sampling.

The findings of Questionnaire on Your Home (1999) are as follows.

Bathroom

1. Most respondents would not accept their bathrooms/toilets with only mechanical ventilation and no windows.
2. Most respondents preferred the gas water heater in the bathroom.

Kitchen

1. Most respondents would not accept their kitchens with only mechanical ventilation and no windows.
2. Most respondents preferred an open kitchen.

Pipe Duct

Most respondents preferred pipes concealed in the pipe duct.

Windows

1. Most respondents would not prefer to have one window added to the living room if one window was to be deducted from the kitchen.
2. Most respondents found bay windows more useful compared with ordinary flush windows.
3. Most respondents preferred genuine 'bay windows' (i.e. projected windows on the same floor level as the rest of the room) if they were to count as a portion of the GFA.
4. Most respondents thought that 1.1m was the appropriate height of the base of a bay window from the finished floor level.
5. If it is less than 1.1m, window grilles should be added. Most respondents would not like to have the grilles.
6. Most respondents would like to have flower boxes on the external walls of the windows.
7. Most respondents would like the windows to extend from floor to ceiling. Most of them wanted them to be in the living room.

Balcony

1. Most respondents found the balcony useful.
2. Most respondents thought the balcony should be discounted in GFA calculation.

Common Areas

1. Most respondents thought that facilities for the disabled should be provided in the common areas of residential buildings.
2. Most respondents thought that the common areas (e.g. lift lobbies, corridors, etc.) in their buildings were comfortable.
3. Most respondents would like to have larger common areas, with a certain discount in GFA.

Ceiling Height

Most respondents regarded 3m as the most comfortable clear ceiling height, including finishes (e.g. false ceiling), for a residential building.

Environmental Control Devices

1. Most respondents would like sunshading devices to be incorporated into their buildings.
2. Most respondents would like noise shields to be incorporated into their buildings.

OBSERVATIONS ON THE TWO SURVEYS ON THE DESIGN OF THE BUILT ENVIRONMENT

The survey conducted among HKIA members (architects) represented a professional view, but the one conducted among the users was of no less concern as it provided certain views of the community, which buildings are meant to serve. Expectation for a better quality of life was common among the two surveys. The main issues are discussed as follows.

Improvement can be made by releasing the restriction on plot ratio or GFA calculation. Architects took this point by making reference to the external wall, environmental protection devices, building services and amenity facilities. Users favoured an open kitchen, concealed pipes in ducts and environmental control devices – items for better-quality living. In the present conditions, most of these have to be exchanged for the precious GFA and cannot be implemented easily.

The balcony was considered useful by the users, who would like it to have a discount in GFA calculation. However, architects considered the balcony should totally account for GFA. From a technical point of view, the balcony can be seen as a sheer cantilevered floor slab, an extension of the living space. However, from an environmental point of view, the balcony can be designed as a positive environmental protection device as well as an additional amenity facility to the living space. The key consideration here may lie in proper management to avoid unauthorized or illegal conversion.

Both users and architects agreed that facilities for the disabled should be provided in the common area. It should be further noted that the common area as an important part of the floor plan is now designed as 'efficient' as possible to give more precious saleable GFA. However, if GFA is discounted but more amenity facilities can be added, it will enhance the quality of the floor plan, giving more design alternatives.

Tall buildings are an attraction for tourists. For example, in New York City, both the World Trade Centre and the Empire State Building have an observation deck and a related independent route for public access and enjoyment. Such provision can be encouraged through discount in or exemption of GFA, similar to the public thoroughfare on the first floor or ground level.

Users were concerned about windows. They would like to have full height windows. They would also like to have 'bay windows' which are projected windows not included in GFA calculation. This is understandable as any additional space in small-sized apartments would be useful for the users. Bay windows have to conform to design criteria established in the PNAP. Whether such compliance results in an environmentally positive or aesthetically pleasing design is debatable, but the fact is that no alternative form of such 'bonus' in GFA allows for alternative design.

Architects were concerned about the classification of a site – the proportion of site boundary abutting streets determines its development potential. It was their opinion that the class of site for a small site should be upgraded when combining with adjacent sites, and that open spaces next to a site should be considered as a street when determining the classification of the site. Regarding this issue, open spaces used as parks are environmentally much more pleasing than a 4.5m-wide street filled with vehicular traffic.

▮ LIST OF ENVIRONMENTALLY-RELATED BUILDING (PLANNING) REGULATIONS

Regulation 4 Buildings not to obstruct, endanger or cause nuisances

(c) No building or fixture thereon shall be so constructed that it permits the escape into or over any adjacent footpath or street at a height of less than 2.5m of any noxious gases or exhaust from any ventilating system.

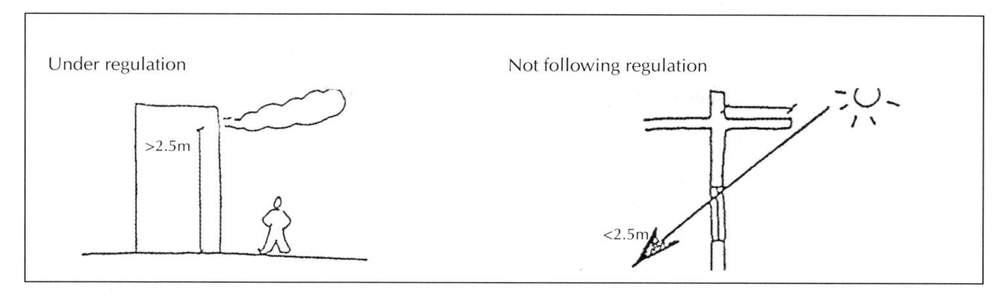

Commentary:

Control on water dripping from the condensation of air-conditioning units should be included in the regulation. Safety control should be applied to objects that are hung on the external wall.

Regulation 7 Eaves, cornices and mouldings

(2) No eaves, cornices or mouldings shall so project for a distance of more than 750mm or at a height of less than 2.5m above the level of the ground.

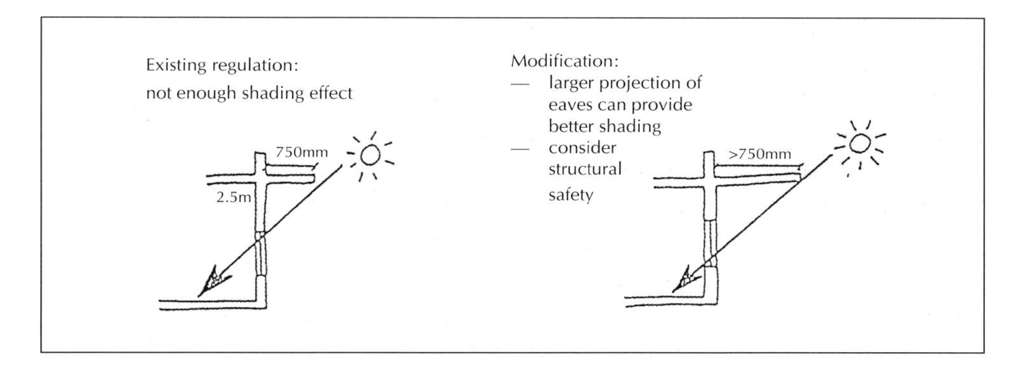

Commentary:

The limit of the projecting distance can be extended to provide better shading. However, structural safety should also be incorporated in the regulation.

Regulation 24 Height of storeys

(1) Every room used or intended to be used for the purpose of an office or for habitation in any building shall have a height of not less than 2.5m measured from floor to ceiling: provided that there shall be not less than 2.3m measured from the floor to the underside of any beam.

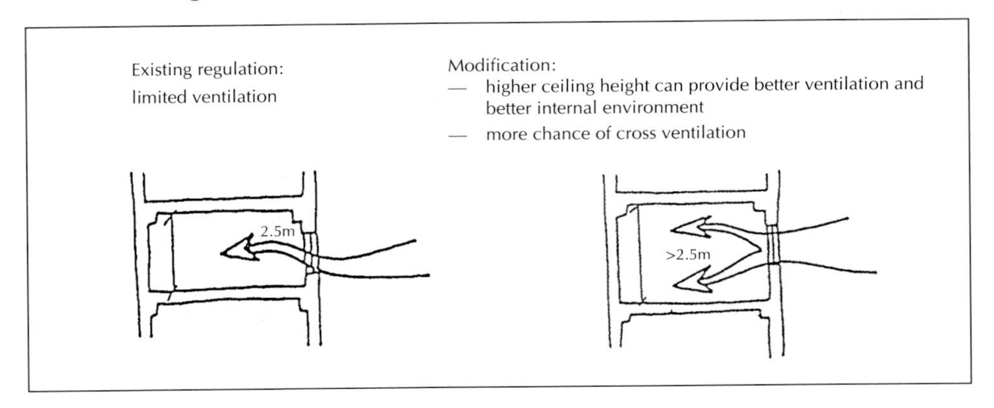

Commentary:

The ceiling height can be extended to provide better ventilation and better internal environment. Cross ventilation can be encouraged by allowing the use of smaller window sizes in the cross-ventilated room.

Regulation 25 Space about domestic buildings

(1) (a) Every domestic building on a class A or B site or on a class C site shall have within the site an open space at the rear, or partly at the rear and partly at the side.

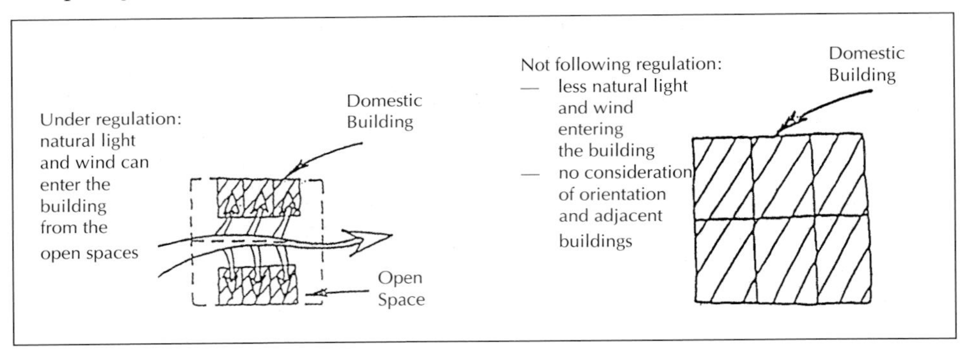

Commentary:

The regulation should take building orientation and adjacent buildings into account.

Regulation 26 New buildings on existing street less than 4.5m wide to be set back from centre line of street

Where the width of an existing street in front of any new building is less than 4.5m, no part of such a building shall be nearer to the centre line of the street than 2.25m.

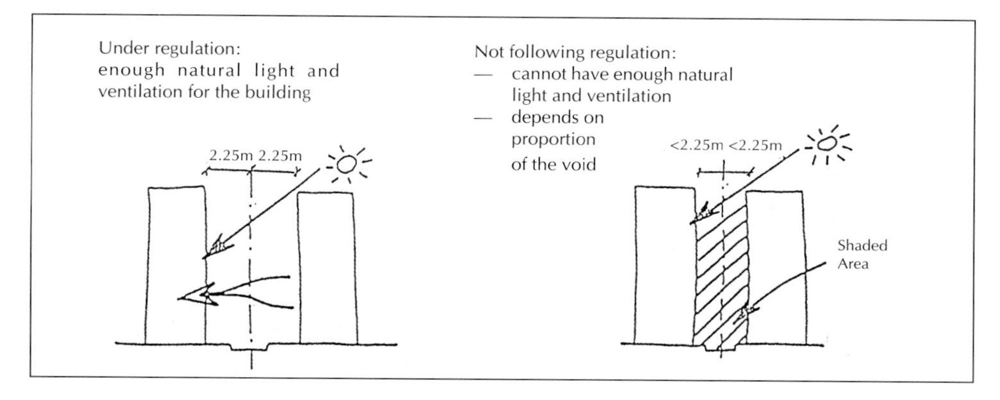

Commentary:

The distance should depend on the proportion of the void instead of a rigid dimension.

Regulation 29 Lighting and ventilation

Every storey of every building used or intended to be used for the purpose of an office or for habitation shall be provided with effectual means of lighting and ventilation.

Regulation 30 Lighting and ventilation of rooms used or intended to be used for habitation or as an office or kitchen

(1) Every room used for habitation or for the purposes of an office or as a kitchen shall be provided with natural lighting and ventilation.

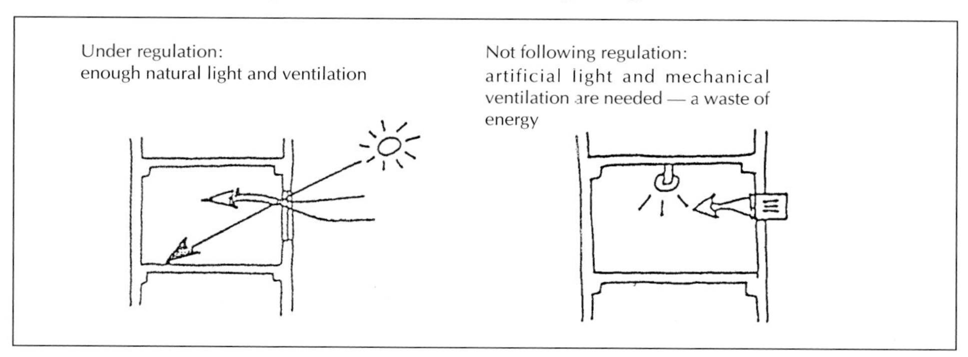

Commentary:

Building (Planning) Regulations 29 to 37 all consider lighting and ventilation, but they are too general. There is no consideration of building orientation. Cross ventilation is not encouraged. Also, these regulations usually do not apply after internal partitioning is added. Besides, the regulations control only the buildings for office and habitation use. Other uses such as shops, banks, industrial, etc., are not under control. These regulations should include these building types too.

Regulation 32 Restriction on distance that any part of room may be from a prescribed window
No part of any room used for habitation shall be more than 9m, measured within the room, from a prescribed window which faces directly into the external air or, where, under and in accordance with regulation 33, a window opens on to an enclosed verandah or balcony or on to a conservatory or on to any similar enclosed place or is, under regulation 71, permitted to open on to an unenclosed verandah or balcony or any other unenclosed place, from the outer edge of the verandah, balcony, conservatory or enclosed or unenclosed place, as the case may be.

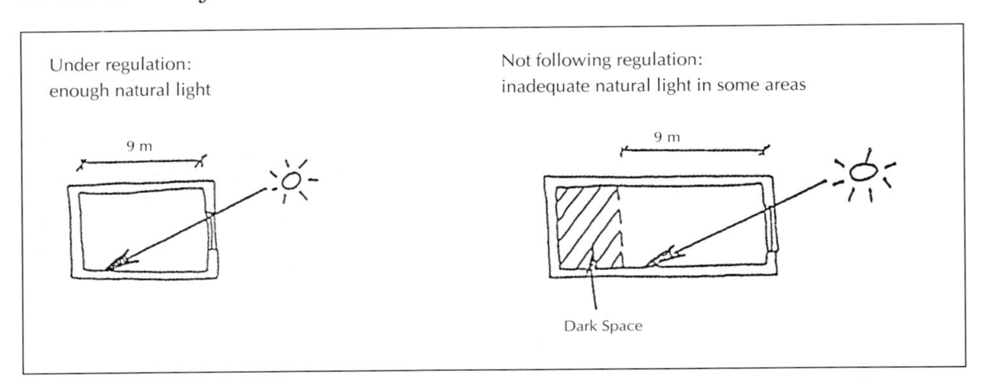

Commentary:

This regulation should also consider the height and width of the space. It should have a three-dimensional consideration. Modification of this regulation can be considered.

Regulation 36 Rooms containing soil fitments
(4) No room containing a soil fitment shall open directly into a room used or intended to be used for the manufacture, preparation or storage of food for human consumption.

In general, there is not enough encouragement for environmental considerations in design to create better-quality living. To achieve a better quality of life, there should be extra incentives for designers and developers to come up with innovative solutions for better habitation in buildings as well as an improved urban fabric.

Under the Building (Energy Efficiency) Regulations (1995), energy efficiency is required to be achieved under Regulation 4:

- A building to which this Regulation applies shall be so designed and constructed as to achieve energy efficiency to the satisfaction of the BA.
- Without prejudice to the generality of subsection (1), a building shall be regarded as being designed and constructed in compliance with that subsection if its external walls and roofs have a suitable OTTV.

This applies to commercial buildings and hotels in respect of appropriate overall insulation for the building enclosure of external walls and roofs. The application of this requirement to residential buildings will need certain adjustment as the occupancy hours and environmental control provisions will be quite different from those for commercial buildings and hotels. It should also be noted that residential buildings under the 'domestic' column of plot ratio has in general less bulk of development than 'non-domestic' buildings.

The Building Enclosure as stated by Centuori (1992) has the mission of acting a barrier against the outside – an appropriate building skin against natural elements. The Building Code assumes certain standards of comfort to be met by a particular surface wrapping. These standards of comfort address concern for protection against fire, water, heat, noise and force of wind as well as natural light provision. These are then quantified in the building code. However, standards of comfort vary from place to place. The basic subtropical climate in Hong Kong will need special specifications.

Quoting from Wong (1998), the enclosure has 'to control the influence of the external environment like solar heat, light, air and noise on the interior space, thus providing comfort'. A few buildings in Hong Kong have been designed to give very positive environmental concern, but part of the design effort is wasted due to the difficulty in overcoming the existing building code and the increase in initial economic outlay.

Besides making an 'energy-efficient' enclosure, the design of building with respect to orientation and cross ventilation is of no less importance. Choosing materials with low embodied energy and minimal maintenance is also part of the consideration in the broad area of sustainable building design. Isolated consideration of environmental factors will lead to unbalanced situations to be rectified only by more expensive means. Such consideration is not encouraged under the present building code.

▮ REVIEW OF PNAP IN EFFECTING QUALITY DESIGN

PNAP are issued by the BA to authorized persons and registered structural engineers. PNAP no. 1 was issued in 1983, entitled 'Practice Note in Force'.

The following is a discussion on how PNAP affect aspects of technical design.

External Wall

No. 59: Cladding

This PNAP explains cladding as facing or decoration onto the external wall and requires attention to structural safety.

If this is just 'additional to the external wall', GFA calculation should not comply with this PNAP. In such case, additional insulation materials on the external wall can be exempted from GFA calculation.

No. 106: Curtain Wall Systems

Technical and maintenance aspects are stated in this PNAP.

However, enforcing maintenance and repair is not a statutory requirement. In view of recent damages caused by Typhoon York (1999), such enforcement should be made obligatory to ensure public safety.

Environmental Aspects

No. 169: Natural Lighting To Staircases

This PNAP states that windows are required for staircases, with possible modification for non-domestic buildings. The introduction of Building (Planning) Regulation 40 is also made here.

Staircase windows will compete for the limited frontage on a residential building due to the 'efficient' design of buildings. The recommendation for improvement by dividing the GFA into saleable and common parts will help to resolve such competition.

No. 172: Energy Efficiency Of Buildings

OTTV calculation is required for hotels, offices and assembly areas, under this PNAP. Calculation is now too elaborate and difficult to be checked.

Application of this PNAP to assembly areas should take account of frequency of usage. Consideration should be given to alternate provision of natural cross ventilation which require alternative more windows in contrast to general provision for satisfaction of OTTV.

No. 219: Lighting And Ventilation For Bathrooms In Domestic Buildings

This PNAP allows modifications to and artificial lighting for domestic-use bathrooms.

This gives design alternatives for bathrooms which are without windows, and is welcomed as an improvement.

No. 230: Water Seepage

This PNAP addresses water-seepage problems that are associated with piping installed in the structure.

Waterproofing of construction cannot be a practical means to solve embedded piping leakage unless a 'double slab' is constructed with a non-structural slab embedding the pipes and another slab for waterproofing.

Amenity Provisions And Better Facilities

No. 112: Buildings To Be Planned For Use By Persons With Disabilities – Building (Planning) Regulation 72

For improvement of this PNAP, buildings with good access for disabled persons should be labelled prominently for public notice.

Public spaces like restaurants, assembly halls, etc., should have, on the same level, toilet facilities for the disabled. Enforcement of proper management is required to prevent unauthorized conversion of these facilities.

No. 116: Amenity Features

In this PNAP, the objectives for the provision of better management, environment and quality of life are general terms for exemption of amenity features from GFA calculation.

In practice, more flexibility should be given. Amenity features taking up floor areas but not intensifying the development can be excluded from GFA calculation. Sky terraces can be encouraged through this consideration.

No. 229: Exclusion Of Floor Area For Recreational Use

This PNAP allows certain recreational facilities in domestic buildings.

Proper management should again be emphasized.

Room Sizes And Calculations

No. 13: Calculation Of Gross Floor Area And Non-accountable Gross Floor Area – Building (Planning) Regulation 23(3)(a) and (b)

Voids, buildings services and curtain walls are exempted from GFA calculation under this PNAP.

Building services can be extended to cover other supporting facilities, such as pipe ducts and meter cabinets, that require management and operation.

No. 27: Height Of Storeys

This PNAP states 2.5m as the minimum height. A greater height should be encouraged by exemption of additional staircases (as a result of the increase in building height) from GFA calculation.

No. 68: Projections In Relation To Site Coverage And Plot Ratio

This PNAP gives definition to certain projections that are exempted from the plot ratio, and also defines the 'bay window'.

The relaxation of the 'bay window' on plot ratio has turned out to be defining the elevation of many residential buildings in Hong Kong, affecting the urbanscape and city skyline. Alternatives should be provided for diversity in elevation design.

No. 107: Bridges Over Streets

This PNAP gives authority to the Advisory Committee on the Appearance of Bridges and Associated Structures (ACABAS) to approve bridges.

Approval time should be coordinated within a statutory period for plan processing.

No. 118: Streets In Relation To Site Area

Transfer of plot ratio is not allowed in large sites under this PNAP.

To give better planning, this transfer should be permitted by working out the lease conditions with Lands Department. One consideration is to view internal streets as 'tunnels' of road space through the site without affecting site integrity.

No. 179: Service Lanes

This PNAP still requires the provision of service lanes.

Now not useful for scavenging reasons, this can be replaced by regulations for fire break or access footpaths if really necessary.

A comprehensive refuse disposal system initiated through the planning level can replace these service lanes.

No. 223: Podium Height Restriction Under Building (Planning) Regulation 20(3)

This PNAP allows possible provision of a 20m-high podium.

In a high-density urban environment, can an increase in podium height create a better aesthetic or environmental effect on the city?

If not, the space occupied by the additional podium height should be compensated by setback or opening up of ground-level space for the public.

SURVEY ON THE DEFINITION OF FLOOR AREA

To investigate the possibility for improvement, a survey was conducted during the second quarter of 1999 among members of the Hong Kong Institute of Architects (HKIA), Hong Kong Institute of Engineers (HKIE), Hong Kong Institute of Surveyors (HKIS) and Hong Kong Institute of Planners (HKIP), targetting architecture-related professionals (see Figure 2). Over 200 questionnaires were returned. Most respondents from this database were employed in the private sector and had at least five years' practical experience.

An outline of the survey results is as follows.

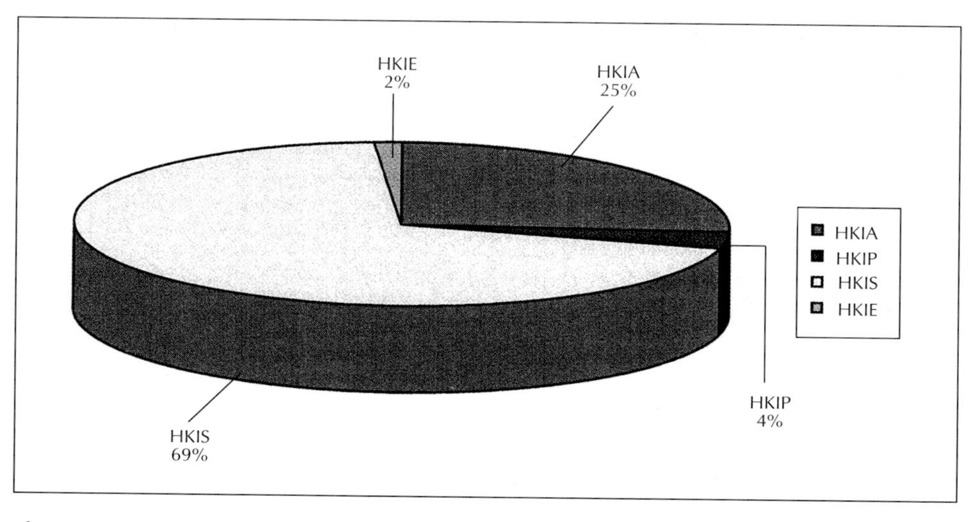

Figure 2 Distribution of participants in the 'Definition of Floor Area' survey from HKIA, HKIP, HKIS, HKIE.

GFA (See Figure 3)

Building (Planning) Regulation 23(3)(a) defines GFA as the area contained within the outer surface of external walls of a building measured at each floor level.

Most respondents agreed to have the following items discounted in GFA calculation under the Building (Planning) Regulation:

1. External wall surfacing and finishes
    - External wall finishes such as tiles, plaster
    - Cladding and the space between the external wall and the cladding
2. Environmental protection devices
    - Sunshading device
    - Noise shield
    - Heat insulating material for the external wall
3. Internal parts relating to building services
    - Lift shaft
    - Services areas e.g. pipe ducts (electrical cable ducts, water pipe ducts, A/C pipe ducts, etc.)
4. Amenity facilities
    - Window flower box which projects more than 500mm
    - Observation deck and related independent transportation route for the public
    - Podium garden
    - Air conditioner hood which projects more than 500mm

Most respondents disagreed to have the following items discounted in GFA calculation under the Building (Planning) Regulation:

1. Lift lobby
2. Staircase
3. Exposed structure such as columns and beams on the external side of the building
4. Incorporated Owners' Committee meeting room
5. True bay window with same floor level as the rest of the room
6. Balcony

Usable Floor Area

The Building (Planning) Regulations define 'usable floor area' (UFA) as any floor space other than staircase, staircase halls, lift landings, the space used in providing water-closet fitments, urinals and lavatory basins, and the space occupied by machinery for any lift, air-conditioning system or similar service.

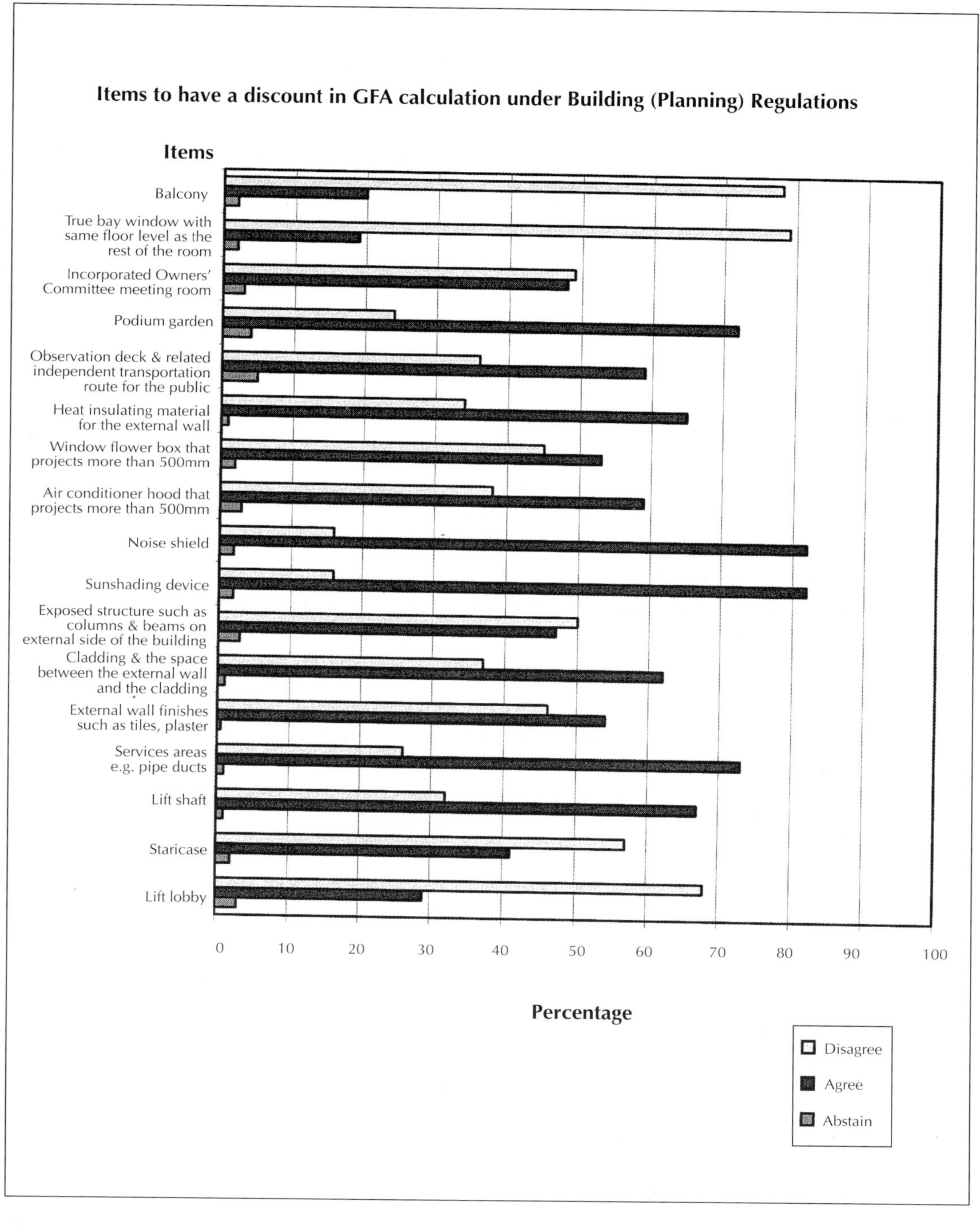

Figure 3 Survey on the Definition of Floor Area: GFA

Most respondents agreed to have the following item discounted in UFA calculation under the Building (Planning) Regulations:

Structural elements, e.g. structural walls, structural columns, etc.

Most respondents disagreed to have the following items discounted in UFA calculation under the Building (Planning) Regulations:

1. Non-structural partition walls including finishes
2. Water-closet fitments
3. Bath-tubs

Construction Floor Area

Most respondents agreed to have the following items included in the calculation of the construction floor area:

1. G/F lobby
2. Transformer room
3. Water tanks

Most respondents agreed that construction floor area should be clearly defined, and that law should be made on the use of construction floor area in selling the units.

Saleable Floor Area (See Figure 4)

Most respondents disagreed that the enclosing walls separating a unit from a lightwell, a lift shaft or any similar vertical shaft, or a common area, should be regarded as external walls which would mean their full thickness should be included (see Figure 4a).

Most respondents agreed to have the following items included in the calculation of saleable floor area (see Figure 4b):

1. Balcony
2. External wall enclosing the unit including the external wall finishes
3. Half of the wall thickness if it is a separating wall between adjoining units
4. Internal partitions within the unit
5. Columns within the unit

Most respondents disagreed to have the following items included in the calculation of the saleable floor area:

1. Staircase
2. Lift shaft
3. Lobby
4. Communal toilet

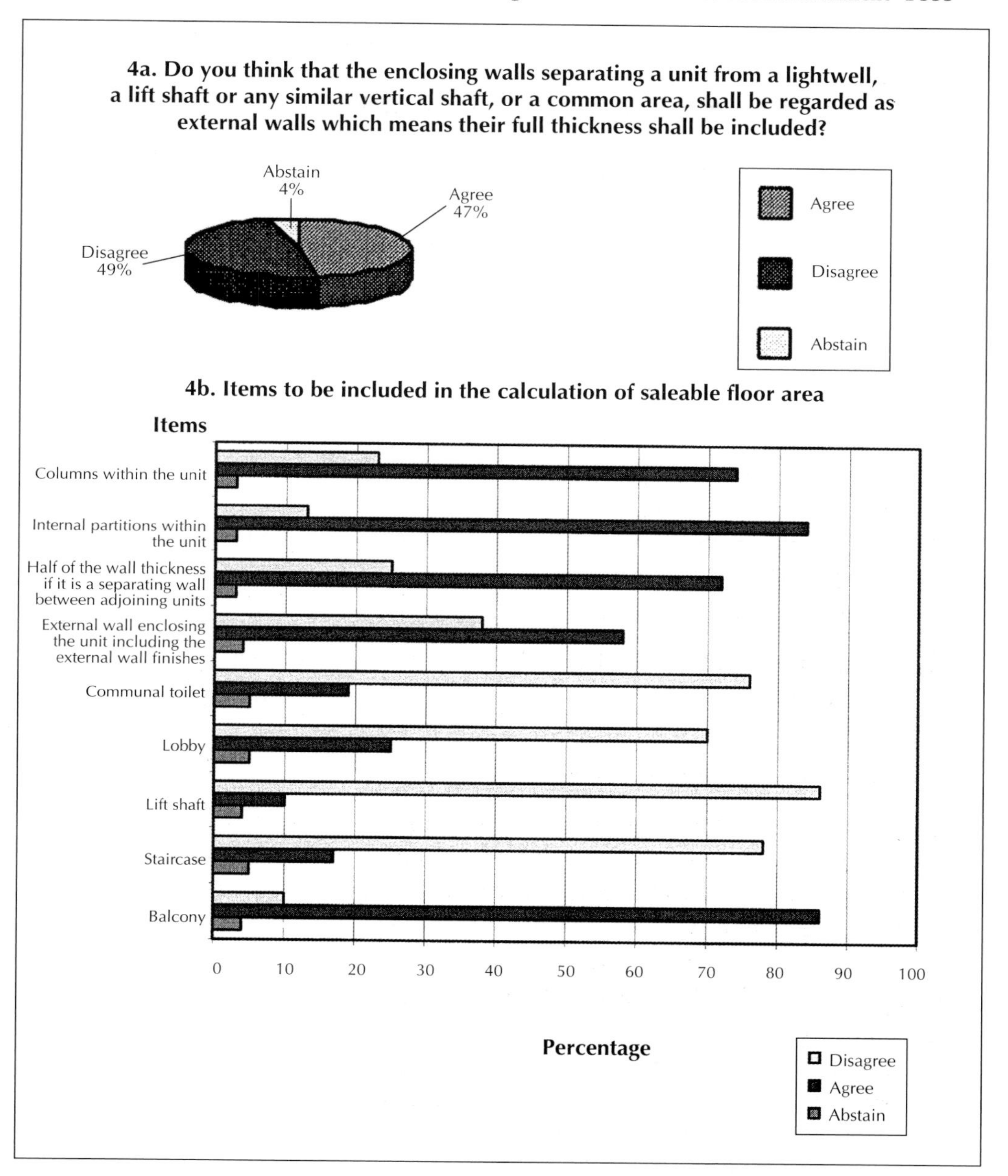

Figure 4 Saleable Floor Area

Lettable Area

Most respondents agreed that the lettable area of a unit occupying an entire floor should be the floor area exclusively allocated to that unit including toilets and lift lobbies but excluding common areas such as staircases, smoke lobbies, lift shaft and plant rooms.

Most respondents agreed that the lettable area of a unit which is one of several units making up an entire floor should be the floor area exclusively allocated to that unit plus a proportionate share of the communal toilets, lift lobbies and passageways among the units on that floor.

Internal Floor Area

Most respondents agreed to have the following items included in the calculation of the internal floor area:

1. Balcony
2. Toilets
3. Internal partitions and columns within the unit

Most respondents disagreed to have the lift lobby included in the calculation of the internal floor area.

Tolerance (See Figure 5)

Most respondents preferred that the tolerance applied to the area of the completed building for sale purposes should be 1%.

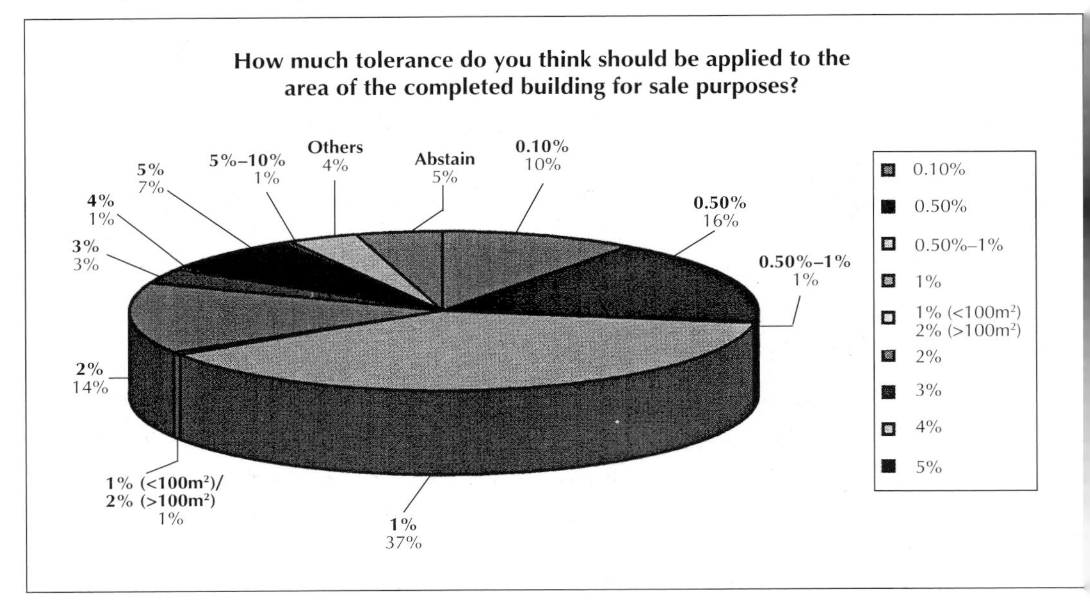

Figure 5 Tolerance

Floor Area For Sale Purposes (See Figure 6)

Most respondents preferred that saleable floor area should be used when selling the units.

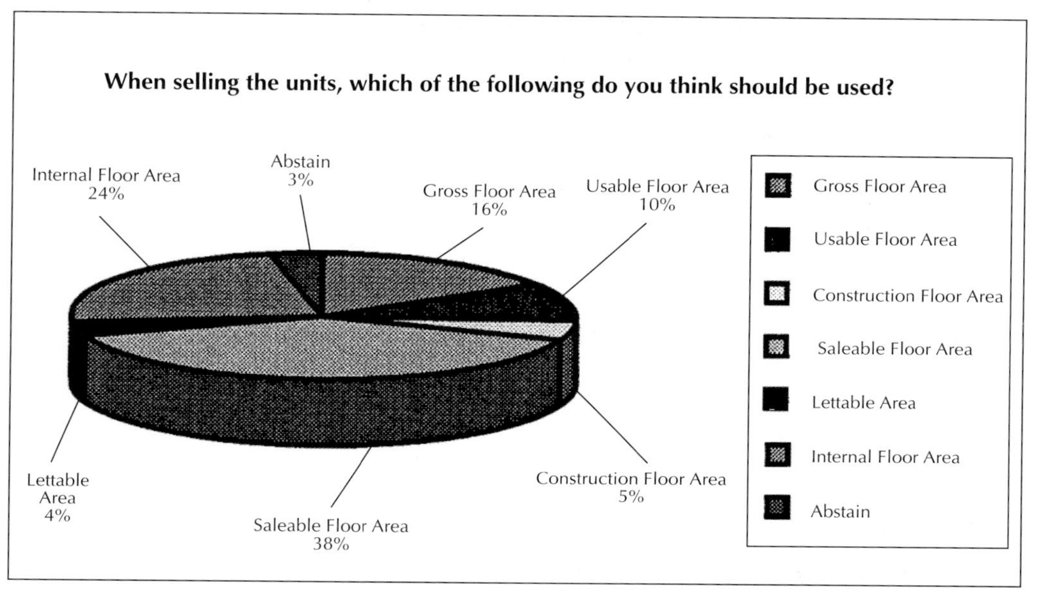

Figure 6 Floor Area for Sale Purposes

OBSERVATIONS ON THE SURVEY ON FLOOR AREA

This survey received generous response from HKIA but only lukewarm response from HKIE and HKIP. However, the questionnaires from HKIE were returned by the chairmen of Building, Building Services and Structural divisions.

GFA (See Figure 7)

GFA is the 'floor area' related to the development potential of the site. The area of the site multiplied by the maximum plot ratio gives the maximum GFA that a site can develop. So the type of functions or the building components that have to be calculated into GFA will be crucial to property developers and end-users.

The survey results regarding GFA are very close to the results of the first survey, which targetted members of HKIA. Components like external wall

Definition from Building (Planning) Regulation 23(3)(a):
'Gross floor area' (GFA) is the area contained within the outer surface of external walls of a building measured at each floor level.

This includes:
- lift lobby
- staircase
- lift shaft
- services areas e.g. pipe ducts (electrical cable ducts, water pipe ducts, A/C pipe ducts, etc.)
- external wall finishes such as tiles, plaster
- exposed structure such as columns and beams on external side of the building
- true bay window with same floor level as the rest of the room
- balcony

The following need not be included in GFA calculation:
- cladding and the space between the external wall and the cladding
- genuine sunshading device
- genuine noise shield
- individual air-conditioner boxes and platforms which have a built-in system for condensate disposal
- window flower boxes provided they are small, individual and non-continuous

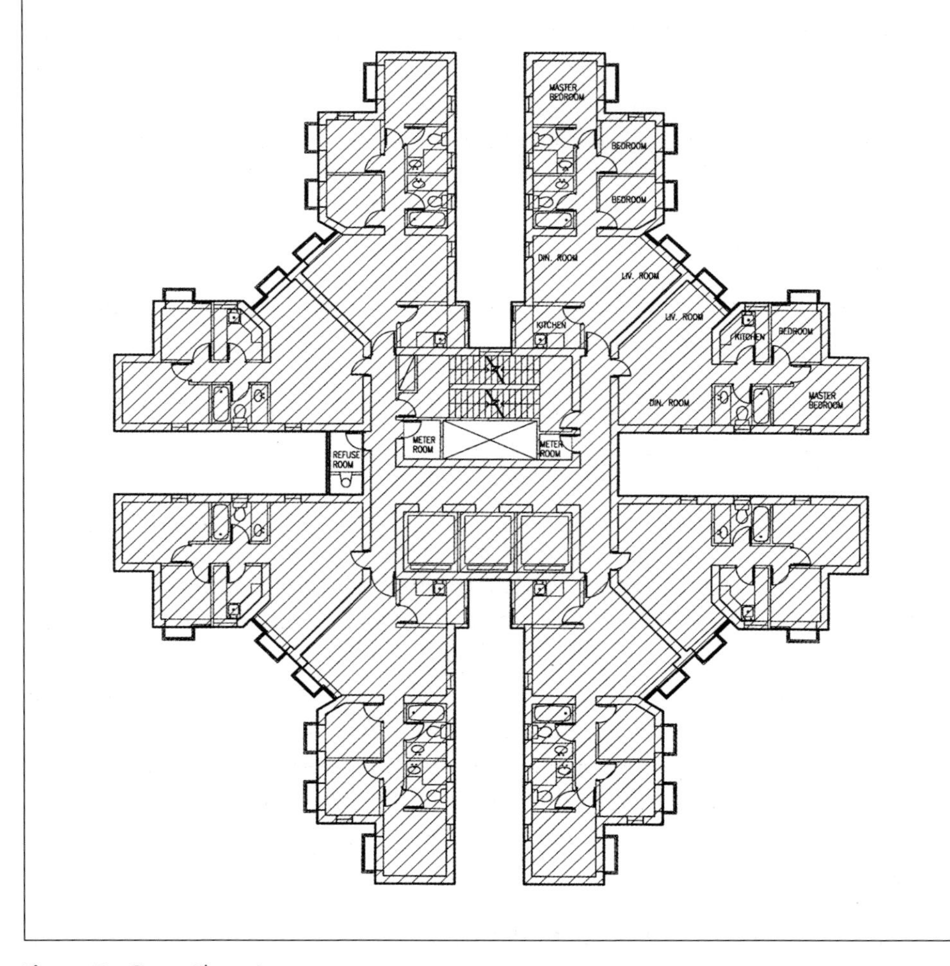

Figure 7 Gross Floor Area

surfacing, environmental protection devices, internal parts relating to building services, and amenity facilities were generally agreed to be exempted from or discounted in GFA calculation.

Another observation is that no conclusion could be drawn as to whether the incorporated owners' committee meeting room should have a discount in GFA calculation, which is contradictory to the previous survey results from HKIA. An examination of the survey figures reveals that the percentages of those who agreed and disagreed were very close. If this is sheer GFA consideration of an additional room, this will be included in GFA calculation. However, if this can be related to amenity facilities for better management, alternative consideration should be given.

Another component to be noted is the exposed structure, such as columns and beams on the external side of a building. The survey results for agreement and opposition were fairly close. If such projections exhibit environmental protection benefits, discount of GFA can be encouraged. In fact, based on the survey statistics, more engineers and surveyors agreed to this exemption than architects.

UFA (See Figure 8)

UFA is used to determine the number of people using the part of the building, which in turn shows the standard requirement for the provision of fire escape staircases, escape route dimensions and the number of sanitary fitments.

Here, the survey results were quite clear in showing that structural elements should be discounted in such UFA calculation.

Construction Floor Area

Besides GFA, function areas are also included in the calculation of the construction floor area. This term is often stated in sale brochures from property developers, but not indicated in the approval process by the BA.

Items like the G/F lobby, transformer room and water tanks should be included, according to this survey. Other possible items may include bay windows, air-conditioning hoods, planter boxes, or even recreational facilities and car parks. However, clear definition should be given to avoid doubts.

Saleable Floor Area (See Figure 9)

This term is used for sale purposes to show the sole ownership for the particular unit. While most respondents can agree to the 'individual' part and 'common'

Definition from Building (Planning) Regulations:
'Usable floor area' means any floor space other than staircase, staircase halls, lift landings, the space used in providing water-closet fitments, urinals and lavatory basins, and the space occupied by machinery for any lift, air-conditioning system or similar service. In common practice, this excludes:

- structural elements
- non-structural partition walls

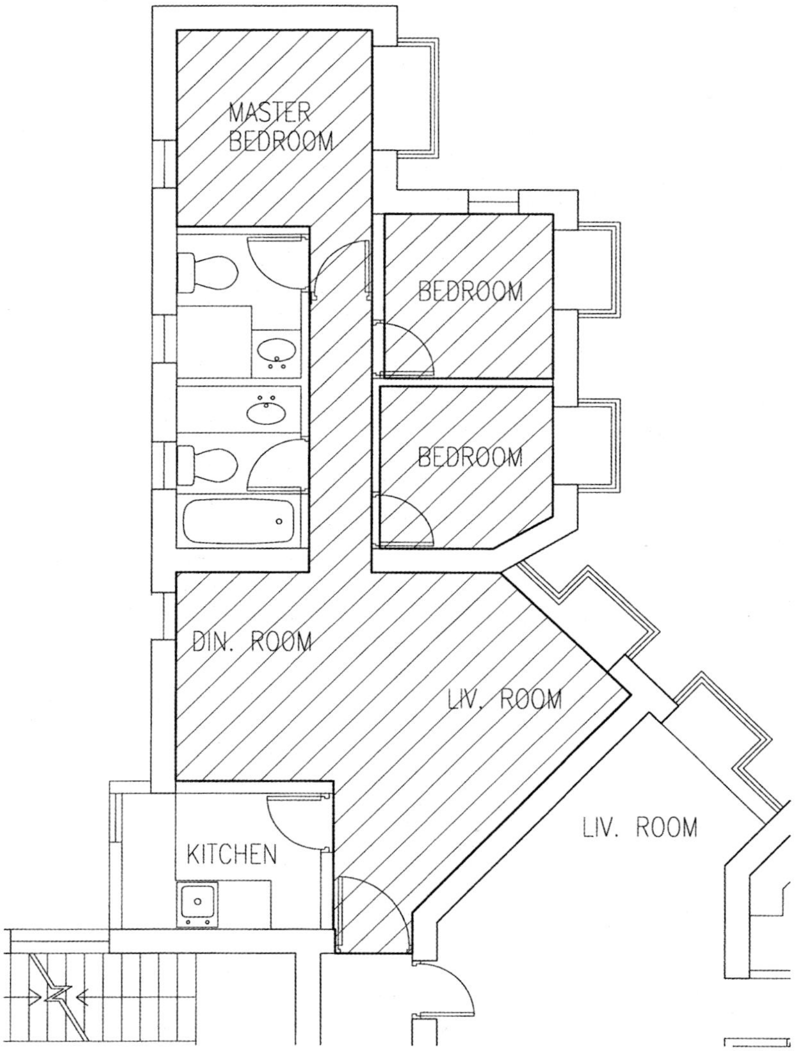

Figure 8 Usable Floor Area

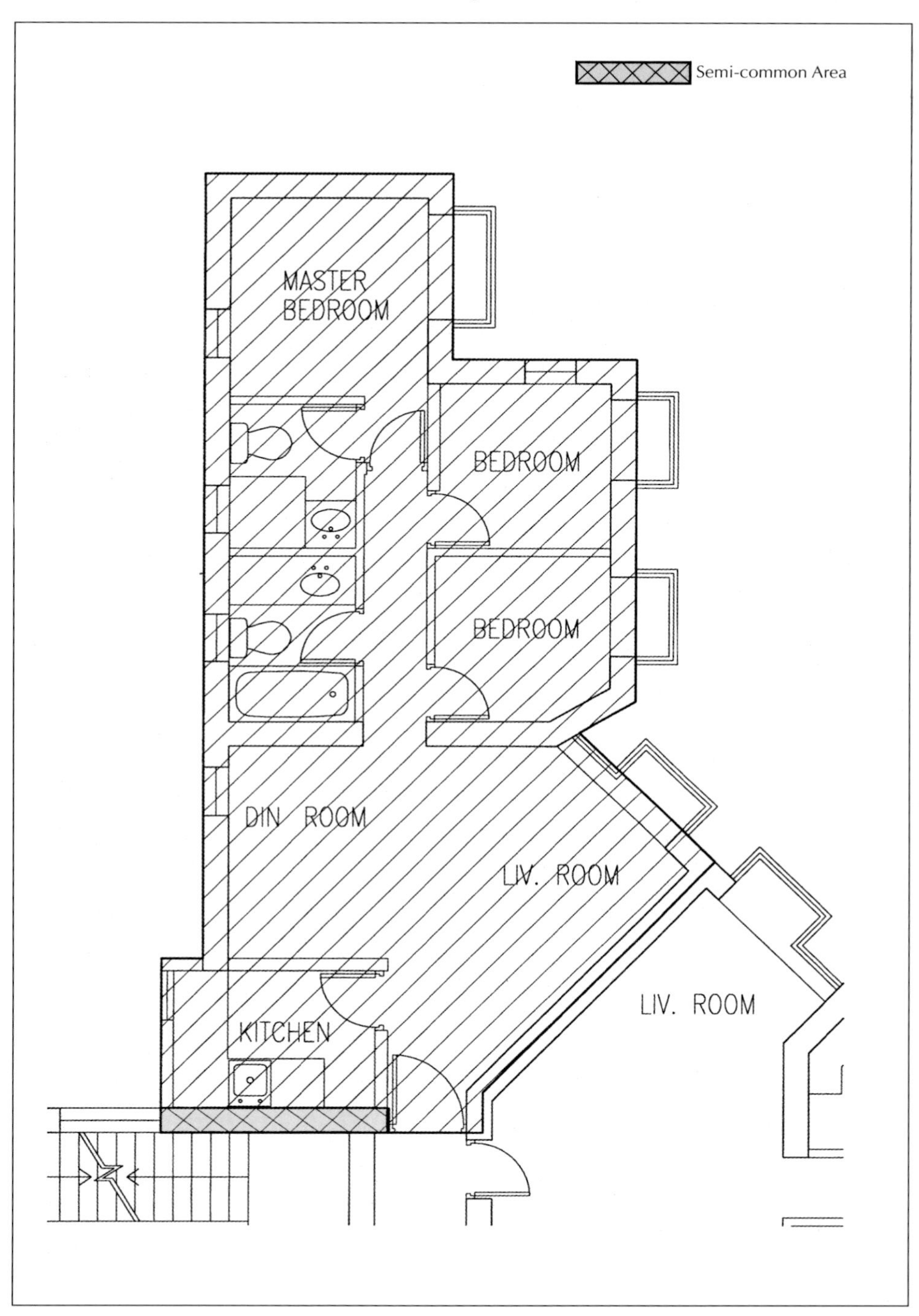

Figure 9 Saleable Floor Area

part of a unit for inclusion into or exclusion from Saleable Floor Area calculation, the grey area lies in the boundary between the 'individual' part and 'common' part. This will be the enclosing walls separating a unit from a lightwell, a lift shaft or any similar vertical shaft, or common area.

In this survey result, the inclusion of these walls into Saleable Floor Area is disagreed. This can be understood since the exclusion of these semi-common areas will allow easier measurement of the actual area.

Lettable Area (See Figure 10)

The lettable area is used for rental purposes. Views were clearly expressed in the survey results.

Internal Floor Area

According to the survey, the internal floor area should measure the enclosed internal floor and include the balcony, toilets, internal partitions and columns within the unit.

Tolerance

Tolerance can be seen as a result of the conventions of deviations in measurement and variation of actual site work. The range of tolerance chosen by the respondents ranged from 0.1% to even 10%, but the majority chose 1%.

Floor Area For Sale Purposes

When selling the units, survey results show that the majority agreed on using saleable floor area as the unified practice.

▮ ALTERNATIVE VIEWS ON FLOOR AREA FROM REAL ESTATE ADMINISTRATION

The same survey (1999) on floor area was also conducted among the Hong Kong Institute of Real Estate Administration (HKIREA). Forty-five questionnaires were returned.

A summary of the survey results is as follows.

Lettable area of a unit shall be the saleable floor area of that unit plus a proportionate share of lift lobbies and passageways among the units on that floor.

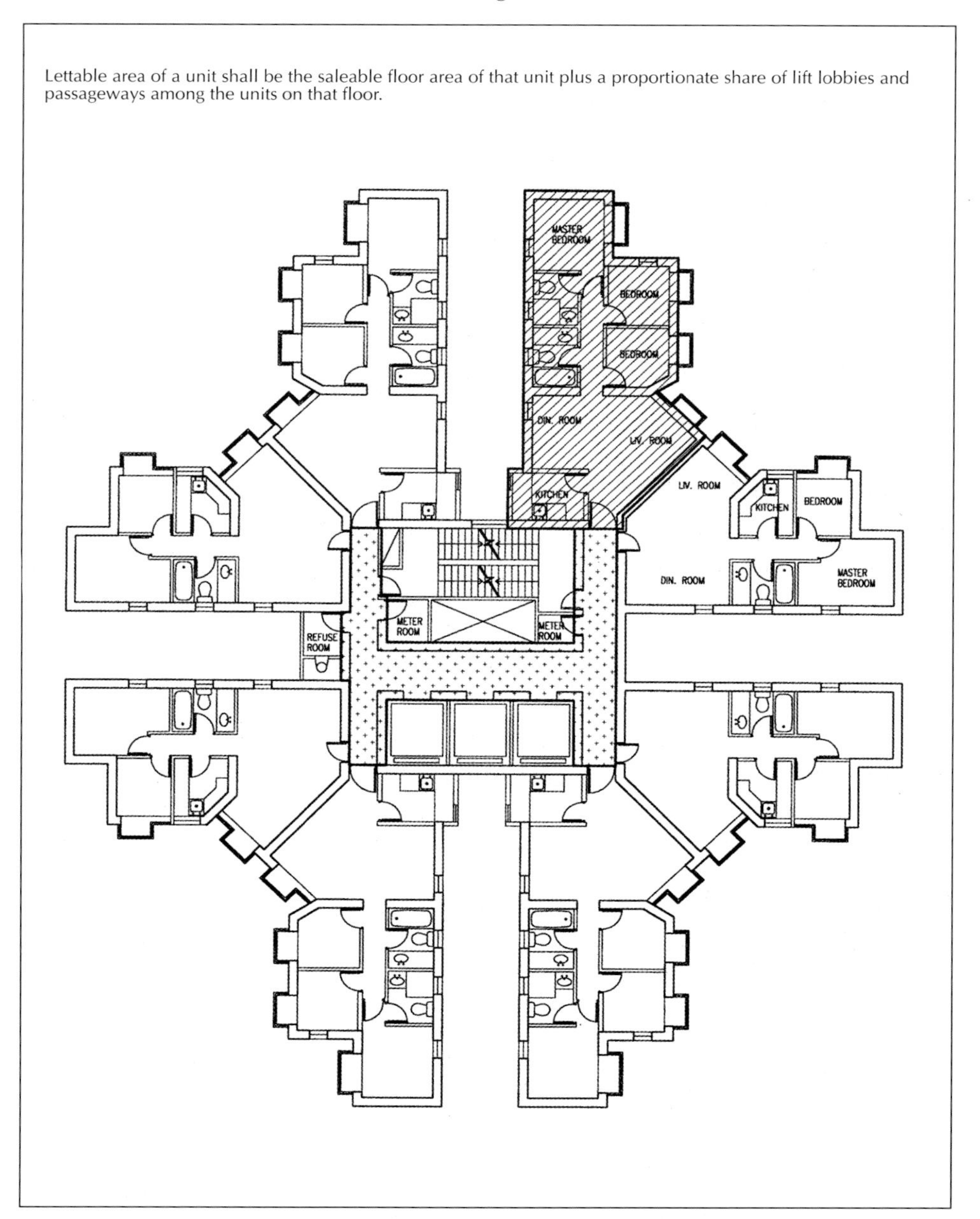

Figure 10 Lettable Area

GFA (See Figure 11)

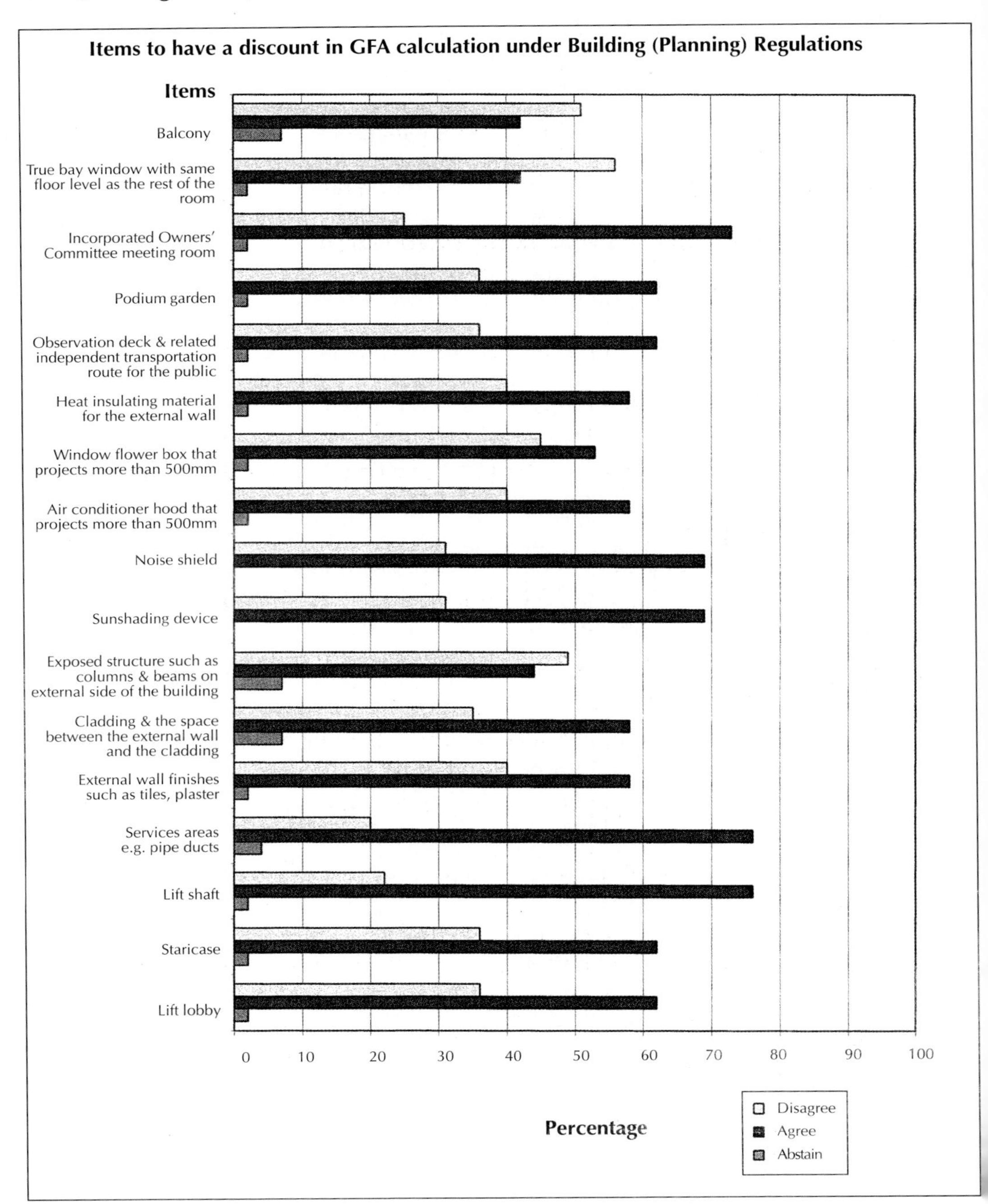

Figure 11 Survey on the Definition of Floor Area: GFA

Most respondents agreed to have the following items discounted in GFA calculation under the Building (Planning) Regulations:

1. External wall surfacing and finishes
    - External wall finishes such as tiles, plaster
    - Cladding and the space between the external wall and the cladding
2. Environmental protection devices
    - Sunshading device
    - Noise shield
    - Heat insulating material for the external wall
3. Internal parts relating to building services
    - Lift shaft
    - Services areas e.g. pipe ducts (electrical cable ducts, water pipe ducts, A/C pipe ducts, etc.)
4. Amenity facilities
    - Window flower box which projects more than 500mm
    - Observation deck and related independent transportation route for the public
    - Podium garden
    - Incorporated Owners' Committee meeting room
    - Air conditioner hood which projects more than 500mm
    - Common parts
    - Lift lobby
    - Staircase

Most respondents disagreed to have the following items discounted in GFA calculation under the Building (Planning) Regulations:

1. Exposed structure such as columns and beams on the external side of the building
2. True bay window with same floor level as the rest of the room
3. Balcony

The HKIREA survey has similar results as the previous joint-institute survey, with the exception that the former advocated giving a discount to the incorporated owners' committee meeting room, lift lobby and staircase in GFA calculation. This can be understood as an increased quality of amenity facilities and safety provisions.

UFA

Most respondents agreed to have the following items discounted in UFA calculation under the Building (Planning) Regulations:

1. Structural elements, e.g. structural walls, structural columns etc.
2. Water-closet fitments
3. Bath-tubs

Most respondents disagreed to have the following item discounted in UFA calculation under the Building (Planning) Regulations: non-structural partition walls including finishes.

This HKIREA survey shows that water-closet fitments and bath-tubs should be discounted in UFA calculation, in addition to the structural elements. This can be understood as these items occupy genuine usable area.

Construction Floor Area

This survey has identical indications as the previous joint-institute survey.

Saleable Floor Area

This HKIREA survey has similar results as the previous joint-institute survey, except for the different opinion on the 'semi-common' area. The former shows that most respondents agreed that the enclosing walls separating a unit from a lightwell, a lift shaft or any similar vertical shaft, or a common area, should be regarded as external walls which would mean their full thickness should be included.

Lettable Area

This HKIREA survey shows identical preferences as the previous joint-institute survey.

Internal Floor Area

Again, this survey indicates identical preferences as the previous joint-institute survey.

Tolerance (See Figure 12)

Most respondents preferred that the tolerance applied to the area of the completed building for sale purposes should be 1%.

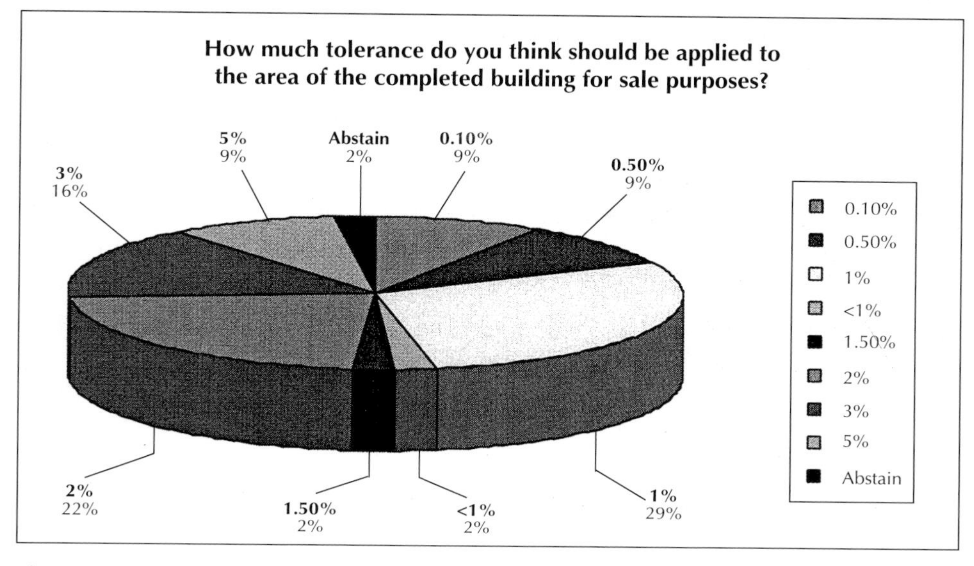

Figure 12 Tolerance

Floor Area For Sale Purposes (See Figure 13)

Most respondents preferred that saleable floor area should be used when selling the units.

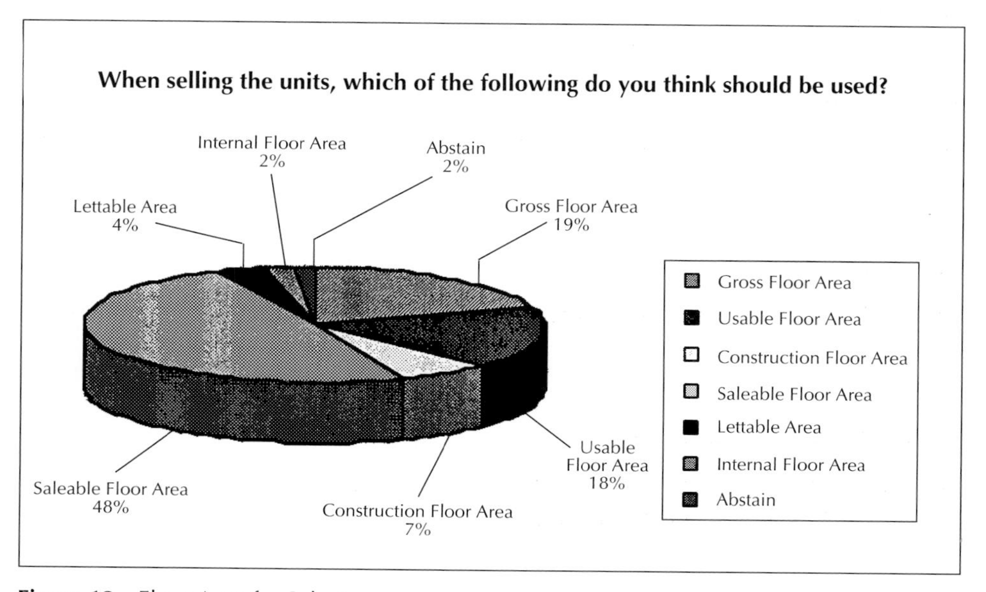

Figure 13 Floor Area for Sale Purposes

▮ RECOMMENDATIONS FOR IMPROVEMENT OF BUILDING CONTROL

It is common consensus to aim for better-quality living in the future. In terms of the psychological aspect, the aesthetic aspects of buildings (shapes, forms, colours and texture) will matter. In terms of the physical aspect, environmental design will take effect and address basic health and safety concerns. In a city, buildings do not exist in isolation; they affect each other and contribute to form the urban fabric. The combined aesthetic aspects of buildings will impart urban forms to the city, giving it a certain sense of cultural identity. The environmental effects of buildings, besides achieving desired indoor conditions, can turn the city into either a genuine green city or a sheer concrete jungle. Thus buildings developing towards a positive direction of quality living will bring about sustainability to a city. With such considerations in mind, the following recommendations are made.

GFA

Hong Kong buildings in the 1950s were usually finished with plastered walls and paint finish, but now, buildings can be finished with tiles, cladding, curtain walls, heat-insulating materials, or other environmentally positive materials. Society advances with improvement in building technology to cater for better-quality living. Building services are now more sophisticated to help maintain efficient and well-managed buildings. Ductworks are necessary. IT pipe ducts may be required. Separated ducts for recycled waste will be needed. More common areas will be required as amenity facilities.

In the 1950s, early buildings were built with minimal building services, often with exposed pipe work on external walls. Buildings were finished with simple materials like plaster and paint. So it is obvious that the GFA applicable to early buildings cannot contribute positively to the quality of life if applied nowadays to modern buildings. Building regulations with better incentives for quality have to be formulated. In this respect, GFA should be interpreted as a means of controlling density – density in terms of population – which will not be affected if spaces for building services, external cladding or environmental protection devices are increased.

Similar thoughts can be applied to amenity areas. However, complication arises due to the overlap of floor area with common area. A calculation method to solve this is to divide GFA into saleable floor area and common area, the proportion of which can be worked out by a statistical survey on existing plans and adding certain amount of amenity facilities. With this method, the 'saleable floor area' portion of GFA is limited so as to control density, but the 'common area' portion is allowed a flexible range so that amenity facilities can be added to improve the quality of life. Through this method, sky gardens are possible, naturally lit and ventilated corridors and lobbies can be provided, and safer escape staircases can be built. This flexibility of common areas can also release stringent floor plan layouts to allow more design freedom for better orientation of living or working units, instead of the currently adapted 'efficient' floor plans.

To implement this system successfully, the building has to be 'watched' to avoid unauthorized use of common areas, by a central management system which can also function as a check on fire services protection as well as wear and tear of the building.

Saleable Floor Area

Different floor areas are meant for different uses, but the common issue is to establish a uniform interpretation for the benefit of the community.

Saleable floor area is described by the HKIS (1999) as 'the ownership and occupation of the premises in relation to the building structure. As such it also reflects to a certain extent the rights and liabilities appurtenant to the premises.' This is a principle component of GFA and is crucial to owners.

Saleable floor area is preferred to be used when selling the units. The grey area in measurement is the 'semi-common' walls between the unit and common areas. This is indicated by the fact that the joint-institute survey advocated for the exemption of such walls from the calculation of the saleable floor area, whereas the HKIREA survey showed the opposite view. A closer look at the 'semi-common' walls reveals that these walls are usually structural elements, the size of which can change with the height of the building. Such change will usually result in more internal space for the unit. So in taking measurements, it will be easier, for the purpose of uniformity, to include such walls.

To make the saleable floor area meaningful to consumers, floor plans should be accompanied by such area announcements showing variations in structural sizes for different ranges of floors. Then the consumer can understand the exact size and 'usefulness' of the floor area. Similarly, other common areas added into the calculation of the saleable floor area to make up the construction floor area should be clearly stated for the benefit of the consumer.

The 1% limit of tolerance was the most preferred figure for measurement. In order to agree on the 'floor area', this has to be worked out with a precise measurement convention, alongside with other limits of tolerance due to site workmanship.

Environmental Issues

The environment is an interdependent world where effects of an object or action are multiple, and affect one another. Buildings have internal and external environments, both determined by the initial design of buildings. To achieve control of the internal environment without damaging the external environment, certain mechanisms or devices will be needed. These are usually external wall features or internal networks. As discussed previously under GFA, the building code has to be modified to give positive encouragement to new projects.

The Building Regulations were not written initially with environmental considerations. However, under the present Building Regulations, the main environmental concern is lighting and ventilation through windows. The method of calculation shows some deficiencies. For example, any open space adjacent to a site is not considered as providing enough room for lighting and ventilation, but a 4.5m-wide street is considered enough. Such prescribed-window calculation should be rewritten with proper consideration of space around the building and whether the building can conform to the breezeway in the urban fabric or proper cross ventilation is planned internally. With the objective of improving the quality of life, such prescribed-window calculation has to be rewritten to allow for flexibility and for merits to be considered in individual projects. Also, the exemption of projected windows from the calculation of GFA has a great impact on the building envelope of residential buildings in Hong Kong. Sheer conformity leads to monotony in design and lack of local character in different districts. An alternative for projected windows should be allowed under the Building Regulations as an amenity feature to give variation in both building appearance and internal use.

The scope of existing energy-efficiency regulations is rather limited. For example, an office building that complies with OTTV requirement may have used highly reflective glazing to prevent heat and light from entering the building. However, this will induce intensive glare in the surroundings, which is a negative environmental effect on urban spaces. Another case is the OTTV for assembly halls, where the frequency of use should be considered together with adequate provision of natural lighting and cross ventilation.

Planning And Site Matters

The peripheral areas around a site often affect its planning, including its development potential. Smaller sites often can increase their development potential by combining with large sites. An open space adjacent to a site, especially when used as a park, should be treated like an 'open space' as a street when establishing the class of site and windows for lighting and ventilation.

Certain regulations not compatible with the modern society have to be modified. The control of buildings based on similarity with buildings in the vicinity should be replaced by certain planning control for the neighbourhood area so as to facilitate a well-planned urban setting with visual corridors, urban breezeway and city skyline. Also, the service lane requirement is outdated. A servicing strategy in the neighbourhood with centralized refuse collection can be an alternative which should also be worked out at the planning level of the city. To create a modern metropolis, building and urban design are inseparable issues. Sustainable planning should be achieved together with formulation of legislative control for quality buildings.

Internal Layout

The size of a room, especially its height, is a basis on which to build up the internal environment. The 3m ceiling height favoured by most respondents is not common among existing residential buildings such as public housing estates. Such provision may mean additional staircase area which adds up GFA. Another restriction may come from limitations set in lease conditions. If a 3m ceiling height is required for the unit to qualify as a quality house, since it achieves spatial airiness and environmental benefits, then building regulations and planning control should be set for its positive encouragement, so that density of population can still be controlled while the quality of life is enhanced.

Modern kitchen facilities now provide safe and healthy cooking. The method of cooking and the type of fuel affect safety. According to the survey, consideration should be made for open kitchens to be used safely in domestic buildings.

Conclusion

Lynch (1984) mentioned that places are modified to fit human activities and that control of space can be a serious obstacle to adaptability. In the high-density city of Hong Kong, it is essential to allow for flexibility and efficiency.

Lynch also stated that flexible provision to 'fit' for the future is a puzzling criterion. He also suggested ease of manipulation and resilience. All these imply a certain degree of changes to be built in with a sense of continuity (response) and an ability to restore after an assumed disaster (recover). Applying this to Hong Kong, such adaptive character should be built in with building control to allow for sustainability and quality. Thus not only the writing, but the interpretation, of the Building Code is important for building control.

Though the Buildings Department (1999) states that 'the design of building forms and internal functions [is] left to the free will of the public', the restrictions imposed by the Building Code as well as the economic pressure for maximum development are overwhelming. To achieve a quality working and living environment, the built environment has to be equipped with better facilities. Changes of existing design have to be introduced to enhance the building envelope, provision of internal services, and amenity facilities. Without increasing population density nor endangering health and safety, such building design has to be encouraged through a reform of existing building regulations; otherwise little achievement can be made in the future. Also of equal emphasis is the need to provide a well-enforced management system to keep the built environment in high quality. The statutory scheme of preventive maintenance of buildings to be proposed (2000) by the Secretary for Planning, Environment and Lands can aim at a combined management system that ensures both building maintenance as well as compliance with the Building Code and fire safety measures.

▌ ACKNOWLEDGEMENTS

This chapter was mainly funded by the University of Hong Kong's Committee on Research and Conference Grants. The title of the research was 'Reviewing building regulations affecting the high density residential development in Hong Kong'.

This is also part of a doctorate research, entitled 'Control of Building Design in the High Density Urban Environment of Hong Kong'.

▌ REFERENCES

Building Authority, Hong Kong. 1995. Building (Efficiency) Regulations.

——. 1998. Building (Planning) Regulations, Parts I to VIA, First to Third Schedules.

——. October 1998. Building Ordinance, cap 123.

Buildings Department, Hong Kong. 1999. *Building development and control in Hong Kong*. 27–35. Hong Kong: Printing Department.

Buildings Ordinance Office, Hong Kong. January 1990. Streets in relation to site area, Building (Planning) Regulation 23(2)(a), PNAP no. 118.

——. November 1993. Projections in relation to site coverage and plot ratio, PNAP no. 68.

——. May 1994. Cladding, PNAP no. 59.

——. May 1994. Height of storeys, Building (Planning) Regulations 3(3) and 24, PNAP no. 27.

——. September 1994. Bridges over streets, Building Ordinance section 31(1), PNAP no. 107.

——. November 1994. Natural lighting to staircases, Building (Planning) Regulation 40, PNAP no. 169.

——. May 1995. Energy efficiency of buildings, Building (Energy Efficiency) Regulation, PNAP no. 172.

——. November 1995. Service lanes, PNAP no. 179.

——. August 1996. Amenity features, PNAP no. 116.

——. December 1997. Lighting and ventilation for bathrooms in domestic buildings, PNAP no. 219.

——. April 1998. Podium height restriction under Building (Planning) Regulation 20(3), PNAP no. 223.

——. June 1998. Calculation of gross floor area and non-accountable gross floor area, Building (Planning) Regulation 23(3)(a) and (b), PNAP no. 13.

——. August 1998. Buildings to be planned for use by person with disabilities, Building (Planning) Regulation 72, PNAP no. 112.

——. January 1999. Exclusion of floor area for recreational use, PNAP no. 229.

——. October 1999. Curtain wall systems, PNAP no. 106.

——. October 1999. Water seepage, PNAP no. 230.

Centuori, Jeanine. 1992. Building codes as dress codes for the protective clothing of buildings. *Architronic*, Vol. 1, No. 1.

Chan, Anson. 1999. *Building development and control in Hong Kong*, Buildings Department, p. 6. Hong Kong: Printing Department.

Le Corbusier. 1927. *Towards a new architecture*. London: Architectural Press.

Downward, Alan. 1992. *Building control: a guide to the law*. 1–6. Reading, Berks.: College of Estate Management.

Greenstreet, Robert C. 1996. The impact of Building Codes and legislation upon the development of tall buildings. *Architronic*, Vol. 5, No. 2.

Hall, A.C. 1996. *Design control: towards a new approach*. 1–19. Oxford: Butterworth Architecture.

Hong Kong Institute of Surveyors. 1999. *Code of measuring practice*. 1–32.

Lung, David. 1998. Investigating Hong Kong architecture and local development. *HKIA* 17: 74–80.

Lynch, Kevin. 1984. *Good city form.* 151–86. Cambridge, Mass.: MIT Press.

Siu, Gordon. 1999. *Building development and control in Hong Kong,* Buildings Department, p. 7. Hong Kong: Printing Department.

Tung, Chee-hwa. 1999. *The 1999 Policy Address: Quality People, Quality Home.* 1, 29–45. Hong Kong: Printing Department.

Wong, Wah Sang. 1998. *Building enclosure in Hong Kong.* 1–54. Hong Kong: Hong Kong University Press.

SURVEYS CONDUCTED

Definition of Floor Area (1999), distributed to HKIA, HKIE, HKIS, HKIP and HKIREA.
Buildings Regulations Affecting Design of Buildings (1999), distributed to HKIA.
Questionnaire on Your Home (1999), collected from 100 citizens of Hong Kong.

Index